NUREG-1567

Standard Review Plan
for Spent Fuel Dry Storage Facilities

I0488110

Final Report

Manuscript Completed: February 2000
Date Published: March 2000

Spent Fuel Project Office
Office of Nuclear Material Safety and Safeguards
U.S. Nuclear Regulatory Commission
Washington, D.C. 20555-0001

ABSTRACT

The Standard Review Plan for Spent Fuel Dry Storage Facilities (FSRP) provides guidance to the staff of the U.S. Nuclear Regulatory Commission for reviewing applications for license approval or renewal for commercial independent spent fuel storage installations (ISFSIs). An ISFSI may be co-located with a reactor or may be away from a reactor site. These installations may be designed for the storage of irradiated nuclear fuel and associated radioactive materials.

The U.S. Code of Federal Regulations (CFR), Title 10, Chapter 1, Part 72, Subpart B, specifies information required to be submitted in applications for license approval and renewal for ISFSIs. Regulatory Guide (RG) 3.48, "Standard Format and Content for the Safety Analysis Report for an Independent Spent Fuel Storage Installation" provides an outline and specific guidance regarding the information to be included in an applicant's safety analysis report (SAR). This standard review plan is intended to ensure the quality and uniformity of the NRC staff reviews by establishing the review scope and requirements.

The FSRP uses a basic outline defined by RG 3.48, modified based on staff experience with SAR reviews. The modified outline will be used for the related safety evaluation report (SER) prepared by the NRC staff in response to the applicant's SAR. The FSRP includes regulatory requirements, staff positions, references to applicable national and other industry standards and codes, acceptance criteria, guidance on preparation of the SER, and other guidance.

In conjunction with the FSRP, the Spent Fuel Project Office (SFPO) developed several SFPO Director's Interim Staff Guidance (ISG) documents. These ISGs were developed to address emerging issues for which interim guidance was needed. Current ISGs are available on the NRC website. Although the FSRP was revised to incorporate most of these ISGs, ISG guidance will continue to be developed when required. The FSRP will be revised periodically to reflect current guidance to the staff.

Comments are solicited on this document and applicable ISGs. The comments will be considered and incorporated into updates to the FSRP, as appropriate. Comments, errors or omissions, and suggestions for improvement should be sent to the Director, Spent Fuel Project Office, Office of Nuclear Material Safety and Safeguards, Mail Stop O-13D13, U.S. Nuclear Regulatory Commission, Washington, D.C. 20555-0001.

TABLE OF CONTENTS

TABLE OF CONTENTS
(Continued)

TABLE OF CONTENTS
(Continued)

TABLE OF CONTENTS
(Continued)

TABLE OF CONTENTS
(Continued)

TABLE OF CONTENTS
(Continued)

TABLE OF CONTENTS
(Continued)

TABLE OF CONTENTS
(Continued)

TABLE OF CONTENTS
(Continued)

TABLE OF CONTENTS
(Continued)

TABLE OF CONTENTS
(Continued)

TABLE OF CONTENTS
(Continued)

TABLE OF CONTENTS
(Continued)

TABLE OF CONTENTS
(Continued)

TABLE OF CONTENTS
(Continued)

TABLE OF CONTENTS
(Continued)

TABLE OF CONTENTS
(Continued)

ACRONYMS AND ABBREVIATIONS

ACI   American Concrete Institute
AISC   American Institute of Steel Constructions
ALARA   as low as is reasonably achievable
ALI   annual limit on intake
ANS   American Nuclear Society
ANSI   American National Standards Institute
API   American Petroleum Institute
ASCE   American Society of Civil Engineers
ASD   allowable stress design
ASME   American Society of Mechanical Engineers
ASTM   American Society for Testing and Materials
AWS   American Welding Society
AWWA   American Water Works Association

B&PV   boiler and pressure vessel (ASME B&PV Code)
BWR   boiling water reactor

CDE   committed dose equivalent (see definition)
CERCLA   Comprehensive Environmental Response, Compensation and Liability Act
CFR   *Code of Federal Regulations*
Ci   curie
C   degrees Celsius (°C)
CAA   Clean Air Act
CAM   continuous air monitor
CP   construction permit
CSRP   Cask Standard Review Plan (NUREG-1536)
CWA   Clean Water Act

D&D   decontamination and decommissioning
DAC   derived air concentration
DBE   design-basis earthquake
DE   design-basis earthquake
DBF   design-basis fire
DBFL   design-basis flood
DBT   design-basis tornado
DBW   design-basis wind
DCSS   dry cask storage system
DHR   decay heat removal
DOE   Department of Energy

ACRONYMS AND ABBREVIATIONS
(Continued)

EA   environmental assessment
EAL   emergency action level
EIS   environmental impact statement
EPA   Environmental Protection Agency
ER   environmental report
ESF   engineered safety feature

F   degrees Fahrenheit (°F)
FAM   financial assurance mechanism
FE   floating earthquake
FEMA   Federal Emergency Management Agency
FONSI   finding of no significant impact
FPP   fire protection program
FSRP   Facilities Standard Review Plan

g.   Gravitational unit (1 g = force exerted on the mass vertically by gravity)
GDC   general design criteria
GULF-GA   Gulf-General Atomic

HTGR   high-temperature gas-cooled reactor
H/U   heatup
HVAC   heating, ventilation, and air conditioning

ICBO   International Conference of Building Officials
ICRP   International Commission on Radiological Protection
IDLH   immediately dangerous to life and health
IEEE   Institute of Electrical and Electronics Engineers
ISFSI   independent spent fuel storage installation

k_{eff} effective multiplication factor
kgf   kilogram-force
Km   kilometers

LANL   Los Alamos National Laboratory
LLNL   Lawrence Livermore National Laboratory
LRFD   load and resistance factor design
LWR   light-water reactor

ACRONYMS AND ABBREVIATIONS
(Continued)

mi	miles
MofS	margin of safety
mrem	millirem
MRS	monitored retrievable storage
MSL	mean sea level
mSv	milliSievert (1 mSv is equivalent to 100 mrem)
MT	magnetic particle test
NA	not applicable
NAS	National Academy of Science
NCRP	National Council on Radiation Protection
NDE	non-destructive examination
NDT	nil ductility temperature
NEPA	National Environmental Policy Act
NESHAP	National Emissions Standard for Hazardous Air Pollutants
NFPA	National Fire Protection Association
NMSS	NRC Office of Nuclear Material Safety and Safeguards
NOAA	National Oceanic and Atmospheric Administration
NPDES	National Pollutant Discharge Elimination System
NQA	nuclear quantity assurance
NR	not required
NRC	U.S. Nuclear Regulatory Commission
NSC	National Safety Council
NTIS	National Technical Information Service
NWPA	Nuclear Waste Policy Act of 1982
OEL	occupational exposure limit
OFA	optimized fuel assembly
OL	operating license
ORNL	Oak Ridge National Laboratory
OSHA	Occupational Safety and Health Administration
PHA	peak horizontal ground acceleration
P&ID	piping and instrumental diagrams
PMF	probable maximum flood
PMP	probable maximum precipitation
PNL	Pacific Northwest Laboratory
PT	dye penetrant test
PWR	pressurized-water reactor

ACRONYMS AND ABBREVIATIONS
(Continued)

QA   quality assurance

RC   reinforced concrete
RCRA   Resource Conservation and Recovery Act
RG   regulatory guide
RT   radiographic test
Rule   Unless used generically, a requirement stated in the Code of Federal Regulations

SAIC   Science Applications International Corp.
SAR   safety analysis report
SER   safety evaluation report
SFPO   Spent Fuel Project Office
SNL   Sandia National Laboratory
SRP   Standard Review Plan
SSCs   structures, systems, and components
SSE   safe shutdown earthquake

UBC   Uniform Building Code, published by the International Conference of Building Officials
USGS   U.S. Geological Survey
USQ   Unreviewed Safety Question. For the FSRP, as defined at 10 CFR 72.48.
UT   ultrasonic test

VT   visual test

WPS   welding procedure specification

GLOSSARY

The following terms are defined here by the staff for the purpose of this FSRP. Many terms are taken from 10 CFR 20.1003 or 10 CFR 72.3. The definitions from these CFR sections have not been changed in the list below, but are repeated for convenience.

Accident-Level. A term used to include both design basis accidents and design basis natural phenomenon events and conditions. See "Design Basis." Resistance, response limit, and functional capability requirements apply for conditions and events that exceed "off-normal" or "Design Event II" as described in ANSI/ANS 57.9.

Annual limit on intake (ALI), means the derived limit for the amount of radioactive material taken into the body of an adult worker by inhalation or ingestion in a year. ALI is the smaller value of intake of a given radionuclide in a year by the reference man that would result in a committed effective dose equivalent of 5 rems (0.05 Sv) or a committed-dose equivalent of 50 rems (0.5 Sv) to any individual organ or tissue. (ALI values for intake by ingestion and by inhalation of selected radionuclides are given in Table 1, Columns 1 and 2, of Appendix B to 20.1001-20.2401). (10 CFR 20.1003)

As low as is reasonably achievable (ALARA), means making every reasonable effort to maintain exposures to radiation as far below the dose limits in 10 CFR 20 as is practical and consistent with the purpose for which the licensed activity is undertaken, taking into account the state of technology, the economics of improvements in relation to state of technology, the economics of improvements in relation to benefits to the public health and safety, and other societal and socioeconomic considerations, and in relation to utilization of nuclear energy and licensed materials in the public interest." (10 CFR 20.1003)

Basic, or fundamental, safety criteria. The following are considered the basic nuclear safety criteria for design of the ISFSI installation:

> Maintain subcriticality
> Prevent release of radioactive material above acceptable amounts
> Ensure radiation rates and doses do not exceed acceptable levels
> Maintain retrievability of the stored radioactive materials

Benchmarking. Validation of the accuracy of a computer code by comparison of results with results of relevant experiments.

Committed dose equivalent (H_T, 50) means the dose equivalent to organs or tissues of reference (T) that will be received from an intake of radioactive material by an individual during the 50-year period following the intake. (10 CFR 20.1003)

GLOSSARY
(Continued)

Confinement Barrier. A structure, system, or component that prevents the release of radioactive substances from areas containing radioactive substances to areas not containing radioactive substances and ultimately, to the environment.

Construction. Includes materials, design, fabrication, installation, examination, testing, inspection, and certification (as required in the manufacture and installation of components).

Controlled Area. Any area to which access is controlled to protect individuals from exposure to radiation and radioactive materials.

Damaged Fuel. Spent nuclear fuel with known or suspected cladding defects greater than a hairline crack or a pinhole leak.

Derived air concentration (DAC) means the concentration of a given radionuclide in air which, if breathed by the reference man for a working year of 2,000 hours under conditions of light work (inhalation rate 1.2 cubic meters of air per hour), results in an intake of 1 ALI. DAC values are given in Table 1, Column 3, of Appendix B to 10 CFR 20.1001-20.2401. (10 CFR 20.1003)

Design Basis. The extreme level of an event or condition for which there is a specified resistance, specified limit of response, or requirement for a specified level of continuing capability. Compares with "Design Events" III and IV of ANSI/ANS 57.9.

Design Event (I, II, III, or IV). Conditions and events as defined and used for ISFSI in ANSI/ANS 57.9.

Docketed. Formal submissions made to the NRC by an applicant. A docket number is assigned to the facility by the NRC and is used for the application and subsequent submissions and other correspondence on the facility. Except when the NRC concurs in a request that material be protected as being "proprietary data" docketed material becomes available for public.

Emergency Power. The power supply that is selected to furnish electric energy to safety-related structures, systems and components when the preferred power supply is not available.

Exclusion Area. [Applies to sites with a reactor only] That area surrounding the reactor, in which the reactor licensee has the authority to determine all activities, including exclusion or removal of personnel and property from the area.

Exemption. As used in the FSRP, an exemption to application of a specific regulatory requirement that must be approved by the NRC.

GLOSSARY
(Continued)

Important Confinement Features. Term used in ANSI/ANS 57.9 but not acceptable to the NRC (per RG 3.60). "Important to safety" should be substituted for "important confinement features" in the standard.

Important to Safety, also "Important to Nuclear Safety." Terms used synonymous in the FSRP. "Important to nuclear safety" is used where there may be a misinterpretation that the classification "important to safety" may also include SSCs which do not have a nuclear safety role but may be important for life safety, fire prevention, prevention or mitigation of property loss, or protection of the environment (from other than radioactive material or radiation). Important to safety can include "safety-related" and "nonsafety-related" SSCs (see definitions). "Structures, system, and components important to safety" mean those features of the ISFSI whose function is: (1) To maintain the conditions required to store spent fuel or high-level radioactive waste safely, (2) To prevent damage to the spent fuel or the high-level radioactive waste container during handling and storage, or (3) To provide reasonable assurance that spent fuel or high-level radioactive waste can be received, handled, packaged, stored, and retrieved without undue risk to the health and safety of the public. (10 CFR 72.3)

In-place radioactive material. Radioactive material that has not escaped through the closest confinement barrier (or liquid containment).

ISFSI. Independent Spent Fuel Storage Installation. ISFSI may be operated by public or private utilities, commercial entities, and governmental agencies.

k_{eff} *"k" effective.* Measure of reactivity. Multiplication factor including all biases and uncertainties at a 95 percent confidence level for indicating the level of subcriticality relative to the critical state. At the critical state $k_{eff} = 1.0$.

Mixed waste. Waste material which is hazardous because of both radioactive material and other hazard(s), such as chemical, toxic, incendiary.

Non-Mechanistic Event. An event, such as cask tip-over, that should be analyzed for acceptable system capability, although a cause for such an event is not identified in the analyses of off-normal and accident-level events and conditions.

GLOSSARY
(Continued)

Nonsafety-Related Electrical Equipment. Equipment whose failure under postulated environmental conditions could prevent satisfactory accomplishment by "safety-related electrical equipment" of prevention or mitigation of the consequences of design basis events. For this definition, design basis events are defined as conditions of normal operation, including anticipated operational occurrences, design basis accidents, external events, and natural phenomena for which the facility must be designed to ensure accomplishment of the stated safety requirement. [Based on description at 10 CFR 50.49(b)(1).] [Also see "Important to Safety" and "Safety-Related Electrical Equipment."]

Normal. The maximum level of an event or condition expected to routinely occur. The ISFSI is expected remain fully functional and to experience no temporary or permanent degradation from normal operations, events, and conditions. Compares to "Design Event I" of ANSI/ANS 57.9. Events and conditions that exceed the levels associated with "normal" are considered to be, and to have the response allowed for, "off-normal" or "accident-level" events and conditions.

Off-Normal. The maximum level of an "off-normal" event or condition, for which there is a corresponding maximum specified resistance, specified limit of response, or requirement for a specified level of continuing capability. Similar to "Design Event II" of ANSI/ANS 57.9. ISFSI SSCs are expected to experience off-normal events and conditions without permanent deformation, and without degradation of capability to provide their full functional capability (although operations may be suspended or curtailed during off-normal conditions) over the full license period. Off-normal is considered to include "anticipated occurrences" as used in 10 CFR 72.

Part. A subdivision of the CFR. References to a "Part" number in this FSRP are to parts of Title 10 of the CFR unless a different title is specified.

Radwaste. Waste which is hazardous as a result of its containing nuclear materials; may be high- or low-level.

Rem. Is the special unit of any of the quantities expressed as dose equivalent. The dose equivalent in rems is equal to the absorbed dose in rads multiplied by the quality factor (1 rem=0.01 sievert). (10 CFR 20.1004)

Restricted Area. Any area, access to which is controlled by the licensee for purposes of protection of individuals from exposure to radiation and radioactive materials. (10 CFR 20)

Retrievability. Capability to retrieve the stored radioactive material without the release of radioactive materials to the environment or radiation exposures in excess of 10 CFR 20 limits

GLOSSARY
(Continued)

(10 CFR 72.122(h)(5)). ISFSI storage systems must be designed to allow ready retrieval of the stored spent fuel for compliance with 10 CFR 72.122(l).

Safety Analysis Report. In the context of this FSRP, the report submitted by the license applicant in compliance with 10 CFR 72, Subpart B. The fundamental contents of the report are described at 10 CFR 72.24. Guidance on content of the report is provided by Regulatory Guide 3.48, "Standard Format and Content for the Safety Analysis Report for an Independent Spent Fuel Storage Installation," October 1981. The SAR is considered to be the submitted application, and supplemental data and responses submitted to the NRC staff to resolve questions arising during the staff's review. Only docketed material is considered to form part of the submission. The effective SAR is considered by the staff to be that submitted, as amplified and/or modified by the supplemental and later submissions.

Safety Evaluation Report. In the context of this FSRP, the report prepared by the NRC staff to document the acceptability of the applicant's safety analysis and other required submissions.

Safety-Related Electrical Equipment. Equipment relied upon to remain functional during and following a design basis event to ensure the capability to prevent or mitigate the consequences of accident-level events. For this definition, design basis events are defined as conditions of normal operation, including anticipated operational occurrences, design basis accidents, external events, and natural phenomena for which the facility must be designed to ensure accomplishment of the stated safety requirement.

Sievert (Sv). is the SI unit of any of the quantities expressed as dose equivalent. The dose equivalent in sieverts is equal to the absorbed dose in grays multiplied by the quality factor (1 Sv=100 rems). (10 CFR 20.1004)

Source Material. (1) Uranium or thorium, or any combination thereof, in any physical or chemical form or (2) ores containing by weight one-twentieth of one percent (0.05%) or more of (I) uranium, (ii) thorium or (iii) any combination of thereof. Source material does not contain SNM. (10 CFR 72.3)

Special Nuclear Material. (1) plutonium, uranium 233, uranium enriched in the isotope 233 or in the isotope 235, and any other material which the Commission, pursuant to the provisions of Section 51 of the Act, determines to be SNM, but does not include source material; or (2) any material artificially enriched by any of the foregoing but does not include source materials. (10 CFR 72.3)

Standby power. The power supply that is selected to furnish electric energy when the preferred power supply is not available. Often used interchangeably with emergency power.

GLOSSARY
(Continued)

Storage confinement cask. Cask, vessel, or other sealed container providing the principal confinement barrier for subject radioactive material while in dry storage. Term includes internal and integral external components unless otherwise specified at the point of use.

Subject radioactive material. The material whose storage is the principal function of the ISFSI. Term includes power reactor spent fuel and other radioactive material associated with spent fuel storage for an ISFSI.

Unrestricted Area. An area to which access is not controlled by the licensee for purposes of protection of individuals from exposure to radiation and radioactive materials. 10 CFR 20

Volume %. The percentage of a mole of the material that is present in a volume that is equal to the standard volume for the material as a gas.

INTRODUCTION

This document is the Facility Standard Review Plan (FSRP). It is intended to provide guidance to the NRC staff who will be conducting a safety review of a site-specific license application for an independent spent fuel storage installation (ISFSI). The objective of this introduction to the FSRP is to give an overview of the entire ISFSI Safety Analysis Report (SAR) review process and to assist the Project Manager who is responsible for coordinating and managing the overall safety review effort. It is also intended to help individual technical reviewers understand how their specific review must be coordinated with other reviews to produce an integrated review.

Review Process

The review process of an ISFSI application involves six major phases: (1) site evaluation, (2) operations systems evaluation, (3) criteria and technical design evaluation, (4) evaluation of proposed programs that support protection of worker and public health and safety, (5) evaluation of accidents, and (6) evaluation of proposed technical specifications.

These six major review phases are illustrated in Figure 1. The figure shows that all reviews proceed primarily from information in the license application (primarily the SAR) and any responses to Requests for Additional Information (RAIs). The review of health and safety programs can proceed independently of the other review efforts (site, operations systems, and design review), but the results of all these efforts must be considered in the accident analysis. The last phase of the safety review addresses the proposed technical specifications and draws on all of the previous review results. Additional details on how the design review is conducted are also presented in a later section of this introduction and in specific chapters of the FSRP.

Site Evaluation. This phase of the review evaluates site characteristics to determine if the applicant has properly identified and quantified natural phenomena such as floods, high winds, high temperatures, and seismic events, and has included them in the ISFSI design bases. The review also determines if the applicant has identified and quantified the site characteristics related to contaminant transport and potentially exposed individuals and population. Specific guidance for conducting this review is presented in Chapter 2 of this FSRP.

Operation Systems Evaluation. This phase of the review evaluates the overall description of the proposed ISFSI, the identification of the major components, and the description of the major spent fuel or high-level waste handling operations and post-storage inspection and monitoring operations.

Criteria and Technical Design Evaluation. This phase of the review is a large part of the ISFSI review effort and must be performed in a coordinated manner with several technical disciplines. Individual components and ISFSI system performance are reviewed for normal conditions, off-normal conditions, and design basis accidents. Selection of the components for the design review should be made after consultation between the various review disciplines (structural, thermal, nuclear criticality safety, shielding, etc.) and the Project Manager. The

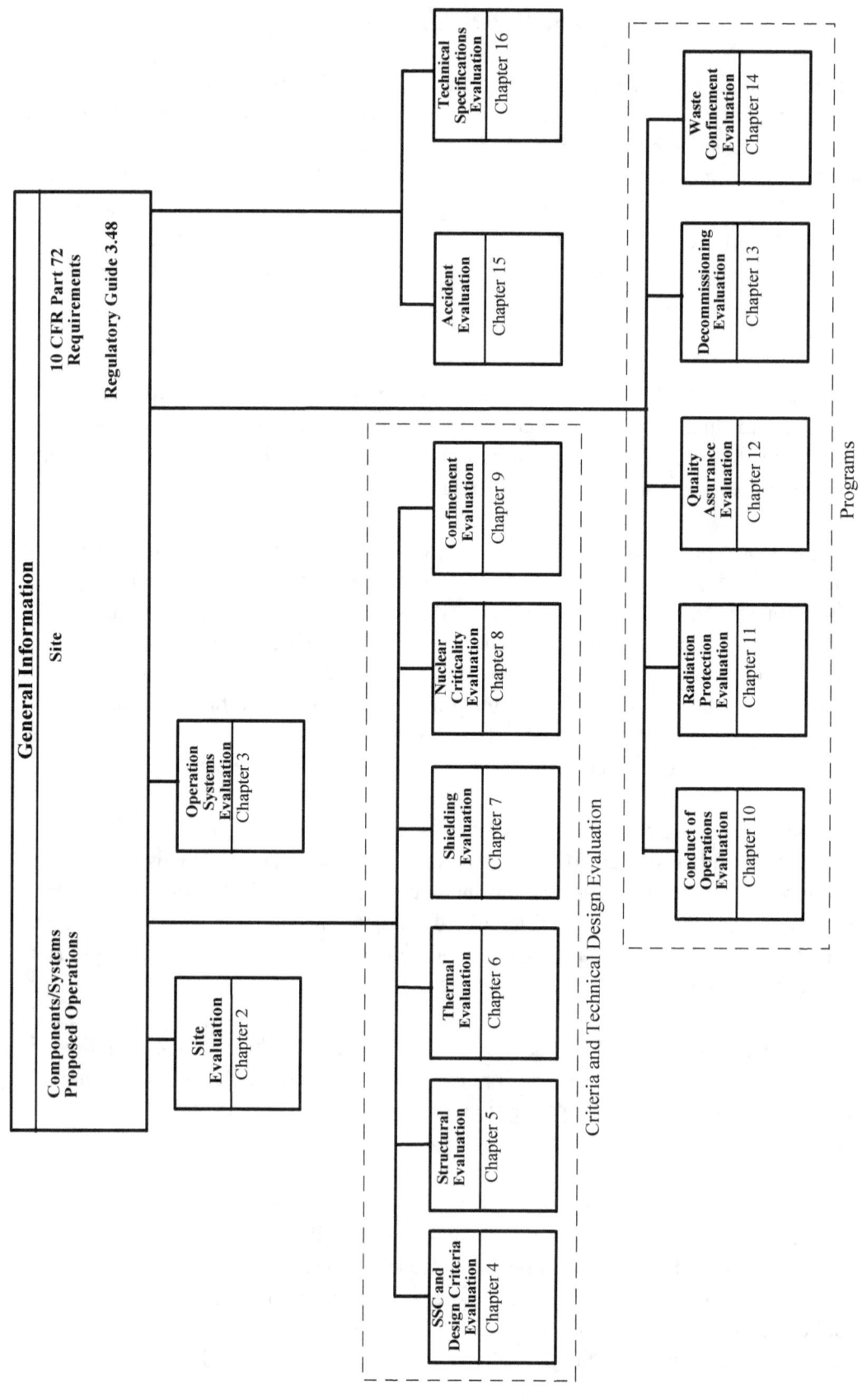

Figure 1 Major Review Phases

system-level performance review focuses on radiological impacts and involves estimates of any material released under normal conditions, off-normal conditions, and design basis accident conditions.

Evaluation of Proposed Programs that Support Protection of Worker and Public Health and Safety. This phase of the review provides assurance that facility operations will not have adverse impacts on public health and safety. These programs include radiation protection, conduct of operations, quality assurance, decommissioning, and waste confinement. Each of these programs can be reviewed independently of the other programs or the design review effort.

Evaluation of Accident Analysis. This phase of the review involves the evaluation of accidents under off-normal events and conditions, as well as accident or design basis events. The results are documented in a separate chapter of the Safety Evaluation Report (SER) and are part of the basis for determining technical specifications.

Proposed Technical Specifications. The previous reviews and analyses have established the bases for the identification of safety limits, limiting conditions, and surveillance requirements deemed necessary to ensure safe operation of the ISFSI.

The criteria and design evaluation effort (Phase 3) involves the detailed review of proposed design criteria, design codes, and the resulting designs. This effort is similar to the design reviews conducted for ISFSI casks certified under 10 CFR Part 72, Subpart L, and conducted according to the guidance of NUREG-1536, "Standard Review Plan for Dry Cask Storage Systems." This FSRP builds on the guidance in NUREG-1536, and the chapters related to criteria and design evaluation often reference specific sections of NUREG-1536.

The evaluation of proposed programs that support protection of worker and public health and safety (e.g., radiation protection, conduct of operations, quality assurance) is a review area for specific licenses rather than the certificate of compliances. Such programs are not developed for review as part of a cask certification application that would be reviewed by using the guidance of NUREG-1536; thus, the review guidance for these areas does not refer to NUREG-1536.

Material in the FSRP Chapters

Each chapter of the FSRP gives six types of information: review objective, areas of review, regulatory requirements, acceptance criteria, review procedures, and evaluation findings.

Review Objective. This section gives an overview to the chapter and establishes the major review objectives of the chapter.

Areas of Review. This section identifies topics and their sequence in the discussion of acceptance criteria and review procedures subsections of each chapter.

Regulatory Requirements. This section summarizes the regulatory requirements from 10 CFR Part 72 expected to be applicable. The reviewer should read the complete language of the current version of 10 CFR Part 72 and independently determine whether the proper set of applicable regulations for the design is being reviewed.

Acceptance Criteria. This section is organized according to review areas established in Section 2 of the specific chapter, and identifies the type and level of information that should be in the application. Specific criteria are identified based on (a) specific language in 10 CFR Part 72, (b) specific language in Regulatory Guide 3.48, and (c) clearly established precedent in Part 72 licensing.

Review Procedures. This section presents a step by step procedure of what to check in the application for each review area. As an aid to the reviewer, this section also provides information on what has been found acceptable in past reviews. Standards that have been found acceptable in specific licensing reviews, or are desirable but not specifically identified in existing regulatory documents, are identified in this section.

Evaluation Findings. This last section of each chapter provides guidance on how findings might be worded for the SER and gives suggested language for findings that indicate compliance with regulations and regulatory guidance.

Safety Evaluation Report Outline

The review results are documented in an SER. The final determination of the organization of an SER is determined by the review Project Manager.

The chapters are presented in an order intended to help a reader of the SER understand:

- Evaluation of the site
- Evaluation of proposed operations
- Evaluation of structures, systems, and components important to safety; and evaluation of design criteria and bases
- Results of specific technical reviews
- Reviews of proposed programs intended to promote protection of worker and public health and safety
- Assessment of potential accidents
- Evaluation of proposed technical specifications.

1 GENERAL DESCRIPTION

1.1 Review Objective

The objective of this chapter is to ensure that the applicant has provided a non-proprietary description of major components and operations that is adequate to familiarize reviewers and other interested parties with the pertinent features of the installation. Figure 1.1 presents an overview of the evaluation process.

1.2 Areas of Review

The following outline shows the areas of review addressed in Section 1.4, Acceptance Criteria, and Section 1.5, Review Procedures:

Introduction
General Description of Installation
General Systems Description
Identification of Agents and Contractors
Material Incorporated by Reference

1.3 Regulatory Requirements

This section identifies and presents a high-level summary of Title 10 of the Code of Federal Regulations (CFR) Part 72 relevant to the review areas addressed by this chapter. The NRC staff reviewer should read the exact regulatory language. A matrix at the end of this section matches the regulatory requirements identified in this section to the areas of review identified in the previous section.

72.22 Contents of application: General and financial information

72.24 Contents of application: Technical information [Contents of SAR]
(b) "A description and discussion of the [Independent Spent Fuel Storage Installation] ISFSI or monitored retrievable storage (MRS) structures"
(f) "Features of ISFSI or MRS design and operating modes to reduce ... radioactive waste volumes."
(l) "A description of the equipment ... to maintain control over radioactive materials in gaseous and liquid effluent"

72.44 License Conditions

A matrix showing the primary relationship of these regulations to the specific areas of review in this chapter is given in Table 1.1. The reviewer should independently verify the relationships in this matrix to ensure that no requirements are overlooked because of unique applicant design features.

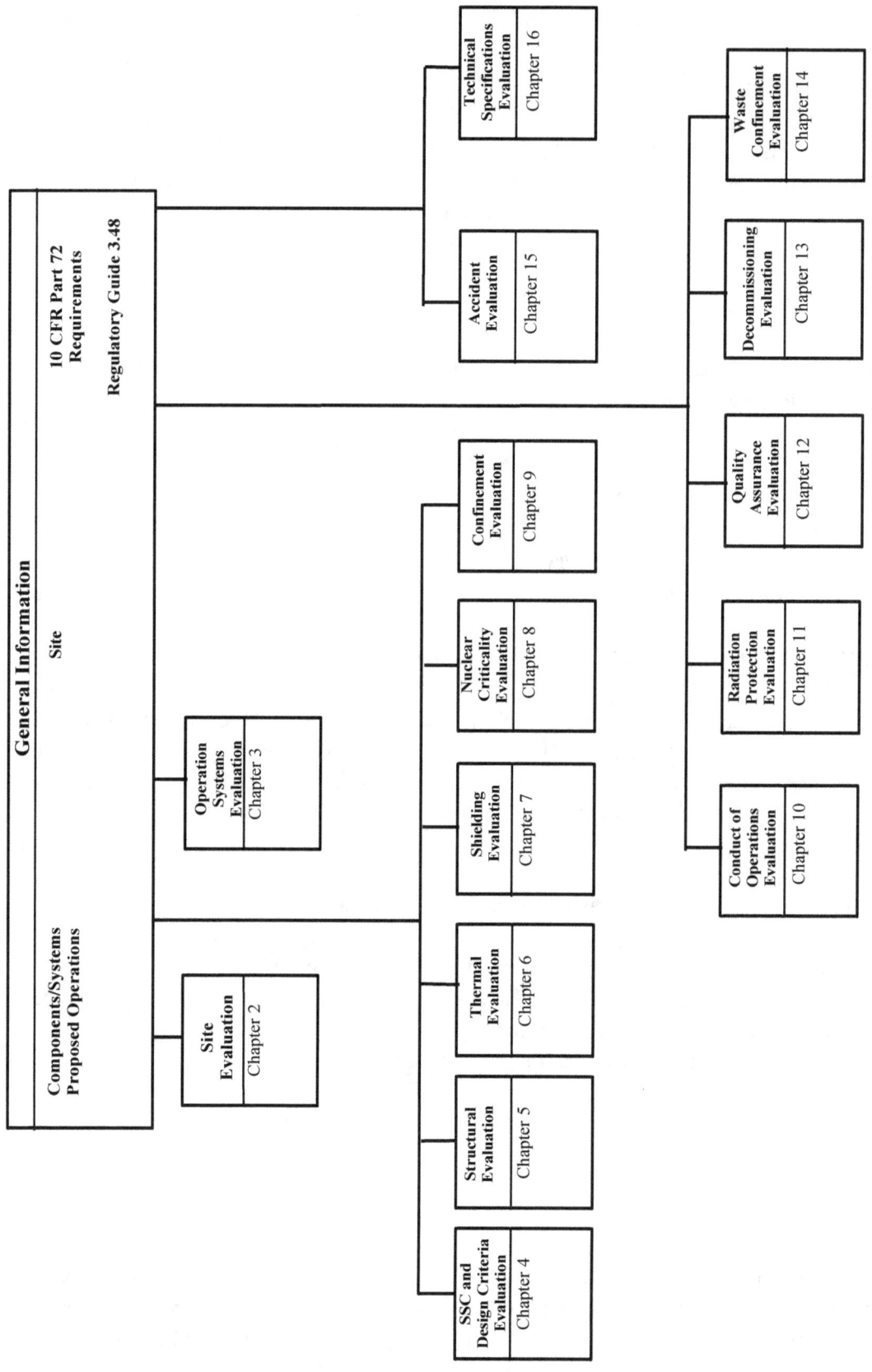

Figure 1.1 Overview of Safety Evaluation

1.4 Acceptance Criteria

This section identifies the acceptance criteria for the material provided in the introduction. The regulatory requirements relevant to the introductory chapter of the Safety Analysis Report (SAR) are found in 10 CFR 72.22, 72.24, and 72.44. The general description should enable all reviewers, regardless of their specific review assignments, to obtain a basic understanding of the principal function and design features of the proposed installation. Regulatory Guide 3.48, "Standard Format and Content for the Safety Analysis Report for an Independent Spent Fuel Storage Installation, (Dry Storage)," provides guidance regarding information that should be included in the general description. Because much of the information relevant to this initial aspect of the review is presented in more detail in other chapters of this Standard Review Plan (SRP), this chapter focuses on familiarization with the system and should be consistent with the remaining sections of the SAR.

Table 1.1 Relationship of Regulations and Areas of Review

Areas of Review	10 CFR Part 72 Regulations		
	72.22	72.24	72.44
Introduction	●	●	
General Description of Installation		●	
General System Description		●	
Identification of Agents and Contractors	●	●	
Material Incorporated by Reference			●

1.5 Review Procedures

The following provides guidance relevant to the review of the general description chapter of the SAR.

1.5.1 Introduction

The reviewer should verify that the principal function and design features of the installation have been presented. The reviewer should verify that the location of the ISFSI and schedules for construction and operation have been presented.

1.5.2 General Description of Installation

The reviewer should verify that the applicant has provided a broad overview of the installation that is non-proprietary and may be used as a tool to familiarize interested parties with the features of the proposed ISFSI. This description should present the principal characteristics of the ISFSI, including its dimension, weights, and construction materials. The reviewer should compare sketches and diagrams, if presented in this section, with the detailed drawings presented elsewhere in the SAR. If the application includes proprietary drawings and descriptions that will remain proprietary upon approval of the license, the sketches, drawings, and diagrams that provide the general description need not show the proprietary features. This may be achieved by depicting less detail or by illustrating generic components that fulfill the design functions that differ from the actual design. However, these representations should show the operational concept and safety-related features in sufficient detail to form an acceptable basis for public review and comment, as necessary for public hearings.

To verify compliance with 10 CFR 72.122(l) *Retrievability*, the reviewer should verify that the facility description demonstrates that the facility is designed to allow for ISFSI decommissioning, and will be used for interim storage, not permanent disposal. 10 CFR 72.122(l) applies to normal and off-normal design conditions and not to accidents. Sections 4,10, and 15 of this SRP describe the staff's recommendations for post-accident recovery with regard to retrievability.

The reviewer should verify that the applicant has presented a general description of the fuel or other contents proposed for storage in the ISFSI. Because a very detailed description of the proposed contents is typically provided in the SAR in Section 3, "Principal Design Criteria," the information presented in Section 1, "General Description," is important only to the extent that it permits overall familiarization with the ISFSI. Key parameters for spent fuel include the type of fuel (i.e., pressurized water reactor [PWR], boiling water reactor [BWR]), number of fuel assemblies, and conditions of the fuel assemblies (i.e., intact, consolidated). This section often includes additional characteristics, such as maximum burnup, initial enrichments, heat load, and cooling time, as well as the assembly vendor and configuration (e.g., Westinghouse 17x17). These characteristics may also be repeated in the principal design criteria. The cover gas, if any, should be identified.

1.5.3 General Systems Description

The reviewer should verify that a summary description of the storage mode and arrangement of the storage structures has been provided. The reviewer should verify that a brief description of the operating systems, including fuel handling, decay heat removal, site-generated waste treatment, and auxiliary systems has been provided. The reviewer should determine if sufficient detail has been provided to result in an understanding of the systems involved.

1.5.4 Identification of Agents and Contractors

The reviewer should verify that the prime agents or contractors for the design, construction, and operation of the installation have been identified. The reviewer should verify that all principal consultants and outside service organizations, including those providing quality assurance (QA) services, have been identified. The reviewer should ensure that the application clearly defines the division and assignments of responsibilities among those parties.

1.5.5 Material Incorporated by Reference

The reviewer should verify that a tabulation of all topical reports incorporated by reference has been provided. The reviewer should verify that any documents submitted to the Commission in other applications and incorporated in whole or in part have been tabulated and a summary included in the appropriate section of the SAR.

1.6 Evaluation Findings

NRC staff reviewers prepare evaluation findings regarding satisfaction of the regulatory requirements related to the introduction and general description. If the documentation submitted with the application fully supports positive findings for each of the regulatory requirements, then the findings should substantially be stated as follows (finding numbering is for convenience in referencing within the SRP and SER):

F1.1   The staff concludes that the information presented in this section of the SAR satisfies the requirements for the general description under 10 CFR Part 72. This finding is reached on the basis of a review that considered the regulation itself; Regulatory Guide 3.48 and accepted practices.

F1.2   Agents and contractors responsible for the design, construction, and operation of the installation have been identified.

F1.3   A tabulation of all topical reports and docketed material, incorporated by reference, has been provided in the SAR.

2 SITE CHARACTERISTICS

2.1 Review Objective

The purpose of the site characteristics review is to make three determinations. The first is whether the applicant has properly identified the external natural and man-induced phenomena for inclusion in the design basis and whether the design basis levels are adequate. The second is whether the applicant has adequately characterized local land and water use and population so that important individuals and populations likely to be affected can be identified. The third is whether the applicant has adequately characterized the transport processes which could move any released contamination from the facility to the maximally exposed individuals and populations.

The results of this review determine the acceptability of site-derived design bases. The determination whether the design basis events were properly incorporated into the proposed design is made in the design review sections of the Standard Review Plan (SRP). The results also determine the location of maximally exposed individuals and populations and the dilution/dispersion parameters to be used by the radiation protection reviewer in determining the impacts of normal operations and accidents.

Because the site characteristic information required of the Safety Analysis Report (SAR) and of the Environmental Report (ER) is similar, common information is normally presented in one document which is referenced by the second document. Thus, the reviewer may need copies of the relevant sections from both documents.

An overview of the site characteristics review process is shown in Figure 2.1 which shows that the site review process draws upon information from the application. Some results are documented in the NRC staff-prepared Safety Evaluation Report (SER); others are used in other technical review areas.

2.2 Areas of Review

The following outline shows the six areas of review addressed in Section 2.4, Acceptance Criteria, and Section 2.5, Review Procedures:

Geography and Demography
>   Site Location
>   Site Description
>   Population Distribution and Trends
>   Land and Water Use

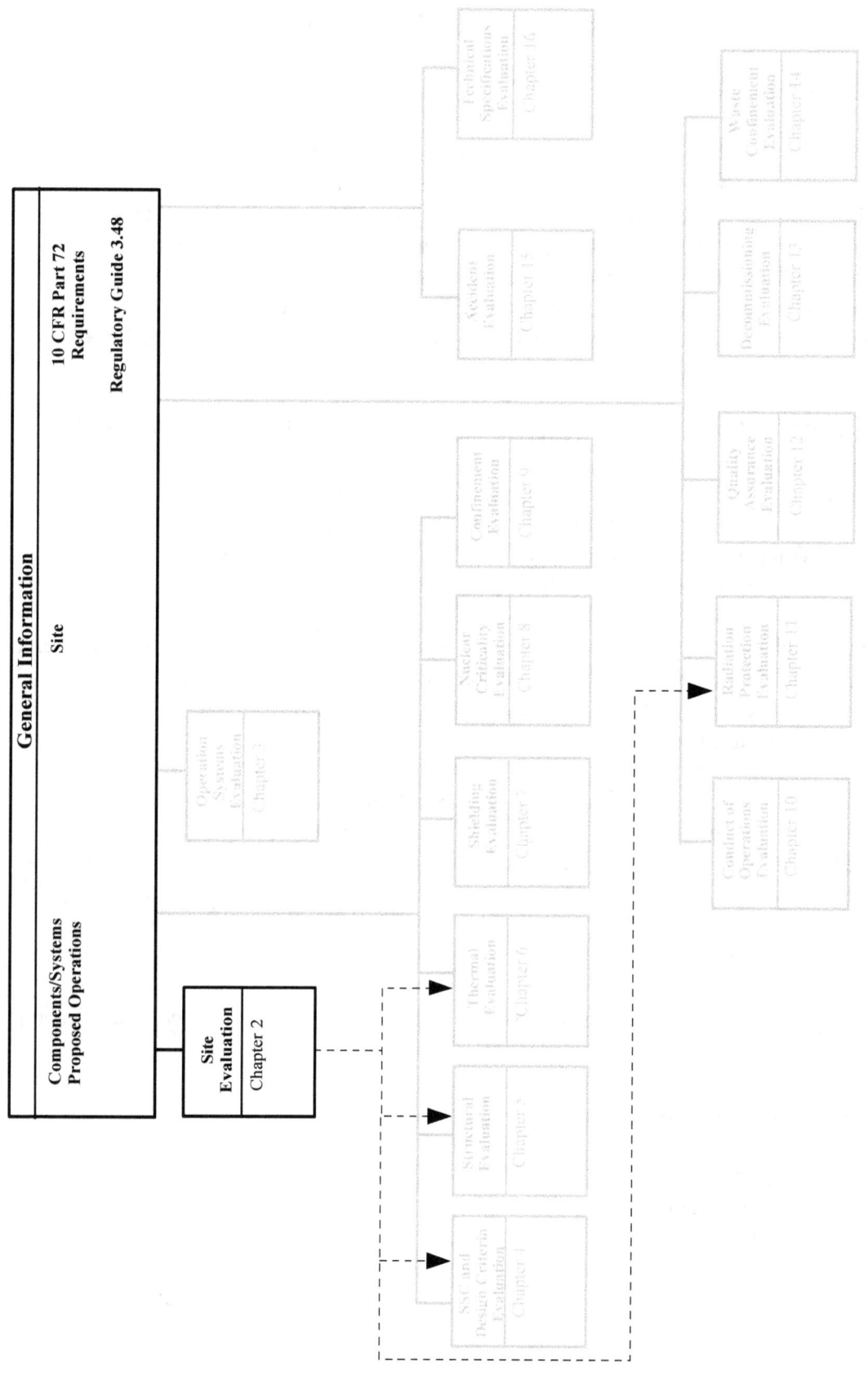

Figure 2.1 Overview of Site Evaluation

Nearby Industrial, Transportation, and Military Facilities

Meteorology
> Regional climatology
> Local meteorology
> Onsite meteorological measurement program

Surface Hydrology
> Hydrologic description
> Floods
> Probable Maximum Flood (PMF) on streams and rivers
> Potential dam failures (seismically induced)
> Probable maximum surge and seiche flooding
> Probable maximum tsunami flooding
> Ice flooding
> Flood protection requirements
> Environmental acceptance of effluents

Subsurface Hydrology

Geology and Seismology
> Basic geologic and seismic information
> Vibratory ground motion
> Surface faulting
> Stability of subsurface materials
> Slope stability

2.3 Regulatory Requirements

This section identifies and presents a high-level summary of Title 10 of the Code of Federal Regulations (CFR) Part 72 relevant to the review areas addressed by this chapter. The NRC staff reviewer should read the exact regulatory language. A matrix at the end of this section matches the regulatory requirements identified in this section to the areas of review identified in the previous section.

72.24 Contents of application: Technical information
(a) "A description and safety assessment of the site...with appropriate attention to the design bases for external events."

72.40 Issuance of license
(a)(2) "The proposed site complies with the criteria in Subpart E;"

72.90 General considerations
(a) "Site characteristics...must be investigated and assessed."
(b) "Proposed site...must be examined with respect to the frequency and the severity of external natural and man-induced events that could affect ... safe operation."
(c) "Design basis external events must be determined for each combination of proposed site and proposed ... design."
(d) "Proposed sites with design basis external events for which adequate protection cannot be provided through ... design shall be deemed unsuitable."
(e) "[T]he potential for radiological and other environmental impacts on the region must be evaluated."
(f) "The facility must...avoid...adverse impacts associated with the occupancy and modification of floodplains."

72.92 Design basis external natural events
(a) "Natural phenomena...that can occur in the region of a proposed site must be identified and assessed."
(b) "Records of the occurrence and severity of...natural phenomena must be collected...and evaluated."
(c) "Appropriate methods must be adopted for evaluating the design basis external natural events."

72.94 Design basis external man-induced events
(a) "The region must be examined for both past and present man-made facilities and activities that might endanger the proposed ISFSI or MRS."
(b) "Information concerning the potential occurrence and severity of such events must be collected and evaluated."
(c) "Appropriate methods must be adopted for evaluating the design basis external man-induced events."

72.96 Siting limitations
(a) "An ISFSI which is owned and operated by DOE must not be located at any site within which there is a candidate site for a [High-Level Waste] HLW repository."
(b) "An MRS must not be sited in any State in which there is any site approved for site characterization for a HLW repository."
(c) "If an MRS is located...within 50 miles of the first HLW repository, any Commission decision approving the first HLW repository application must limit the quantity of spent fuel or high-level radioactive waste that may be stored."
(d) "An MRS authorized by section 142(b) of [Nuclear Waste Policy Act] NWPA ... may not be constructed in the State of Nevada."

72.98 Identifying regions around an ISFSI or MRS site

(a) "The regional extent of external phenomena ... that are used as a basis for the design ... must be identified."

(b) "The potential regional impact due to the construction, operation or decommissioning...must be identified."

(c) "Those regions identified pursuant to paragraphs (a) and (b) of this section must be investigated with respect to:

> (1) The present and future character and the distribution of population,
>
> (2) Consideration of present and projected future uses of land and water within the region, and
>
> (3) Any special characteristics that may influence the potential consequences of a release of radioactive material during the operational lifetime of the ISFSI or MRS."

72.100 Defining potential effects of the ISFSI or MRS on the region

(a) "The proposed site must be evaluated with respect to the effects on populations."

(b) "Each site must be evaluated with respect to the effects on the regional environment."

72.102 Geological and Seismological characteristics

(a)   (1) "East of the Rocky Mountain Front...except in areas of known seismic activity...sites will be acceptable if investigations show no unstable geological characteristics ... for vibratory ground motion ... in excess of an appropriate response spectrum anchored at 0.2 g."

> (2) "For those sites that have been evaluated under paragraph (a)(1) of this section that are east of the Rocky Mountain Front, and that are not in areas of known seismic activity, a standardized design earthquake (DE) described by an appropriate response spectrum anchored at 0.25 g may be used. Alternatively, a site-specific DE may be determined by using the criteria and level of investigations required by appendix A of part 100 of this chapter."

(b) "West of the Rocky Mountain Front ... and in other areas of known potential seismic activity, seismicity will be evaluated by the techniques of appendix A of part 100 of this chapter."

(c) "Sites other than bedrock sites must be evaluated for their liquefaction potential or other soil instability due to vibratory ground motion."

(d) "Site-specific investigations and laboratory analyses must show that soil conditions are adequate for the proposed foundation loading."

(e) "In an evaluation of alternative sites, those which require a minimum of engineered provisions to correct site deficiencies are preferred."

(f) "The ... DE for use in the design of structures must be determined as follows:

> (1) For sites that have been evaluated under the criteria of appendix A of 10 CFR Part 100, the DE must be equivalent to the safe shutdown earthquake (SSE) for a nuclear power plant.
>
> (2) Regardless of the results of the investigations anywhere in the continental U.S., the DE must have a value for the horizontal ground motion of no less than 0.10 g with the appropriate response spectrum."

72.122 Overall requirements

(b) "Protection against environmental conditions and natural phenomena.

   (1) Structures, systems, and components (SSCs) important to safety must be designed to accommodate the effects of ... site characteristics and environmental conditions associated with normal operation, maintenance, and testing of the ISFSI or MRS and to withstand postulated accidents.

   (2) SSCs important to safety must be designed to withstand the effects of natural phenomena.... The design bases for these SSCs must reflect:

      (i) Appropriate consideration of the most severe of the natural phenomena reported for the site and surrounding area

      (ii) Appropriate combinations of the effects of normal and accident conditions and the effects of natural phenomena.

      The ISFSI or MRS should also be designed to prevent massive collapse of building structures or the dropping of heavy objects on the spent fuel or high-level radioactive waste or on to structures, systems, and components important to safety."

   (4) "If the ISFSI or MRS is located over an aquifer which is a major water resource, measures must be taken to preclude the transport of radioactive materials to the environment through this potential pathway."

A matrix showing the primary relationship of these regulations to the specific areas of review in this chapter is given in Table 2.1. The reviewer should independently verify the relationships in this matrix to ensure that no requirements are overlooked because of unique applicant design features.

Table 2.1 Relationship of Regulations and Areas of Review

Areas of Review	10 CFR Part 72 Regulations									
	72.24	72.40	72.90	72.92	72.94	72.96	72.98	72.100	72.102	72.122
Geography & Demography	•		•			•	•	•		
Nearby Facilities	•	•	•		•	•	•	•		•
Meteorology	•	•	•	•			•			•
Surface Hydrology	•	•	•	•			•			•
Subsurface Hydrology	•						•			•
Geology & Seismology	•	•	•	•			•		•	•

2.4 Acceptance Criteria

The specific acceptance criteria for methods used to identify design criteria are presented in the appropriate parts of this section. No specific acceptance criteria for factors such as atmospheric dispersion or population location are applied in assessing the impacts. Rather, the applicant must supply accurate information so that realistic impacts can be estimated.

2.4.1 Geography and Demography

10 CFR 72.90, 72.98, 72.100, and 72.122 require that the SAR contain information about the site geography, population, and water and land uses. The criteria given here indicate the kind and degree of detail of information required in an application before a reviewer can validate its adequacy for use in an impact analysis.

2.4.1.1 Site Location

Information on site location of the proposed ISFSI and nearby facilities should clearly describe the location by stating the site's host State and county, and its latitude, longitude, and Universal Transverse Mercator coordinates. Maps and aerial photographs of the site should be presented with radial coverage extending a minimum of 8 km (5 mi) from the site. A detailed map of the site area should clearly show adjacent buildings, roads, railroads, transmission lines, wetlands, and surface water bodies. The reviewer should be aware of the limitations on ISFSI and MRS siting which are listed in 10 CFR 72.96, and the potential changes to these limitations which may have been enacted by Congress.

2.4.1.2 Site Description

A site map should clearly indicate the site boundary and the controlled area (if different from the site boundary), controlled area access points, and the distances from the boundary to significant features of the installation. The SAR should discuss the applicant's legal responsibilities for the properties described, such as ownership, lease, or easements. Topographic maps should reveal the site topography and surface drainage patterns as well as roads, railroads, transmission lines, wetlands, and surface water bodies on the site. Vegetative cover and surface soil characteristics should be described to facilitate evaluation of fire hazards and erosion. Other activities conducted by the applicant within the controlled area should be identified, and potential interactions with ISFSI operation discussed.

2.4.1.3 Population Distribution and Trends

Current population data and projections should be presented. A sector map of population should divide the area within a 8-km (5-mi) radius of the site by concentric circles with radii of 1.5, 3, 5, 6.5 and 8 km (approximately 1, 2, 3, 4, and 5 miles), and by 22.5-degree segments, each segment centered on one of the 16 compass points. Current and projected populations in each sector should be given. The population data should be overlain on a base map which shows any cities or towns. The maximally exposed individual(s) should be specifically identified and a rationale for their selection (e.g., nearest well, closest person downwind in the predominant wind direction) presented.

2.4.1.4 Land and Water Use

Use of land and water within an 8-km (5-mi) radius should be described. Residential, farming, dairy, industrial, and recreational uses of land and water should be presented in sufficient detail to allow estimates of concentrations of radionuclides to populations from any airborne or liquid effluents.

2.4.2 Nearby Industrial, Transportation, and Military Facilities

10 CFR 72.94 requires that the region be examined for man-made facilities that might endanger the proposed ISFSI or MRS. The SAR should indicate the locations of nearby industrial, transportation, military, and nuclear installations on a map which clearly shows their distance and relationship to the ISFSI. All facilities within an 8-km (5-mi) radius and all relevant facilities at greater distances should be included. For each facility, the products or materials produced, stored, or transported should be described, and any potential hazards to the ISFSI from activities or materials at the facilities should be discussed. Any effect of these facilities on the specific ISFSI design basis should also be discussed.

2.4.3 Meteorology

10 CFR 72.90 requires that site characteristics affecting the safety of the proposed ISFSI or MRS must be assessed. The SAR should describe the meteorological conditions at the site and vicinity. Conditions which influence the design and operation of the facility should be identified, and sources of all information should be stated. Enough information should be provided to permit NRC staff to independently evaluate atmospheric diffusion characteristics of the site area. Enough information should also be provided to permit NRC staff to determine the basis for the high winds (either straight line or tornado winds) and high temperature used in the design basis.

2.4.3.1 Regional Climatology

The SAR should describe the climate of the region, including temperature, precipitation, relative humidity, general airflow, pressure patterns, cloud cover, average wind speeds, and prevalent wind direction. Ranges and seasonal variations of these parameters should be discussed. Climate characteristics attributable to terrain should be mentioned. Data on the frequency, intensity, and duration of severe weather should be presented. For example, the SAR should address: temperature, wind, and precipitation extremes; hurricanes, tropical storms, tornadoes, lightning strikes; and snow, ice, and hail storms. Data sources and reliability should be discussed. The rationale for the design basis winds and temperature should be stated in the application.

2.4.3.2 Local Meteorology

The description of local meteorology should summarize data on temperature, wind speed and direction, and relative humidity collected onsite as well as at nearby weather stations. The representativeness of data collected offsite should be discussed. If offsite data adequately represent onsite conditions, then onsite data may not be necessary. For the purpose of evaluating atmospheric diffusion, topographic maps at two different scales should be provided. One should show detailed topographic features, as modified by the facility, within an 8-km (5-mi) radius of the site. A smaller-scale map should show topography out to a 16-km (10-mi) radius. This map should be accompanied by profiles of maximum elevation over distance from the center of the installation out to 16-km (10-mi) for each of the 22.5 degree compass-point sectors.

2.4.3.3 Onsite Meteorological Measurement Program

The meteorological data collected onsite should be reviewed to ensure its adequacy for NRC staff to conduct independent atmospheric dispersion estimates for both postulated accidents and expected routine releases of gaseous effluents. The meteorological data should be provided in the form of joint frequency distributions of wind speed and wind direction by atmospheric stability class. The SAR should state the measurements made, the locations and elevations of measurements, descriptions of the instruments used, instrument performance specifications, calibration and maintenance procedures, and data analysis procedures. Any onsite program and any programs to be used during operations to estimate offsite concentrations of airborne effluents should be described in conformity to Regulatory Guide 1.23, "Onsite Meteorological Programs," criteria for an acceptable onsite meteorological measurements program, and its format for presenting stability class data. If no onsite measurement program exists, the applicant should provide justification for using data from nearby stations.

2.4.4 Surface Hydrology

10 CFR 72.98 requires that the present and future uses of land and water within the region be investigated. The SAR should contain adequate information for an independent review of all surface hydrology-related design bases, performance requirements, and operating procedures important to safety.

2.4.4.1 Hydrologic Description

The SAR should characterize the surface hydrologic features of the region, area, and site, because this information is the basis for hydrologic engineering analyses. The location, size, and hydrologic characteristics of all streams, rivers, lakes, and adjacent shore regions which influence or may influence the site or facilities under severe hydrologic conditions, should be described. Topographic maps of the area and the site should be provided to give a clear understanding of these features. A map of the site area should indicate any proposed change to the natural drainage features. If the site is vulnerable to river flooding, any river control structures upstream or downstream of the site should be identified.

The SAR should identify the sources of the hydrologic information, the types of data collected, and the methods and frequency of collection. The SAR should also list the structures important to safety, including their exterior accesses, and equipment and systems which may be affected by hydrologic features. The SAR should note any surface waters which could potentially be affected by normal or accidental effluents from the site. A listing of any population groups which use such surface waters as a potable water supply should be provided, as well as the size of these population groups, location, and water-use rates.

2.4.4.2 Floods

The SAR should adequately support any claim that the proposed site is flood-dry, that is, with structures important to safety so high above potential sources of flooding that safety is obvious or can be documented with little analysis, as indicated in American National Standards Institute/American Nuclear Society (ANSI/ANS) 2.8-1981.

If the site is not flood-dry, then the SAR should identify the design basis flood (DBF) and provide a rationale for this specific design basis. Such a rationale should contain a synopsis of the flood history of the site, including dates and maximum water levels. Causes of past and potential future flooding, such as river or stream floods, surges, tsunami, dam failures, ice jams, etc., should be provided. A detailed analysis of the flooding potential of the site is required, as discussed in SRP Sections 2.4.4.3 through 2.4.4.9. This information should be detailed enough for NRC staff to perform an independent flood analysis of the site, as described in NUREG-0800, "Standard Review Plan for the Review of Safety Analysis Reports for Nuclear Power Plants," Section 2.2.2(II).

2.4.4.3 Probable Maximum Flood on Streams and Rivers

The SAR must consider the PMF on adjacent streams and rivers in its detailed flood analysis. If the SAR did not follow the approach for assessing PMFs in ANSI/ANS 2.8-1981, then it should describe the alternative approach used. The steps taken to derive the probable maximum precipitation (PMP) over the applicable drainage area, the precipitation losses, the amount of runoff, and the PMF should be shown. Drainage basins should be identified on a topographic map. The estimated discharge hydrograph for the PMF at the site and, if applicable, a similar hydrograph without the effects of an upstream reservoir should be included. The conversion of the PMF peak discharge into water elevation at the site should be described. Wind-wave activity which could coincide with the PMF should be discussed. Finally, the locations and associated water levels for which PMF determinations have been made should be summarized.

2.4.4.4 Potential Dam Failures (Seismically Induced)

If potential dam failures are necessary to identify flood design bases, then the SAR should discuss the effects of potential seismically induced dam failures (both upstream and downstream) on the water levels of streams and rivers. Descriptions of existing or proposed dams and reservoirs which could influence conditions at the site should be provided and include seismic design criteria for dams. The potential dam failure modes which lead to the most critical consequences for the site (flood or low reservoir level) should be described. Domino-type dam failures from floodwaves should be considered where applicable. Finally, the reliability of the water level estimate should be addressed.

2.4.4.5 Probable Maximum Surge and Seiche Flooding

If the site is at risk of inundation from surge or seiche flooding, these hazards should be described. Water bodies which could impact the site should be described, and the surge and seiche history of the site should be provided. The frequency and magnitudes of potential causes of surges, such as hurricanes, wind storms, squall lines, and other mechanisms should be described. A graph of the calculated maximum surge hydrograph should be provided. The potentially coincident wind-generated waves and the possibility of wave oscillation at natural frequencies should be described. Estimates of potential wave runup, erosion, and sedimentation, and any site facilities designed to guard against these processes, should be described.

2.4.4.6 Probable Maximum Tsunami Flooding

If the site abuts a coastal area, the hazards posed by tsunami should be analyzed. The history of tsunami in the region--be it recorded, translated, or inferred from the geologic record--should be analyzed. The analysis should include all potential tsunami generators, such as specific faults, fault zones, volcanoes, and potential landslide areas. The maximum tsunami height from these causes should be estimated at the source, in deep water offshore from the site, and onshore. A probable maximum tsunami should be derived from these analyses. Near-shore routing, wave breaking, bore formation, and resonance effects of this tsunami should be discussed. Any structures designed to protect against tsunami flooding should be described.

2.4.4.7 Ice Flooding

If the site is not subject to flooding caused by ice jams, a brief statement of explanation should be provided. If the site is subject to ice-jam flooding, an analysis of this hazard should be provided. The history of ice jam formation in the region and the location of ice-generating mechanisms relative to the facility should be described. Any structures designed to protect against flooding from ice jams should also be described.

2.4.4.8 Flood Protection Requirements

The static and dynamic consequences of all types of flooding on each storage structure and component important to safety should be described if the previous flooding analyses indicate that the structure or component is subject to flooding. The design bases required to ensure that all structures and components can survive all design flood conditions should be included.

2.4.4.9 Environmental Acceptance of Effluents

The ability of the surface water and ground water environment to disperse, dilute, or concentrate normal and inadvertent liquid releases of radioactive effluents for the full range of anticipated operating conditions, including accident scenarios leading to worst-case releases, should be described. All potential surface water and ground water pathways by which radionuclides could reach existing and potential water users should be identified. Any potential for water recirculation, sediment concentration, or hydraulic short-circuiting of cooling ponds should be assessed in anticipation of normal or accidental releases of radionuclides.

2.4.5 Subsurface Hydrology

10 CFR 72.122 requires that measures be taken to preclude the transport of radioactive materials to the environment through subsurface characteristics. The SAR should contain adequate information for an independent review of all subsurface hydrology-related design bases and compliance with dose radiological exposure standards.

If the site is located over an aquifer which is a source of well water, the groundwater aquifer(s) beneath the site, the associated hydrologic units, and their recharge and discharge areas should be described. The results of a survey of groundwater users, well locations, source aquifers, water uses, static water levels, pumping rates, and drawdown should be provided. A water table contour map showing surface water bodies, recharge and discharge areas, and locations of monitoring wells to detect leakage from storage structures should also be provided. Information on monitoring wells should include: wellhead elevation, screened interval, installation method, and representative hydrochemical analyses. An analysis bounding the potential groundwater contamination from site operations should be provided. A graph of time versus radionuclide concentration at the closest existing or potential downgradient well should be included.

2.4.6 Geology and Seismology

10 CFR 72.102 requires that the SAR describe the geological and seismological setting of the site and surrounding region. Conditions which may influence the design and operation of the facility should be identified, and sources of all information should be stated. Enough information for an independent evaluation of the potential ground vibrations and the seismic and fault displacement hazards at the site area should be provided. Design bases for ground vibration, surface faulting, subsurface material stability, and slope stability should also be provided.

2.4.6.1 Basic Geologic and Seismic Information

Basic geologic and seismic characteristics of the site and vicinity should be provided. The geologic history of the area should describe its lithologic, stratigraphic, and structural conditions. A large-scale geologic map of the site area showing the surface geology and the location of major facilities should be provided. A stratigraphic column and cross-sections should also be provided. Planar and linear features of structural significance such as folds, faults, synclines, anticlines, basins, and domes should be identified on a geologic map showing bedrock surface contours. A description of the site geomorphology should include areas of potential landsliding or subsidence, and a topographic map showing geomorphic features and principal site facilities should be provided. The results of pertinent geophysical investigations in the area, such as seismic refraction, seismic reflection, aeromagnetic, or geoelectrical surveys, should also be provided.

The SAR should evaluate geologic features from an engineering geology perspective. Detailed static and dynamic engineering properties of soil and rock underlying the site should be provided, with the results integrated to provide a comprehensive understanding of the surface and subsurface conditions. A small-scale map should show major features of the installation and the locations of all borings, trenches, and excavations. Small-scale cross-sections should demonstrate relationships between major foundations and subsurface materials, structures, and the water table. Finally, any physical evidence concerning the behavior of surficial site materials during previous earthquakes should be presented.

2.4.6.2 Ground Vibration

The design basis ground vibration and a rationale for its selection should be presented and explained. The rationale should list historical earthquakes which could have affected the site, their dates, epicenter locations, and magnitudes. This listing of events is not constrained by distance and may include entries for distant structures, such as the New Madrid fault system. All faults and epicenters should be displayed on maps of appropriate scales. The fault map should include all potentially significant faults or parts of faults within 161 km (100 mi), regardless of capability. All capable faults (as defined in 10 CFR Part 100, Appendix A) which may be of significance in establishing the design basis ground vibration for the site should be identified and adequately described. The maximum ground vibration at the site should be derived from the potential earthquakes from all capable faults and from floating earthquakes (FEs, those not associated with a previously identified structure).

2.4.6.3 Surface Faulting

Surface faulting at the site and underlying tectonic structures which have caused or might cause faulting should be described. The capability of any mapped faults 300 m (1000 ft) or longer within 8 km (5 mi) of the site should be described. Those judged capable should be described in detail, with special attention to their displacement history and their relationship to any regional tectonic structures.

2.4.6.4 Stability of Subsurface Materials

The stability of the rock (defined as having a shear wave velocity of at least 1166 m/s [3500 ft/s]) and soil beneath the foundations of the facility structures while subjected to the design basis ground vibration should be described. The geologic features which could affect the foundations, such as areas of potential uplift or collapse, or zones of deformation, alteration, structural weakness, or irregular weathering, should be described. The static and dynamic engineering properties of the materials underlying the site, as well as the physical properties of foundation materials should be described. A plot plan showing the locations of all borings, trenches, seismic lines, piezometers, geologic cross-sections, and excavations, with all installation structures superimposed, should be provided. Plans and profiles showing the extent of excavations and backfill, as well as compaction criteria, should be provided. The water table history and anticipated groundwater conditions beneath the site during facility construction and operation should be described. Analyses of soil and rock responses to dynamic loading should be provided, and potential liquefaction beneath the site should be discussed. Criteria, references, or methods of design used, along with safety factors, should be discussed.

2.4.6.5 Slope Stability

The stability of all natural and man-made slopes, both cut and fill, the failure of which could adversely affect the site, should be described. Cross-sections of the slopes and a summary of the static and dynamic properties of embankment and foundation soil and rock underlying the slopes, should be provided. The design criteria and analyses used to determine slope stability should be described.

2.5 Review Procedures

2.5.1 Geography and Demography

The reviewer should use the methods stated below to perform the compliance review of the geography and demography information in the SAR.

2.5.1.1 Site Location

Confirm that the site location, its relationship to political boundaries, and the natural and anthropogenic features of the area are properly described. Use U.S. Geological Survey (USGS) topographic maps and aerial photos (obtained either independently or from the applicant) to verify the location described in the SAR.

2.5.1.2 Site Description

Ensure that the site maps clearly delineate boundary and controlled areas. Confirm that distances between the controlled area boundary and the storage location, as well as other possible effluent release points, are accurately reported. These distances should agree with those used in SAR Section 8, accident analyses. Check that access to the controlled area will be adequately restricted to protect individuals outside the area, as required by 10 CFR 72.104. Ensure that the orientation of plant structures with respect to nearby roads, railways, and waterways is shown, and that there are no obvious ways by which transportation routes within the controlled area can interfere with normal ISFSI operations. Use site visits to verify information in the site description.

2.5.1.3 Population Distribution and Trends

Confirm that the source of the population data used in the SAR is appropriate and that the basis for population projections is reasonable. The population data can be compared to other data available from local or State agencies, councils of government, Census Bureau CED tapes and projections, or any Bureau of Economic Analysis special census. Note significant differences from SAR data which may require clarification. Determine whether the rationale for identifying the maximally exposed individual is consistent with local meteorology and patterns of land and water use.

2.5.1.4 Land and Water Use

Compare SAR land use information to existing data on land use, land use controls such as zoning, potential for growth, and other factors which may encourage or retard population growth between the facility and the nearest population. Confirm the identification of any bodies of water or aquifers used by humans, livestock, or farms within 8 km (5 mi) of the site. Compare SAR information with available independent data on water use and any projections of future water use in the vicinity of the site. Consider the level of detail appropriate to the projected distance of the nearest future population center to the site and the level of projected water withdrawal within 8 km (5 mi) of the site.

2.5.2 Nearby Industrial, Transportation, and Military Facilities

Review the potential hazards associated with nearby facilities. In addition to obvious industrial or nuclear facilities in the area, consider other anthropogenic features which could conceivably pose a hazard, such as transportation routes, railroads, and airports. Accuracy of the SAR information can be confirmed by referring to USGS maps, aerial photos, or other documents, such as applications from any nearby nuclear plants. Use contacts with local, State, and other Federal agencies.

Review specific information relating to types of potentially hazardous material expected to be transported in the area including: distance, quantity, and frequency of shipment. The hazards from nearby facilities may include, but are not limited to: explosions of chemicals, flammable material, or munitions; detonation of explosives stored at mines or quarries; structure, petrochemical, brush, or forest fires; and release of toxic gases. Consider aircraft size, velocity, weight and fuel load in assessing the hazards of aircraft crashes on an installation near an airport. Analyze the effects of any airborne pollutants from nearby facilities and the effects of a possible collapse of any discharge stacks on site. Determine if the methods used by the applicant to quantify offsite hazards are consistent with the guidance in Chapter 15, Accident Analysis, of the FSRP. Identify potential accidents which cannot be eliminated from consideration as design basis events because the consequences could affect facility safety features. Ensure that such accidents are adequately considered in the design criteria of SAR Chapter 3.

2.5.3 Meteorology

The reviewer should use the methods stated below to perform the compliance review of the meteorology information in the SAR.

2.5.3.1 Regional Climatology

Review the SAR description of climate parameters against standard references listed in NUREG-0800, Section 2.3.1(II) for verifying meteorological discussions and data. Confirm that the data sources are reliable and that the level of detail in the database is appropriate. Ensure that climate data are based on long-term data gathering at National Weather Service (NWS) stations and other sites with reliable meteorological monitoring equipment. Review the information on severe weather, especially strong wind and wind-borne missiles, and check for consistency with the values used to develop structural design criteria in SAR Chapter 3.

2.5.3.2 Local Meteorology

Use maps and site visits to become familiar with the locations of all primary meteorological stations. Review the topographic maps for the accurate location of features, and confirm accurate portrayal of topography on the topographic profiles. Review summaries of the meteorological data for adequacy and completeness of the database. Whenever possible, review the onsite wind speed and atmospheric stability data which are used to model atmospheric diffusion because airflow and vertical temperature structure can vary substantially over short horizontal distances. If only offsite data are available, determine how well the data represent site conditions. Consult references in NUREG-0800, Section 2.3.2(III), to evaluate the representativeness of weather stations and periods of record. Data summaries from nearby stations with long periods of records should well represent long-term meteorological extremes. Ensure consistency between these extreme values and those used to develop structural and thermal design criteria.

2.5.3.3 Onsite Meteorological Measurement Program

Review two areas in this section, the instruments gathering the meteorological data and the data itself, by examining instrument siting, meteorological sensors, recordings of meteorological sensor output, instrument surveillance, and data acquisition and reduction, as discussed in detail in NUREG-0800, Section 2.3.3.

Review the joint frequency distributions of wind speed, wind direction, and atmospheric stability. Ensure that measurement heights and data recording periods are appropriate. In addition, determine the climatic representativeness of the joint frequency distribution by comparing with data from nearby stations which have collected reliable meteorological data over a long period, such as 10-20 years. Ensure that the meteorological measurement program is consistent with gaseous effluent release structures and systems design. Assume that the effluent release structure and system design are commensurate with the degree of risk to public health and safety.

2.5.4 Surface Hydrology

The reviewer should use the methods stated below to perform the compliance review of the surface hydrology information in the SAR.

2.5.4.1 Hydrologic Description

Ensure that all relevant hydrologic features are addressed and properly described by using USGS topographic maps and available independent hydrologic reports for this verification. Determine whether hydrologic features which influence or may influence the site under severe hydrologic conditions (e.g., a flood) have been adequately described. Review the criteria governing operation of any upstream or downstream river control structures for scenarios of problems in river management. Examine any proposed alterations to the natural drainage pattern of the site. Ensure that the design of any systems, structures, and components important to safety can accommodate the effects of these alterations. Review local hydrologic reports to confirm the identity of population groups getting potable water from the described hydrologic features. Use references in NUREG-0800, Section 2.4.1(II), to verify information provided for this section by the applicant.

2.5.4.2 Floods

Review any claim that the site is flood-dry. Consider that a descriptive statement of circumstances and relative elevations may be enough to complete such a review. Evaluate the bases of any analogy with comparable watersheds for which PMF levels have been determined or approximations of PMF levels used. Require details only to the level required to prove that structures important to safety are safe from flooding. Ensure that conservatism is used in all methods and assumptions. Consult ANSI/ANS 2.8-1981 for descriptions of acceptable procedures to demonstrate flood-dry status.

If the site is not clearly flood-dry, review in detail the analysis called for in Sections 2.5.4.3 through 2.5.4.9. Determine whether SAR Chapter 3 adequately addresses the DBF in Sections 2.5.4.3 through 2.5.4.9.

2.5.4.3 Probable Maximum Flood on Streams and Rivers

Review the SAR derivation of the PMF. Rely on information from actual storms in the region of the drainage basin. Consider storm configurations, maximum storm precipitation amounts (compare these with NWS and Army Corps of Engineers [COE] determinations), time distributions, orographic effects, storm centering, seasonal effects, antecedent storm sequences, and antecedent snowpack. Confirm by calculations that the maximum storm precipitation distribution for the drainage basin is conservative. Review the SAR analysis of the absorption capability of the drainage basin. Ensure that assumptions of initial losses, infiltration rates, and antecedent precipitation are reasonable and justified. Review the SAR model for calculating runoff, as well as the input data such as hydrologic response characteristics of the watershed. Check that subbasin drainage areas and topographic features are mapped properly, and review the tabulation of drainage areas, runoff, and reservoir and channel-routing coefficients. Confirm that the PMF hydrograph represents the flow from the PMP and any possible coincident snowmelt.

Determine whether the PMF analysis considers any existing or proposed upstream dams or river structures and their ability to withstand a PMF. Confirm the maximum water flows from breaches if they are not designed to withstand a PMF. Review the PMF stream course response model and its ability to compute floods of various magnitudes up to the severity of a PMF. Review any reservoir and channel-routing assumptions, and the assessment of initial conditions, outlet works, spillways, coincident wind-wave action, wave protection, and reservoir design capacity. Review the process of translating PMF discharge to peak water level at the site by such means as: topographic profiles, reconstitution of historical floods, standard step methods, roughness coefficients, bridge and other losses, extrapolation of coefficients for the PMF, estimates of PMF water surface profiles, and flood outlines. Review the SAR discussion of the effects on structures from runup and the static and dynamic effects of wave action which may occur coincidentally with the PMF peak water level.

Perform an independent analysis of the PMF by using alternative data and interpretations when available. Require additional justification if the SAR analyses are more than five percent less conservative than independent NRC estimates.

Consult the following documents in reviewing SAR data and analyses: Regulatory Guide 1.59, "Design Basis Floods for Nuclear Power Plants," guidance for estimating the PMF design basis; Regulatory Guide 1.102, "Flood Protection for Nuclear Power Plants," description of acceptable flood protection for safety-related facilities; and NWS and Army COE documents for estimating PMF discharge and water level conditions at the site.

2.5.4.4 Potential Dam Failures (Seismically Induced)

Review whether the applicant considered all relevant dams and reservoirs which could affect the site in the event of failure. Review the drainage areas above reservoirs, and ensure that all dam structures, appurtenances, and ownership are completely described. Review the reservoir elevation/storage relationships and short- and long-term storage allocations. Ensure that the discussion of dam failures considers all factors including: landslides, antecedent reservoir levels, domino-type multiple dam failures, and base river flow coincident with the flood peak, but not necessarily the simultaneous occurrence of the PMF with a seismic dam failure. Ensure that a conservative analysis has been used and that it assumes that the maximum earthquake (based on historic seismicity) coincides with full reservoirs and either a flood half the size of the PMF or a standard-project flood as defined by the Army COE. Review for conservatism the basis for selecting the maximum earthquake which can lead to dam failure.

Review the calculations used to derive the peak flow rate and water level at the site which could result from the worst possible dam failure. Examine all methods and coefficients used in these calculations, and ensure that the analytical methods apply to such artificially large floods. Review the discussion of static and dynamic effects of the floodwave at the site. Examine the assumptions used to attenuate the wave if credit is taken for downstream attenuation of a floodwave. Ensure that wind waves which may coincide with the flood are properly considered.

Conduct a more refined analysis, as described in NUREG-800, Section 2.4.4(III), if this flooding analysis indicates a potential flooding problem. To the extent possible, conduct an independent analysis of the flooding effects from a seismically induced dam failure by using simplified, conservative procedures according to guidance in ANSI/ANS 2.8-1981. Require additional justification if the SAR analyses are more than five percent less conservative than independent NRC estimates.

2.5.4.5 Probable Maximum Surge and Seiche Flooding

Review the descriptions of potential surge and seiche sources, and ensure that they address the most severe combination of reasonable meteorological parameters including: storm track, wind fields, wind fetch, and bottom effects. Use NUREG-0800, Section 2.4.5(III), for its discussion of methods to develop the maximum hurricane parameters for a site, to estimate the maximum surge still water elevations at coastal sites, and to estimate coincident wind-generated waves and runup. Use National Oceanic and Atmospheric Administration (NOAA) Technical Report NWS-23 for its descriptions of the meteorological characteristics of the probable maximum hurricane for the East and Gulf Coasts, the most severe combination of meteorological parameters of moving squall lines for the Great Lakes, and the most severe combination of meteorological parameters capable of producing high storm-induced tides for the West Coast.

Confirm that ambient water levels, including tides and sea-level anomalies, are conservatively estimated. Use NUREG-0800, Section 2.4.5(II), for its discussion of water level estimation methods which follow NOAA and Army COE guidance. Ensure that the method of developing the surge hydrograph from the meteorological, hydrological, and site-specific information is appropriate. Review the information on wave action which may coincide with surges. Ensure that estimates of wave height and runup are adequately conservative and, if appropriate, include breaking waves. Review the analysis of wave resonance within any lakes or harbors near the site.

To the extent possible, conduct an independent analysis of the water level and wave height for surges and seiches by using alternative data and interpretations when available. Require additional justification if the SAR analyses are more than five percent less conservative than independent NRC estimates.

2.5.4.6 Probable Maximum Tsunami Flooding

Review the historical tsunami information for completeness. Review the tabulation of source areas capable of generating tsunami at the site for completeness. Evaluate the seismic characteristics of the tsunami generators, including fault location and orientation, as well as amplitude and areal extent of potential vertical displacement, to ensure that conservative values have been used. Examine this information for consistency with that provided in the SAR geology and seismology section. Review the tabulation of maximum tsunami wave heights which can be generated at each local source and the maximum deep-water heights generated by distant sources. Review the process used to identify the source of the probable maximum tsunami for transparency. Examine the method used to translate tsunami waves from deep water, offshore locations to the site. Review the analysis of local factors which may affect the magnitude of tsunami flooding, such as coastline shape, offshore land areas, hydrography, and stability of the coastal area. Ensure the reasonableness of assumptions and the inclusion of appropriate bathymetric data in the analysis. For the probable maximum tsunami, review the analysis of potential breaking wave formation, bore formation, resonance effects, or other factors which can affect the maximum height of the tsunami water level. Use NUREG-0800, Section 2.4.6(III), for references for evaluating ambient tide and wave conditions, oscillation of waves at natural periodicity, and the adequacy of protection from flooding.

To the extent possible, conduct an independent analysis of the source of the probable maximum tsunami and its resulting water height at the site by using alternative data and interpretations when available. Require additional justification if the SAR analyses are more than five percent less conservative than independent NRC estimates.

2.5.4.7 Ice Flooding

Determine whether ice flooding poses a threat to the site on the basis of a review of the applicable literature describing historical occurrences of icing in the region, and, if so, ensure the adequacy of the SAR historical description. Use NUREG-0800, Section 2.4.7(II), for references in researching the history and potential for ice formation in the region. Ensure that the SAR properly considers all ice-related hazards, such as ice jam floods, wind-driven ice ridges, and ice-produced forces which could affect the site. If feasible, conduct an independent analysis of the ice flooding hazard by using independent data and assumptions.

2.5.4.8 Flood Protection Requirements

Compare the estimated DBF level (both SAR and any independent estimates) with the locations and elevations of safety-related components to confirm whether flood protection is necessary, and if so, to what levels. If flood protection is necessary, review the facility flood design basis for compatibility with the positions in Regulatory Guide 1.59. Use Regulatory Guide 1.102 for guidance on appropriate flood protection measures which must protect against both static and dynamic flooding effects. Review the SAR for flood protection measures based on standard engineering practice, such as that developed by the Army COE, in positive flood control and shoreline protection.

2.5.4.9 Environmental Acceptance of Effluents

Evaluate scenarios for routine and accidental releases to ensure consideration of worst-case releases of radionuclides into surface water or groundwater. Examine the physical parameters used in calculating the transport paths and times of liquid effluent between the release point and receptors downstream or downgradient. Confirm that mathematical models used to analyze flow and transport have been verified by field data and have used conservative input parameters. Site-specific data sources used in modeling the transport of radionuclides through water should be adequately described and referenced.

Use independent data and assumptions to the extent possible to assess the transport capabilities and potential contamination pathways of the surface water and groundwater environments. Focus this independent assessment on transport to existing and possible future water users under both normal and accident conditions. Use NUREG-0800, Section 2.4.13(III) for its descriptions of simplified calculational procedures for models used to assess effluent transport through surface water and groundwater.

2.5.5 Subsurface Hydrology

Review the descriptions of hydrogeologic units beneath the site. For each hydrogeologic unit, ensure proper representation of potentiometric level, hydraulic gradient and conductivity, effective porosity, storage coefficient, recharge and discharge areas, and potential for groundwater flow reversal. For the water table aquifer, ensure that seasonal fluctuations in the water level have been conservatively bounded. Compare the SAR chemical analyses, including major ions, pH-Eh values, and presence of radionuclides, with analyses obtained independently.

Review the information on existing groundwater use, such as withdrawal points, pumping rates, source aquifers, and drawdown. Use reports by the USGS or a State geological survey in reviewing site hydrogeology and water withdrawal downgradient of the site.

Review the analysis of the potential effects of the facility on any groundwater recharge areas within the site, including dewatering during construction. Ensure that this analysis uses conservative assumptions and input values. Confirm that estimated groundwater withdrawal volumes during facility operation are conservative and that drawdown or other effects on the aquifer(s) are addressed.

Review the transport characteristics of aquifers which are subject to radionuclide contamination. Ensure that contamination pathways are adequately described and that models and codes used to predict radionuclide migration are appropriate for the site. Ensure that potential future groundwater uses are conservatively estimated. If warranted, conduct an independent analysis of radionuclide migration by using an alternative transport model or independent data.

2.5.6 Geology and Seismology

The reviewer should use the methods stated below to perform the compliance review of the geology and seismology information in the SAR.

2.5.6.1 Basic Geologic and Seismic Information

Verify the documentation of the results from all independent surveys, geophysical studies, borings, trenches, and other investigations. Consider descriptions of techniques, graphic logs, photographs, laboratory results, and identification of principal investigators. Consider references to published reports, dissertations, and personal communications. Review both the reports cited in the SAR, as well as other relevant reports on local geology.

Review the SAR discussion of basic site characteristics which may be problematic in siting an ISFSI, such as high seismic activity or recent volcanic activity. Scrutinize any SAR statement that the presence of unstable geologic characteristics will not have a deleterious effect on the facility or that their effects are within the design bases of all facility components important to safety.

Examine the geologic maps, cross-sections, and stratigraphic columns provided in the SAR. For each lithologic unit, review the origin, unit thickness, physical characteristics, mineral composition, and degree of consolidation. Use the summary logs of borings, excavations, and trenches in reviewing lithology. Compare the geologic map for the site area with other available published maps. If the SAR interpretations differ substantially from published literature, ensure that the differences are noted and that the SAR interpretations are adequately justified. Review the bedrock contour map to confirm that all relevant structural features are accurately represented. Review the description of the site geomorphology to ensure that all significant landforms, including the geologic processes which engendered them, are properly described. Ensure that all locations of potential landsliding, subsidence, or uplift resulting from natural or anthropogenic processes have been identified, and that any associated hazards have been evaluated.

Review the results of any geophysical surveys, with particular attention to the methods by which the data were gathered. Compare the interpretations of stratigraphy and structures with other cross-sections. Require that discrepancies be explained. Examine any values of compressional and shear wave velocities for reasonableness.

Review the plan showing the locations of all major features of the facility, as well as the locations of all borings, trenches, and excavations. Examine the cross-sections showing the relationships of engineered structures to subsurface material. Ensure that the water table (and fluctuation range) is represented accurately and that groundwater can not have an adverse effect on these structures. Review the profile drawings showing the extent of excavation and backfill, as well as the compaction criteria for the engineered backfill. Ensure that compaction criteria meet appropriate engineering standards. Determine whether the SAR conservatively evaluates the effects of deformation zones such as shears, joints, fractures, faults, or folds on structural foundations. Ensure that alteration zones, irregular weathering profiles, and zones of structural weakness composed of crushed or disturbed materials have been addressed in terms of engineering geology.

Examine the tabulation of the static and dynamic engineering soil and rock properties of the various materials underlying the site, including grain size classification, Atterberg limits, water content, unit weight, shear strength, relative density, shear modulus, Poisson's ratio, bulk modulus, damping, consolidation characteristics, seismic wave velocities, density, porosity, strength characteristics, and strength under cyclic loading. Ensure that the data are substantiated with appropriate representative laboratory test records. Give extra attention to mechanical properties of aquifer materials and any fine-grained materials associated with the uppermost confined or semi-confined aquifer. Scrutinize any site materials which may have an adverse response to seismic shaking, as well as any rocks or soils which may be unstable because of their mineral composition, lack of consolidation, or water content. For those which may respond adversely to seismic shaking, ensure that conservative estimates are used for seismic response characteristics, such as liquefaction, thixotropy, differential consolidation, cratering, and fissuring. Review the SAR for inclusion of available data on the behavior of site geologic materials during previous earthquakes. Review the analytical techniques and safety factors used in evaluating the stability of foundations for all structures and embankments under normal operating and extreme environmental conditions.

2.5.6.2 Ground Vibration

Examine the maps of earthquake epicenters and faults in the region. Confirm that the epicenter map adequately represents the locations of the tabulated historical earthquakes. Ensure that the earthquake tabulation comes from a credible source; compare it to an alternate earthquake catalog if available. Confirm that sound practices are used in estimating the magnitudes of historical earthquakes which pre-dated seismological instrumentation. Consider differences in soil and bedrock properties between the site and the location where earthquake intensity was reported.

Review the descriptions of any capable faults, including length, relationship to regional tectonic structures and the regional stress regime, and the nature and amount of the maximum displacement per event during the Quaternary. Ensure that suitable methods, such as those outlined by Slemmons (1977), determined fault capability. Ensure that fault studies used photogeologic work and field investigations. Compare SAR findings to any alternative published interpretations. Review any justification of non-capability for any fault within 161 km (100 mi) of the site which, if it produced its maximum magnitude earthquake at its closest distance to the site, would produce site ground acceleration greater than or equal to the design value. Confirm that field investigations and conservative assumptions justify the classification of such fault as non-capable. Use trench excavations in determining capability if a fault is overlain by Late Pleistocene sediments.

Review the SAR calculation of the ground motion design basis value as defined by a response spectrum corresponding to the peak horizontal ground acceleration (PHA). A standardized design basis earthquake described by an appropriate response spectrum anchored at 0.25 g may be used for the site if it meets three criteria: 1) located east of the Rocky Mountain front, 2) not in a seismically active region (e.g., New Madrid, MO; Charleston, SC; or Attica, NY), and 3) not subject to ground motion above 0.2 g (per an appropriate response spectrum) as shown by reconnaissance investigation. Alternatively, for sites meeting the three criteria, the procedures of 10 CFR Part 100, Appendix A may be used to identify design basis values. Follow the procedures in 10 CFR Part 100, Appendix A, to derive the ground motion design basis value if the site does not meet these three criteria.

Review the ground motion value derived from Appendix A methods by using the following procedures. Ensure that all capable faults have been considered as seismic sources, with the maximum magnitude earthquake occurring on the fault at its nearest approach to the site. Ensure that the maximum magnitude event is based on an accepted fault length-magnitude relationship, such as Slemmons et al. (1982) or Bonilla et al. (1984). Use a widely accepted attenuation model such as Campbell and Bozorgnia (1994) to ensure that the peak ground acceleration at the site is calculated from the earthquake magnitude and the site-to-source distance. Ensure that the SAR analysis considered an FE, that it based the FE magnitude on the seismological history of the tectonic province, and that it used 15 km as the site-to-source distance for calculating ground acceleration at the site. Ensure that the SAR considered adjacent provinces and their characteristic FEs if the site is near a tectonic province boundary. Ensure that the site-to-source distance for a FE in an adjacent province is 15 km or the closest approach of the province to the site, whichever is greater. (Note: Reviewers should be aware of proposed changes to 10 CFR Part 100. The earthquake which engenders the greatest peak ground acceleration at the site is the design basis earthquake. Presently, Appendix A methodology assigns a site being evaluated a design basis earthquake (DBE) equal to the SSE for a nuclear power plant in the same location.)

Ensure that site-specific response spectrum used to derive PHA from the DBE considers the specific engineering properties of the material underlying the site, including seismic wave velocities, density, water content, porosity, and strength. Ensure that the design criteria in SAR Chapters 3 and 7 consider the design ground motion value.

2.5.6.3 Surface Faulting

Review the evaluation of tectonic structures underlying the site. Consider whether boreholes or geophysical surveys were used to reveal buried structures. Determine the need for geophysical or other studies to establish the presence or absence of such structures if local geology investigations provide some evidence that buried, potentially active structures may underlie the site. Ensure that the evaluation of onsite structures considers the effects of man's activities, such as mining activity, loading effects from dams or reservoirs, and pumping fluids out of or into the subsurface, and the proclivity of faults to slip. Confirm that all faults more than 300 m (1000 ft) long and passing within 8 km (5 mi) of the site have had their capability assessed. Examine these assessments to ensure that the conclusions are based on sound geologic principles and practices, and in cases where capability remains equivocal, a preponderance of the available geologic evidence. Review the information provided on fault length and relationship to regional tectonic structures, the nature and amount of Quaternary displacement, and the magnitude of the maximum Quaternary displacement event for those faults which are deemed capable. Ensure that the outer limits of the fault or fault zone have been identified along the trace 16 km (10 mi) in either direction of the point where the fault makes its closest approach to the site. Ensure that any fault displacement, if the site is subject to surface faulting, does not exceed the design criteria. Require a large safety margin if critical facilities are to be located in areas subject to displacement because fault displacement is a difficult phenomenon to assess.

2.5.6.4 Stability of Subsurface Materials

Review the description of geologic features to ensure that no natural features which could affect foundation stability during ground shaking have been overlooked. Examine the tabulations of the physical and engineering properties for the foundation materials underlying the site. Ensure that foundation material properties include grain size classification, consolidation characteristics, water content, Atterberg limits, unit weight, shear strength, relative density, shear modulus, damping, Poisson's ratio, bulk modulus, strength under cyclic loading, seismic wave velocities, density, porosity, and strength characteristics. Compare selected values against representative laboratory test results to confirm the accuracy of the values of selected properties.

Examine the plans and profiles of the locations of investigative studies and facility structures. Confirm that the plans include all appropriate boreholes, trenches, etc. Ensure that the profiles accurately show the relationships between structure foundations and subsurface materials and the groundwater and engineering characteristics of the subsurface materials. Review the plans and profiles which show excavation and backfill activity to ensure that compaction criteria are substantiated with representative laboratory or field test records. Examine the tables and profiles of the compressional and shear wave velocities in the soil and rock beneath the site. Ensure that these data were gathered by appropriate methods. Examine any graphic logs of boreholes, trenches, or other excavations for accuracy. Ensure that the analyses of the soil and rock responses to dynamic loading are conservative.

Review the discussion of liquefaction potential of material beneath the site. Conduct an independent analysis to verify a claim that liquefaction-susceptible soils are absent beneath the site. The reviewer should ensure that the discussion of soil zones with the potential for liquefaction includes relative density, void ratio, ratio of shear stress to initial effective stress, number of load cycles, grain size distribution, degrees of cementation and cohesion, and groundwater elevation fluctuations.

Ensure that the analysis for soil stability uses the appropriate response spectra in determining the design ground motion from the DBE. Ensure that the static analyses address settlement and lateral pressures and are accompanied by representative laboratory data. Review the specifications for any techniques, such as grouting, vibraflotation, rock bolting, or anchors, required to improve unstable subsurface conditions. Ensure that designs follow proper engineering standards. Examine the safety factors and the criteria, references, or methods of design used in ensuring that the facility can withstand seismic ground motion and surface faulting.

2.5.6.5 Slope Stability

Examine the slope cross-section drawings for accuracy. Review the static and dynamic properties of the embankment and foundation soil and rock beneath the slope to ensure that the values are reasonable and substantiated with representative laboratory test data. Ensure that stability assessments address the potential effects of erosion, deposition, and seismicity, either individually or in combination. Ensure that erosional processes discuss sheet and rill flow, mass wasting, and valley widening. Ensure that the compaction specifications are based on representative laboratory analyses. Review the logs of core borings and test pits taken in these areas for any proposed borrow areas. Ensure that the analyses supporting the slope and erosional stability findings use conservative methods and assumptions.

2.6 Evaluation Findings

Prepare evaluation findings on compliance with SRP Section 2.1 review objectives and Section 2.3 regulatory requirements on site characteristics. Use the following statements (or similarly phrased statements) of findings if the documentation submitted with the application, including the SAR and the ER, fully supports positive findings for each of the regulatory requirements (finding numbering is for convenience in referencing within the FSRP and SER):

F2.1   The SAR provides an acceptable description and safety assessment of the site on which the [ISFSI/MRS] is to be located, in accordance with 10 CFR 72.24(a).

F2.2   The proposed site complies with the criteria of 10 CFR 72 Subpart E, as required by 10 CFR 72.40(a)(2).

Communicate to the NRC project manager the inadequacies in the site characterization, the reasons for an inability to make fully positive findings, and the additional information, analyses, or design changes which must be provided to NRC before the review can continue.

2.7 References

NRC documents referenced in this section are identified in Section 17, Consolidated References.

American National Standards Institute/American Nuclear Society (ANSI/ANS), 2.8-1981, "Determining Design-Basis Flooding at Power Reactor Sites."

Bonilla, M.G., R.K. Mark, and J.J. Lienkaemper, "Statistical Relations among Earthquake Magnitude, Surface Rupture Length, and Surface Fault Displacement," Bulletin of the Seismological Society of America, vol. 74, pp. 2379-2411, 1984.

Campbell, K.W., and Y. Bozorgnia, "Near Source Attenuation Peak Horizontal Acceleration From Worldwide Accelerograms Recorded from 1975 to 1993," Fifth U.S. National Conference on Earthquake Engineering, Chicago, IL, July 10-14, 1994.

National Oceanic and Atmospheric Administration, Technical Report NWS-23, "Meteorological Criteria for the Standard Project Hurricane and Probable Maximum Hurricane Windfields, Gulf and East Coasts of the United States."

Slemmons, D.B., "State-of-the-Art for Assessing Earthquake hazards in the United States: Report 6, Faults and Earthquake Magnitude," Miscellaneous Paper S-73-1, U.S. Army Engineer Waterways Experiment Station, Corps of Engineers, Vicksburg, Mississippi, 1977.

Slemmons, D.B., P. O'Malley, R.A. Whitney, D.H. Chung, and D.L. Bernreuter, "Assessment of Active Faults for Maximum Credible Earthquakes of the Southern California-Northern Baja Region," University of California, Lawrence Livermore National Laboratory publication no. UCID 19125, p. 48, 1982.

3 OPERATION SYSTEMS

3.1 Review Objective

The objective of this chapter is to evaluate for clarity and completeness the description of all operations, including systems, equipment, and instrumentation, particularly as they relate to handling and storage of spent fuel or solidified high-level waste, confinement of nuclear material, and management of expected and potential radiological dose. Sufficient detail should be provided to ensure that reviewers can understand the operations and the operations' effects on the design evaluations. Safety features required to maintain the installation in a safe condition should be described; however, evaluation of those features should be performed in the appropriate technical sections.

Figure 3.1 presents an overview of the operation systems evaluation process. The figure shows that this review process draws on information in the application and the regulatory requirements.

3.2 Areas of Review

The following outline shows the areas of review addressed in Section 3.4, Acceptance Criteria, and Section 3.5, Review Procedures:

Operation Description
Spent Fuel and High-Level Waste Handling Systems
Other Operating Systems
Operation Support Systems
Control Room and Control Area
Analytical Sampling
Shipping Cask Repair and Maintenance
Pool and Pool Facility Systems

3.3 Regulatory Requirements

This section identifies and presents a high-level summary of Title 10 of the Code of Federal Regulations (CFR) Part 72 relevant to the review areas addressed by this chapter. The NRC staff reviewer should read the exact regulatory language. A matrix at the end of this section matches the regulatory requirements identified in this section to the areas of review identified in the previous section.

72.24 Contents of application: Technical information [Contents of SAR]
(b) "A description and discussion of the ISFSI or MRS structures."
(f) "Features of ISFSI or MRS design and operating modes to reduce ... radioactive waste volumes."

> (l) "A description of the equipment ... to maintain control over radioactive materials in gaseous and liquid effluent."

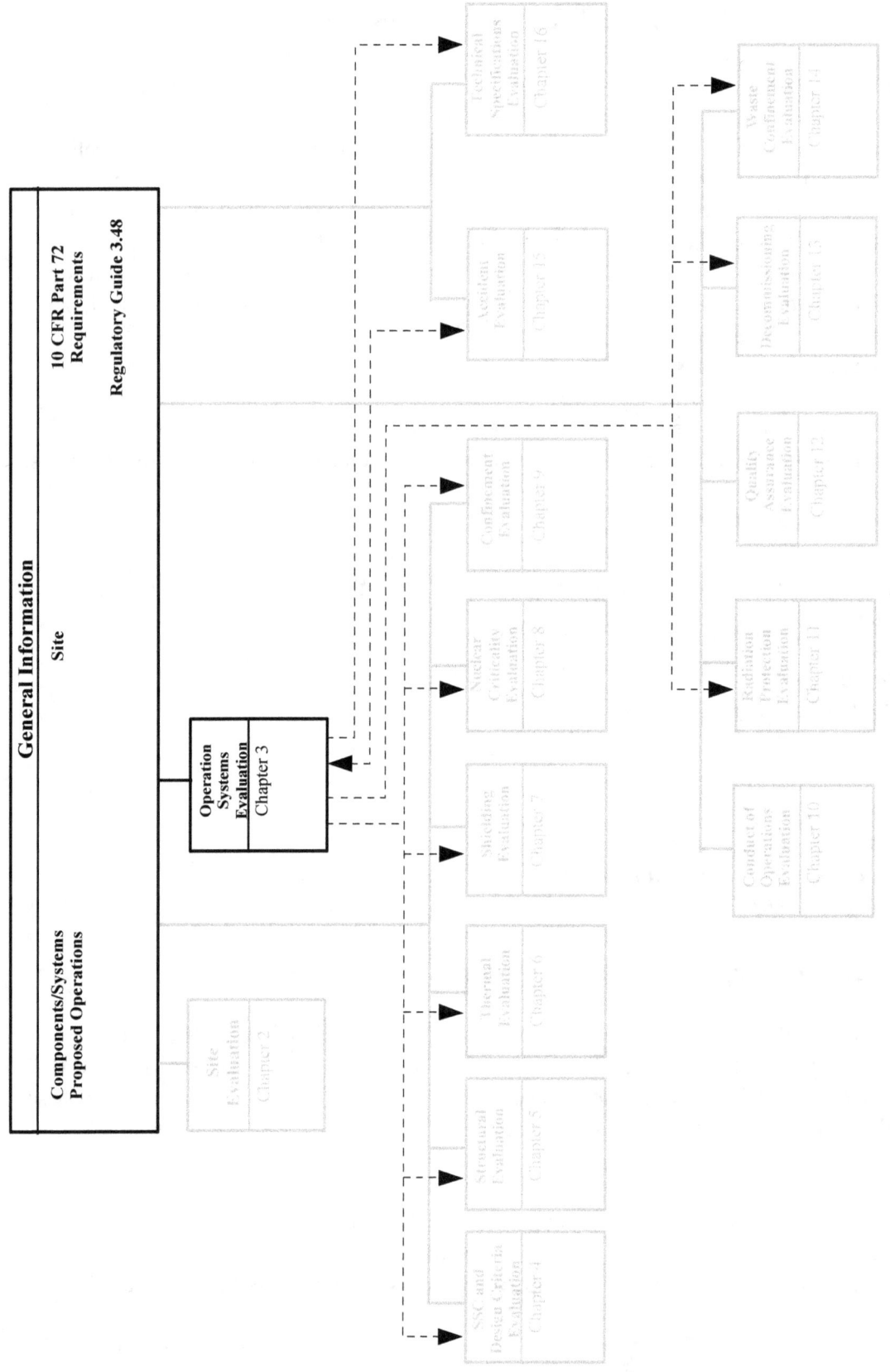

Figure 3.1 Overview of Operational Systems Evaluation

72.40 Issuance of license.

(a) "The application ... meets the ... requirements ... and..."
  (5) "Proposed operating procedures ... protect health and minimize danger."
  (13) "There is reasonable assurance that:
    (i) The activities ...can be conducted without endangering the health and safety ... and
    (ii) these activities will be conducted in compliance with the applicable regulations."

72.44 License Conditions.
(c) "Technical specifications must include..."
(1) "Functional and operating limits and monitoring instruments and limiting control settings."
(2) "Limiting conditions."
(3) "Surveillance requirements."

72.104 Criteria for radioactive materials in effluents and direct radiation...
(b) "Operational restrictions must be established to meet [ALARA] objectives..."
(c) "Operational restrictions must be established for radioactive materials in effluents and direct radiation levels..."

72.122 Overall requirements.
(f) "Testing and maintenance of systems and components."
(h) "Confinement barriers and systems."
(i) "Instrumentation and control systems."
(j) "Control room or control area of the ISFSI or MRS under off-normal or accident conditions."
(k) "Utility or other services."
(l) "Retrievability."

72.124 Criteria for nuclear criticality safety.
(c) "Criticality Monitoring."

72.126 Criteria for radiological protection.
(b) "Radiological alarm systems."
(c) "Effluent and direct radiation monitoring."
(d) "Effluent control."

72.128 Criteria for spent fuel, high-level radioactive waste, and other radioactive waste storage and handling.
(a) "Spent fuel and high-level radioactive waste storage and handling systems."
(1) "A capability to test and monitor components important to safety."
(2) "Confinement structures and systems."

72.150 Instructions, procedures, and drawings.

A matrix which shows the primary relationship of these regulations to the specific areas of review associated with this Standard Review Plan (SRP) chapter is given in Table 3.1. The NRC staff reviewer should verify the association of regulatory requirements with the areas of review presented in the matrix to ensure that no requirements are overlooked as a result of unique applicant design features.

3.4 Acceptance Criteria

This section identifies the acceptance criteria used for the operation systems review. Information on systems may be fully described functionally at the Safety Analysis Report (SAR) chapters oriented on physical design and specific safety-related functions (such as installation and structural design, thermal, criticality, and confinement), since detailed information need only be included once in the SAR. The primary purpose of this chapter is a review of the functional description of the systems operations, flowsheets showing sequences of operations and controls, and drawings showing proper functioning of each system. Additional description of the information that should be in the SAR is provided for each of the review areas.

3.4.1 Operation Description

Operation description relates to the overall storage functions and operation of the installation. The applicant should provide an overview of operations. Acceptable criteria for operation system descriptions are given in NUREG-1536, Chapter 8, Section IV, items 1 through 6.

3.4.2 Spent Fuel and High-Level Waste Handling Systems

The regulatory requirements given in 10 CFR 72.124, 10 CFR 72.128, 10 CFR 72.150, and 10 CFR 72.166 address the information to be included in a license application. The SAR should include information as described in Regulatory Guide 3.48 Section 5.2 on spent fuel (and high-level waste if for an MRS) handling systems. The descriptions of the spent fuel or high-level waste handling systems must be clear. The functions of transfer from transportation vehicles, receipt inspection, and initial decontamination should be addressed if the operations are performed independently of a 10 CFR 50 license review. The transfer facility and its use should be described, including its use during the stages of operation of the ISFSI. Spent fuel and high-level waste handling systems in a pool facility used for wet transfer is addressed in a following section.

3.4.3 Other Operating Systems

The scope of this section is taken to be all operating systems important to safety that are not covered in Sections 3.4.1 (Operation Description) and 3.4.2 (Spent Fuel and High-Level Waste Handling Systems) except that instrumentation and controls are covered in 3.4.4 and analytical sampling is covered in 3.4.6. "Other operating systems" and "auxiliary systems" that are

Table 3.1 Relationship of Regulations and Areas of Review

Areas of Review	10 CFR Part 72 Regulations									
	72.24	72.40	72.44	72.104	72.122	72.124	72.126	72.128	72.150	72.166
Operation Description	●	●	●	●	●	●	●	●	●	
Spent Fuel and High-Level Waste Handling Systems	●			●		●		●	●	●
Other Operating Systems	●		●	●	●	●	●	●	●	●
Operation Support Systems	●		●		●					
Control Room and Control Area					●					
Analytical Sampling	●		●		●					
Shipping Cask Repair and Maintenance								●		
Pool and Pool Facility Systems	●		●	●	●	●	●	●	●	

important to safety should be as described in Regulatory Guide 3.48 Sections 4.3 and 5.3 and noted in the narrative descriptions or flowcharts describing the operation of the ISFSI. 10 CFR 72.122 requires that the SAR include clear descriptions of the systems and system equipment and controls used to assure safety. These items must be consistent with other parts of the SAR.

Examples of "other operating systems" that may be classified as important to safety include ventilation and off-gas systems, electrical systems, air supply systems, steam supply and distribution systems, water supply systems, fire protection systems, air sampling systems, decontamination systems, and systems related to chemical hazards.

3.4.4 Operation Support Systems

10 CFR 72.122 requires that the SAR include information on operation support systems, primarily instrumentation and control (I&C) systems and component spares or alternative equipment. These items should be as described in Regulatory Guide 3.48 Section 5.4. This information should include an analysis or other acceptable basis for determination that operation support systems important to safety remain operational under accident-level conditions. The SAR should include clear descriptions of the operation support systems and descriptions of equipment and controls used to assure safety, which are consistent with other parts of the SAR.

3.4.5 Control Room and Control Area

10 CFR 72.122 requires that the SAR include a discussion of how a control room and control room areas permit the installation to operate safely under normal, off-normal, and accident conditions. The SAR should include clear descriptions of the control room and control area.

The NRC has accepted omission of a control room for ISFSI operations that have not involved control of operations within a pool or use of a powered cooling system for material in storage. A control room and redundancy for control of functions important to safety in a separate control area is acceptable for ISFSI with pool facilities.

3.4.6 Analytical Sampling

The SAR should include a discussion of the provisions for obtaining samples for analysis necessary to ensure that the ISFSI is operating within prescribed limits. The SAR should include a description of the facilities and equipment available to perform the required tests.

3.4.7 Shipping Cask Repair and Maintenance

The SAR should contain a description of the shipping cask repair and maintenance facilities. The operation of these facilities, including provision for contamination control and occupational exposure minimization, should also be included. Note that the ownership, maintenance, and use of a shipping cask for shipping nuclear material by an ISFSI or MRS licensee is governed by the requirements of 10 CFR 71 only.

3.4.8 Pool and Pool Facility Systems

For ISFSI or MRS with a pool, the pool facility and the associated equipment constitute the principal capability for handling the subject radioactive material outside its storage confinement barrier or with that barrier open. The SAR should include clear descriptions of the pool and pool facility systems and descriptions of pool facility equipment and controls used to assure safety, which are consistent with other parts of the SAR. Section 9.1.2 of NUREG-0800 presents pool and pool facility systems requirements for a 10 CFR 50 license and should be used as guidance in the design of a 10 CFR 72 facility. Because a pool facility used only for wet transfer presents unique requirements, specific criteria are not provided.

The NRC accepts pool facilities for licensing under 10 CFR 72 if those facilities meet the requirements for nuclear power plants licensed under 10 CFR 50.

3.5 Review Procedures

The following provides review guidance relevant to the operation systems evaluation. This review is oriented on functions and the compatibility of proposed systems with performance of those functions. Since the NRC does not review and approve procedures, the review of the descriptions of functions constitutes the principal basis for assessing the assurance provided by the submitted documentation. Reviews in other FSRP sections determine quantitative functional performance for functional and structural performances.

3.5.1 General Operating Functions

Review the description of operating functions for completeness. The reviewer should compare the functions with descriptions included in other licensing documentation to confirm acceptability. If a previously certified cask design is used, the functions described in the SAR under review should be checked for compatibility with those functions that were included in the SAR for the certified cask.

Acceptance of the description of general operating functions can be based on information provided in the flowsheets and narrative descriptions of steps. The reviewer should ensure that the applicant has fully described the appropriate procedures, equipment involved, and personnel requirements. Review procedures can be found in the next-to-the-last paragraph of the NUREG-1536, Chapter 8, Section V introduction which precedes subsection 1 (Cask Loading).

3.5.2 Spent Fuel and High-Level Waste Handling Systems

Review procedures for spent fuel handling systems are given in the NUREG-1536, Chapter 8, Section V, items 1, 2, and 3. A review for handling high-level waste follows the same procedure. Because the spent fuel and high-level waster handling systems have many interfaces with other systems of the facility, verify that these interfaces are addressed and that continuity of operations can occur under all operational conditions.

3.5.3 Other Operating Systems

For systems that are important to safety, review the description of the location of the various systems in relationship to their functional objectives. Verify that provisions for coping with unscheduled occurrences have been described so that a single failure within one of the auxiliary systems will not result in a release of radioactive material. The reviewer should evaluate the systems to ensure that the design includes performance under normal operating loads, loading situations resulting from primary failure and/or accident conditions, and loading situations required for the safety of a shutdown operation. If a system requires a technical specification, verify that it has been included in the SAR.

3.5.4 Operation Support Systems

Review the descriptions of the I&C systems for adequacy of definition of their function. Systems that are important to safety should describe all major components, operating characteristics, locations of sensors and alarms, threshold levels for I&C that produce alarms, automatic and manual control actions to be triggered, and safety criteria. The NRC has accepted omission of instrumentation and monitoring for passively cooled welded-closure storage casks if a periodic check for air cooling effectiveness is included as a technical specification.

Consider the projected accident-level and off-normal events (addressed at FSRP Chapter 15) and the roles that the I&C has in avoiding or mitigating significant radiological consequences of those events. Verify that consideration has been given to the redundancy required to ensure safe operation or safe curtailment of operations under accident conditions. Verify that spare or alternative instrumentation, if provided, has been designed to ensure safe functioning.

Review proposed technical specifications that include reliance on an I&C system performance.

3.5.5 Control Room and Control Area

Review the control room and control area functions, equipment, instrumentation and control links, and staffing for consistency and appropriateness for the intended functional control and safety roles. Information on these different aspects of the control room/area may be at various locations within the SAR.

The explanation for omission of a control (and/or monitoring) room/area should be provided. For example, explanations may include: a description of functions and procedures (flowsheets and narrative descriptions) that provide for performance without need for a centralized control room, the acceptability of accident-level and off-normal event and condition analyses that show acceptable levels of maximum response and safety without use of a control room, and/or the desirability that damage avoidance and mitigation be based on passive measures to the extent feasible.

3.5.6 Analytical Sampling

The review of the analytical sampling operation should verify that the types of samples and rate of sampling are appropriate for the condition being monitored. Provisions should be included for obtaining samples during off-normal conditions to ensure that prescribed limits have not been exceeded. The SAR should describe the facilities and equipment that will be available to perform the analyses. Disposition of laboratory wastes should also be described.

The review should compare the proposed analytical sampling operations with those of existing facilities for reasonableness as documented in FSAR that cover similar facilities and prior license applications.

3.5.7 Shipping Cask Repair and Maintenance

The principal concern for review of any shipping cask repair capability incorporated into the ISFSI or MRS is that the applicant recognizes the need for receiving and inspection of loaded shipping casks and shipping cask decontamination. If a repair capability is to be provided on-site for repair of storage confinement and on-site transfer casks, the skills and equipment necessary for shipping cask repair will probably also exist.

The status of the pool facility may be one of several possibilities: (1) the pool facility may be a new facility to be licensed under 10 CFR 72, (2) the pool facility may be licensed under a 10 CFR 50 license which is being terminated and a transfer to 10 CFR 72 is being requested, or (3) the pool facility may exist under DOE ownership, and NRC licensing under 10 CFR 72 is being requested.

If the pool facilities are to be built, the functions and various systems should reflect NRC and industry standards for pool facility design and use. The reviewer should compare the descriptions of functions and the structures, systems, and components (SSCs) to support those functions with descriptions of existing pool facilities and associated FSARs. Bases for significant discrepancies between functions and/or equipment used to perform the various functions should be provided.

If the review is to include approval of an existing licensed pool facility, the reviewer should review and compare descriptions of functions and component SSCs in the FSAR, NRC staff and field inspector reports on operations of the pool facility, and any finding associated with the facility or with the proposed use of the pool facility.

If the pool facility exists but is unlicensed, review the suitability of the design and systems for the proposed functions and the acceptability of the described functions to permit licensing under 10 CFR 72. Review proposed modifications of the facility with the reviewer for structural evaluation. Compare the proposed functions and supporting SSCs with pool facilities used at nuclear power plants to assist confirmation that the design and functions are compatible.

3.6 Evaluation Findings

NRC staff reviewers prepare evaluation findings regarding satisfaction of the regulatory requirements related to operations systems. If the documentation submitted with the application fully supports positive findings for each of the regulatory requirements, then the findings should substantially be stated as follows (finding numbering is for convenience in referencing within the SRP and Safety Evaluation Report [SER]):

F3.1   [If applicable] The [ISFSI/MRS] is to be located on the same site as another facility licensed by the NRC. Potential interactions between these facilities and the [ISFSI/MRS] have been evaluated, in accordance with 10 CFR 72.24(a) and 72.40(a)(3) and have been determined to be acceptable and pose no undue risk to any of the facilities.

F3.2   The SAR includes acceptable descriptions and discussions of the projected operating characteristics and safety considerations, in compliance with 10 CFR 72.24(b).

F3.3   The SAR provides reasonable assurance that the activities to be authorized by the license can be conducted without endangering the health and safety of the public and will be in compliance with the applicable regulations of 10 CFR 72.40(a)(13).

F3.4   [One of the following, as appropriate] The design of the [ISFSI/MRS] provides for an acceptable [control room/control area] as part of the facilities to be built, in compliance with 10 CFR 72.122(j). - or -

The operating procedures and schedule for operations for the [ISFSI/MRS] acceptably provides for control during storage operations to be accomplished from the security/monitoring/surveillance office facility and for control during loading, transfer, and unloading operations to be from temporary control facilities, for which there are acceptable provisions included in the design. This is considered to acceptably comply with 10 CFR 72.122(j). - or -

The [ISFSI/MRS] is to be located on a site with existing facilities suitable and available for control of [ISFSI/MRS] operations under off-normal or accident conditions, whose use will not interfere with other operations on the site important to safety, in compliance with 10 CFR 72.40(a)(3) and 72.122(j).

F3.5   The proposed [ISFSI/MRS] facilities include the following utility service systems [identify]. [If appropriate] The following utility service systems are important to safety: [identify]. The [ISFSI/MRS] design provides for redundant systems to the extent necessary to maintain, with adequate capacity, the ability to perform safety functions assuming a single failure; in compliance with 10 CFR 72.122(k)(1).

F3.6   The proposed design of the [ISFSI/MRS] emergency utility services acceptably permits testing of the functional operability and capacity of each system and permits operation of associated safety systems, in compliance with 10 CFR 72.122 (k)(2).

F3.7   The proposed design of the [ISFSI/MRS] includes the following systems and subsystems which require continuous electric power to permit continued functioning of all systems essential to safe storage: _____ [identify]. The design of the [ISFSI/MRS] acceptably provides for timely emergency power for these systems and subsystems, in compliance with 10 CFR 72.122(k)(3).

F3.8   The descriptions of the proposed [ISFSI/MRS] functions and operating systems with regard to retrieval of stored radioactive material from storage, in normal, off-normal, and accident conditions are acceptable and comply with 10 CFR 72.122(l).

F3.9   Acceptable capability to test and monitor components important to safety are provided in the design and procedures for the [ISFSI/MRS], in compliance with 10 CFR 72.128(a)(1).

If the design of the confinement cask system has been previously certified under 10 CFR 72 Subpart L, the following evaluation finding statements would also be appropriate:

F3.10   The proposed [ISFSI/MRS] uses a cask system that has been previously certified by the NRC.

3.7 References

NRC documents referenced are identified at Consolidated References, Chapter 17.

4 SSC AND DESIGN CRITERIA EVALUATION

4.1 Review Objective

The objective of the review is to ensure that the applicant acceptability defines: (1) the limiting characteristics of the spent fuel or other high-level radioactive waste materials to be stored, (2) the classification of structures, systems and components (SSCs) according to their importance to safety, and (3) the design criteria and design bases, including the external conditions during normal and off-normal operations, accident conditions, and natural phenomena events.

Figure 4.1 presents an overview of the SSC and design criteria evaluation process. The figure shows that this review process draws on information in the application and the regulatory requirements.

4.2 Areas of Review

The following outline shows the areas of review addressed in Section 4.4, Acceptance Criteria, and Section 4.5, Review Procedures:

Materials to be Stored
> Spent Fuel
> High-Level Radioactive Waste

Classification of Structures, Systems, and Components

Design Criteria for SSCs Important to Safety
> General
> Structural
> Thermal
> Shielding and Confinement
> Criticality
> Decommissioning
> Retrieval

Design Criteria for Other SSCs

4.3 Regulatory Requirements

This section identifies and presents a high-level summary of Title 10 of the Code of Federal Regulations (CFR) Part 72 relevant to the review areas addressed by this chapter. The NRC staff reviewer should read the exact regulatory language. A matrix at the end of this section matches the regulatory requirements identified in this section to the areas of review identified in the previous section.

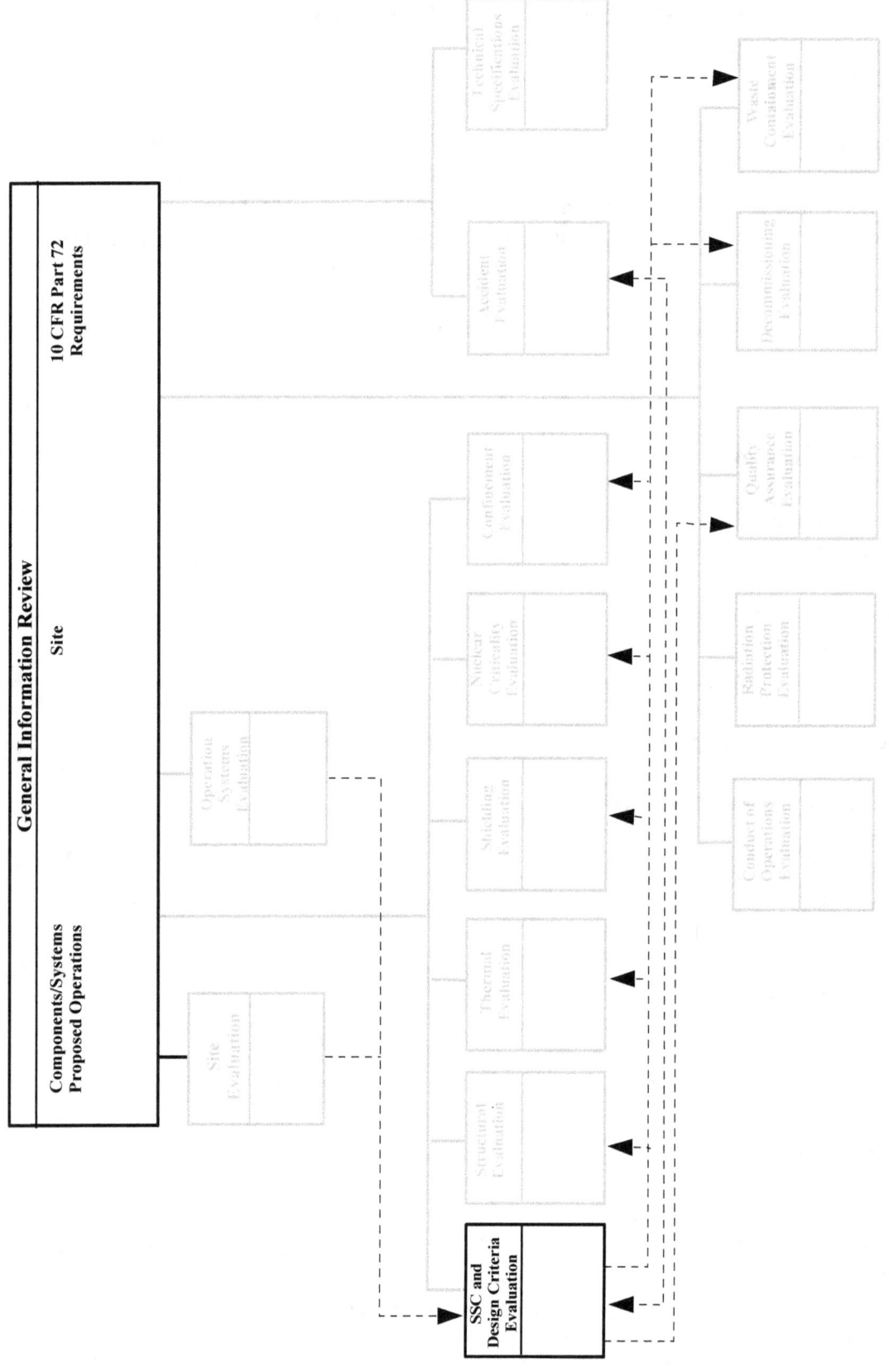

Figure 4.1 Overview of SSC and Design Criteria Evaluation

72.2 Scope

(a) "Except as provided in Section 72.6(b), licenses issued under this part"
    (1) "Power reactor spent fuel to be stored in a complex"
    (2) "Power reactor spent fuel to be stored in an [Monitored Retrievable Storage] MRS owned by DOE"

72.3 Definitions
Structures, systems, and components important to safety
    (1) "To maintain the conditions required to store spent fuel or high-level..."
    (2) "To prevent damage to the spent fuel or high-level radioactive waste..."
    (3) "To provide reasonable assurance that the spent fuel..."

72.6 License required; types of licenses
(b) "A general license...own spent fuel or high-level radioactive waste without regard"

72.24 Contents of application: Technical information
(c) "The design of the [Independent Spent Fuel Storage Installations] ISFSI or MRS in sufficient detail ... including:"
    (1) "The design criteria"
    (2) "The design bases and the relation of the design bases to design criteria"
    (4) "Applicable codes and standards."
(n) "The description must identify the structures, systems and components important to safety"

72.102 Geological and seismological characteristics
(a) "East of the Rocky Mountain Front"
(b) "West of the Rocky Mountain Front"
(c) "Sites other that bedrock sites must be evaluated for their liquefaction"
(d) "Site-specific investigations and laboratory analyses must show that the soil conditions"
(e) "In an evaluation of alternative sites, those which require a minimum of engineered"
(f) "The design basis earthquake (DE) for use in the design of structures must be"

72.104 Criteria for radioactive materials in effluents and direct radiation from an ISFSI or MRS
(a) "During normal operations and anticipated occurrences, the annual dose"
(b) "Operational restrictions must be established to meet as low as is reasonable"
(c) "Operational limits must be established for radioactive materials in effluents"

72.106 Controlled area of an ISFSI or MRS
(a) "For each ISFSI or MRS site, a controlled area must be established."
(b) "Any individual located on or beyond the nearest boundary of the controlled area"
(c) "The controlled area may be traversed by a highway,"

72.120 General considerations
(a) "Pursuant to ... must include the design criteria for the proposed storage installation"
(b) "The MRS must be designed to store either spent fuel or solid high-level radioactive wastes"

72.122 Overall requirements

(a) Quality Standards. "SSCs important to safety must be designed, fabricated"(b) Protection against environmental conditions and natural phenomena.

    (1) "SSCs important to safety must be designed to ...postulated accidents."

    (2) "SSCs important to safety must be designed to ... natural phenomena"

        (i) "Appropriate consideration of the most severe of the natural phenomena"

        (ii) "Appropriate combinations of the effects of normal and accident conditions"

    "The ISFSI or MRS should also be designed to prevent massive collapse of building"

    (4) "If the ISFSI or MRS is located over an aquifer"

(c) Protection against fires and explosions. "SSCs important to safety must be designed"

(d) Sharing of SSCs. "SSCs important to safety must not be shared"

(e) Proximity of sites. "An ISFSI or MRS located near other nuclear facilities"

(f) Testing and maintenance of system and components. "Systems... to permit inspection"

(g) Emergency capability. "SSCs important to safety must be designed for emergencies"

(h) Confinement barriers and systems.

    (1) "The spent fuel cladding must be protected"

    (2) "For underwater storage of spent fuel"

    (3) "Ventilation systems and off-gas systems"

    (4) "Storage confinement systems must have...continuous monitoring"

    (5) "The high-level radioactive waste...that allows handling and retrievability"

(i) Instrumentation and control systems. "Instrumentation and control systems...to monitor systems"

(j) Control room or control area. "A control room or control area, if appropriate"

(k) Utility or other services.

    (1) "Each utility service system must be designed to meet emergency conditions"

    (2) "Emergency utility services must be designed to permit testing"

    (3) "Provisions must be made... in the event of a loss of the primary electric power"

    (4) "An ISFSI or MRS...may share common utilities and services"

(1) Retrievability. "Storage systems must be designed to allow ready retrieval"

72.124 Criteria for nuclear criticality safety

(a) Design for criticality safety. "Spent fuel handling,...be maintained subcritical"

(b) Methods of criticality control. "... the design of an ISFSI or MRS must be based"

(c) Criticality Monitoring. "A criticality monitoring system shall be maintained"

72.126 Criteria for radiological protection.

(a) Exposure control. "Radiation protection systems must be provided for all areas"

    (1) "Prevent the accumulation of radioactive material"

    (2) "Decontaminate those systems to which access is required"

    (3) "Control access to areas of potential contamination"

    (4) "Measure and control contamination"

    (5) "Minimize the time required to perform work in the vicinity of radioactive"

    (6) "Shield personnel from radiation exposure."

(b) Radiological alarm systems. "Radiological alarm systems must be provided"

(c) Effluent and direct radiation monitoring

      (1) "As appropriate for the handling and storage"

      (2) "Areas containing radioactive materials"

(d) Effluent control. "The ISFSI or MRS must be designed to provide"

72.128 Criteria for spent fuel, high-level radioactive waste, and other radioactive waste storage and handling.

(a) Spent fuel and high-level radioactive waste storage and handling systems. "Spent fuel storage"

      (1) "A capability to test and monitor components"

      (2) "Suitable shielding for radioactive protection under normal and accident conditions"

      (3) "Confinement structures and systems"

      (4) "A heat-removal capability having testability"

      (5) "Means to minimize the quantity of radioactive wastes generated."

(b) Waste treatment. " Radioactive waste treatment facilities must be provided"

72.130. Criteria for decommissioning

"The ISFSI or MRS must be designed for decommissioning"

72.144 Quality assurance program

(a) "...licensee shall identify the SSCs to be covered by the quality assurance program"

(c) "The licensee shall base the requirements...on the following considerations"

      (1) "The impact of malfunction or failure of the item on safety;"

      (2) "The design and fabrication complexity"

      (3) "The need for special controls"

      (4) "The degree to which functional compliance can be demonstrated by inspection"

      (5) "The quality history and degree of standardization of the item."

72.182 Design for physical protection

"The design for physical protection must show the site layout and the design features..."

(a) "The design criteria for the physical protection of the proposed ISFSI or MRS;"

(b) "The design bases and the relation of the design bases to the design criteria..."

(c) "Information relative to materials of construction..."

72.236 Specific requirements for spent fuel storage cask approval

(a) "Specification must be provided for the spent fuel..."

(b) "Design bases and design criteria..."

(c) "Spent fuel is maintained in a subcritical condition."

(d) "Radiation sheilding and confinement must be provided."

(e) "The cask must be designed to provide redundant sealing of confinement systems."

(f) "The cask must be designed to provide adequate heat removal capacity without active cooling...."

(g) "The cask must be designed to store the spent fuel safely for a minimum of 20 years...."

(k) "The cask must be conspicuously and durably marked with:..."

(l) "The cask...must be evaluated...to demonstrate...confinement...under normal, off-normal, and credible accident conditions."

A matrix that shows the primary relationship of these regulations to the specific areas of review associated with this FSRP chapter is given in Table 4.1. The NRC staff reviewer should verify the association of regulatory requirements with the areas of review presented in the matrix to ensure that no requirements are overlooked as a result of unique applicant design features.

Table 4.1 Relationship of Regulations and Areas of Review

Areas of Review	10 CFR Part 72 Regulations							
	72.2	72.3	72.6	72.24	72.102	72.104	72.106	72.120
Material to be Stored	●	●	●					●
Classification of SSCs		●		●				
Design Criteria of SSCs Important to Safety				●	●	●	●	●
Design Criteria of SSCs				●				

Table 4.1 Relationship of Regulations and Areas of Review (continued)

Areas of Review	10 CFR Part 72 Regulations							
	72.122	72.124	72.126	72.128	72.130	72.134	72.182	72.236
Material to be Stored								●
Classification of SSCs						●		
Design Criteria of SSCs Important to Safety	●	●	●	●	●	●	●	●
Design Criteria of SSCs								

4.4 Acceptance Criteria

This section identifies the acceptance criteria used for the various review areas. The acceptance criteria are based on regulatory requirements, Regulatory Guides, and staff judgments.

4.4.1 Materials To Be Stored

4.4.1.1 Spent Fuel

The regulatory requirements given in 10 CFR 72.2 (a)(1) and (a)(2) identify power reactor spent fuel as material to be stored. 10 CFR 72.6 (b) states that the general license to store spent fuel or high-level radioactive waste may be issued without regard to quantity. 10 CFR 72.120 (b) discusses the acceptable form, i.e., solid fuel or high-level radioactive waste. The applicant must provide information on the spent fuel to be stored including, but not limited to, reactor type (e.g., Boiling Water Reactor, Pressurized Water Reactor, etc.), fuel manufacturer and model designation and number, fuel physical characteristics, fuel cladding material, thermal characteristics, radionuclide characteristics (e.g., gamma and neutron source terms), and history and census, including burnup, initial enrichment, and cooling time. The applicant must also provide information on the ranges of parameters of the spent fuel to be stored. Bounding parameters for further fuel storage should be listed.

In the SAR, the applicant must specify if damaged fuel is to be stored at the ISFSI. Damaged fuel should be canned for storage and transportation. The purpose of canning is to confine gross fuel particles to a known, subcritical volume during off-normal and accident conditions, and to facilitate handling and retrievability. As proof that the fuel is undamaged, the applicant, at a minimum, should review the fuel records and verify that the fuel was undamaged. Also, the applicant should specify that prior to loading, the fuel assemblies will receive an external visual examination for any obvious damage. For fuel assemblies where reactor records are not available, the applicant should provide alternate information which provides reasonable assurance that the fuel is undamaged or that damaged fuel loaded in a storage or transportation cask is canned in addition to the external visual examination for any obvious damage.

Rod cluster control assemblies, burnable poison (rod) assemblies, thimble plugging assemblies, and primary and secondary source assemblies are materials associated with the storage of spent fuel assemblies. Title 10, Code of Federal Regulations (10 CFR), Section 72.3, "Definitions," states, "...Spent fuel includes the special nuclear material, byproduct material, source material, and other radioactive materials associated with fuel assemblies." The applicant should define the range and types of spent fuel or other radioactive materials that the DCSS [dry cask storage system] is designed to store. For DCSSs that will be used to store activated components associated with a spent fuel assembly, the applicant should specify the types and amounts of radionuclides, heat generation, and the relevant source strengths and radiation energy spectra permitted for storage in the DCSS. Specifically, the applicant should describe:

- The design bases source term (radiological and thermal components). The source term should be based on a saturation value for activation of cobalt impurities or on cobalt activation from a specified maximum burn-up and minimum cool time. The applicant should describe other activation products, as appropriate.

- The effects of gas generation must be considered in the design pressure for the cask, including (1) the release of gas from additional components, and (2) the volume occupied by additional components on the cask internal pressure.

- Additional weight and length of the proposed material must be considered in the structural and stability analyses.

- The thermal analysis must consider (1) the added heat from these components, and (2) the effects of heat transfer within and to/from the fuel assembly by the addition or absence of these components. This would ultimately affect the maximum predicted cladding temperature.

- In terms of a criticality evaluation, absent direct physical measurements, the applicant should not take credit for any negative reactivity from residual neutron absorbing material remaining in the control components. A bounding analysis would assume that no control components are present. Credit for water displacement may be taken provided adequate structural integrity and placement under accident conditions is demonstrated. Also, the applicant may need to consider the effects of displacing borated water, if applicable.

4.4.1.2 High-Level Radioactive Waste

The regulatory requirements given in 10 CFR 72.3 define high-level radioactive waste and 10 CFR 72.120 (b) establish that the spent fuel or solid high-level waste are the acceptable waste forms. Liquid high-level radioactive waste is not acceptable for storage. Furthermore, if a pool type facility is proposed, the solidified waste form shall be a durable solid with demonstrable leach resistance. The applicant must provide information on the waste form, proposed storage package, characteristics of any encapsulation material, radionuclide characteristics, heat generation rate, and history and census. The Safety Analysis Report (SAR) must also include both the ranges of parameters of the known material to be stored and the bounding parameters of any additional materials that may be stored.

4.4.2 Classification of Structures, Systems, and Components

The applicant must identify all SSCs important to safety and provide a rationale for the identification. SSCs are classified into two broad categories: important to safety or not. The NRC review involves both categories; however, SSCs important to safety are reviewed in greater depth. Acceptance criteria for classification of SSCs important to safety are discussed in 10 CFR 72.3, 10 CFR 72.24 (n), and 10 CFR 72.144 (a) and (c).

The chapter on Installation Design and Structural Review discusses five areas of review which generally include SSCs identified as important to safety. These areas of review are: confinement structures, systems, and components; pool confinement facilities; and reinforced concrete structures; other SSCs important to safety; and other SSCs subject to NRC approval. Similarly the chapters on Thermal Evaluation, Radiation Shielding Evaluation, Criticality Evaluation, Confinement Evaluation, Waste Confinement, Radiation Protection, and Decommissioning have review areas that must be considered in identifying SSCs important to safety.

4.4.3 Design Criteria for SSCs Important to Safety

4.4.3.1 General

The regulatory requirements for design bases and general design criteria are given in 10 CFR 72.24 (c)(1), (c)(2), and (c)(4); 10 CFR 72.106 (a) and (c); 10 CFR 120 (a) and (b); 10 CFR 122 (a) through (l); 10 CFR 72.144; and 10 CFR 72.182 (a), and (b). The applicant must identify design criteria and design bases for all SSCs determined to be important to safety. The basic design criteria for SSCs which are important to safety shall: maintain subcriticality, maintain confinement, ensure radiation rates and doses for workers and public do not exceed acceptable levels and remain as low as is reasonably achievable (ALARA), maintain retrievability, and provide for heat removal (as necessary to meet the above criteria). Acceptance criteria for the specific design criteria are discussed in detail in each of the chapters.

The principal design criteria and bases should include the following items:

- Normal design conditions and parameters, including site-specific environmental conditions such as ambient temperature, humidity, and insolation; and operational parameters such as maximum load capacity of cranes and handling equipment; and maximum dimensions of the casks or other critical equipment to be handled

- Off-normal design conditions and parameters, including site-specific environmental conditions such as ambient temperatures and insolation, and operational parameters which do not approach accident conditions

- Accident design events, including site-specific environmental conditions such as tornado wind velocities, tornado pressure drop, maximum wind velocities, design basis earthquake, peak explosive over pressure, peak flood elevation, and accident design events such as maximum dose rates associated with hypothetical accidents including a cask drop or loss of pool coolant

Codes and standards and other detailed criteria applicable for ISFSI and MRS SSCs important to safety are presented or referenced in the Standard Review Plan (SRP) chapters addressing structural evaluation, thermal evaluation, shielding evaluation, nuclear criticality safety, confinement, waste management and decommissioning.

The FSRP chapter on site evaluation addresses review of site characteristics that must be included in design criteria and bases for natural phenomena.

4.4.3.2 Structural

The regulatory requirements for structural aspects of SSCs important to safety are given in 10 CFR 72.24 (c)(1), (c)(2), (c)(3), and (n); 10 CFR 72.102 (a), (b), (c), (d), (e), and (f); 10 CFR 72.120 (a) and (b); and 10 CFR 72.122 (a), (b)(1), (b)(2) and (b)(3), (c), (d), (f), (g), (h), (i), (j), and (k).

The applicant must present the structural design criteria and design bases for the proposed ISFSI or MRS. The structural design criteria and bases presented by the applicant for an ISFSI or MRS must address the design magnitudes of loads and limits derived from site characteristics and analyses of normal, off-normal, and accident-level conditions. The design bases presented by the applicant must include dead load, live load, lateral rail pressure, thermal loads, wind loads, accident loads, earthquake loads, and flood loads. Design bases guidance for tornado protection are given in Regulatory Guides 1.76, "Design Basis Tornado for Nuclear Power Plants," and 1.117, "Tornado Design Classification." Guidance for flood protection is given in Regulatory Guides 1.59, "Design Basis Floods for Nuclear Power Plants," and 1.102, "Flood Protection for Nuclear Power Plants." Guidance for protection against seismic events is given in Regulatory Guides 1.29, "Seismic Design Classification," 1.60, "Design Response Spectra for Seismic Design of Nuclear Power Plants," 1.61, "Damping Values for Seismic Design of Nuclear Power Plants," 1.92, "Combing Modal Responses and Spatial Components in Seismic Response

Analysis," and 1.122, "Development of Floor Design Response Spectra for Seismic Design of Floor-Supported Equipment or Components."

4.4.3.3 Thermal

The regulatory requirements relating to design bases and design criteria for thermal considerations are given in 10 CFR 72.122 (a), (b)(1), (b)(2) and (b)(3), (c), (d), (f), (g), (h), and (i); and 10 CFR 72.128 (a)(4). The applicant must identify thermal design criteria and bases. These criteria and bases must recognize the site temperature range and the specific materials used in ISFSI or MRS components.

Another aspect of thermal design criteria and design bases is fire protection. Guidance for fire protection is given in Regulatory Guide 1.120, "Fire Protection Guidelines for Nuclear Power Plants."

4.4.3.4 Shielding and Confinement

The regulatory requirements for shielding and confinement are given in 10 CFR 72. 24 (c)(1), (c)(2) and (c)(4); 10 CFR 72.104 (a), (b), and (c); 10 CFR 72.106 (a), (b), and (c); 10 CFR 72.122 (a), (b), (c), (d), (e), (f), (g), (h), and (i); 10 CFR 72.126 (a), (b), (c) and (d); and 10 CFR 72.128 (a) and (b). The applicant must identify shielding and confinement design criteria and design bases. These criteria and bases should discuss any proposed compliance with Regulatory Guides 8.5, "Criticality and Other Interior Evacuation Signals;" 8.25, "Air Sampling in the Workplace;" 8.34, "Monitoring Criteria and Methods to Calculate Occupational Radiation Doses;" and 1.143, "Design Guidance for Radioactive Waste Management Systems, Structures, and Components Installed in Light-Water-Cooled Nuclear Power Plants."

4.4.3.5 Criticality

The regulatory design bases and design criteria for criticality safety are given in 10 CFR 72.124 (a), (b), and (c). The application must identify nuclear criticality safety design criteria and design bases. These criteria and bases should discuss any proposed compliance with Regulatory Guides.

4.4.3.6 Decommissioning

10 CFR 72.130 outlines the regulatory requirements for decommissioning considerations. The applicant must identify any decommissioning design criteria and design bases. The application must also discuss compliance with any relevant Regulatory Guides.

Planning for decommissioning and design guidance for facilitating decommissioning are addressed in the FSRP chapter on decommissioning.

4.4.3.7 Retrieval

General regulatory requirements for retrieval capability are given in 10 CFR 72.122 (a), (b)(1), (b)(2), and (b)(3), (c), (f), and (h). Retrievability is specifically outlined in 10 CFR 72.122 (l). The applicant must include design criteria and design bases for retrieval.

The design criteria and bases for the ISFSI or MRS storage system must recognize the need for facilities, equipment, and procedures for the removal of spent fuel or solidified high-level radioactive waste from storage systems, and the transfer of this material into another storage system or a transportation cask. The design developed in compliance with the criteria must be able to retrieve spent fuel or the solidified high-level waste following normal and off-normal design conditions. Specific retrieval facilities, equipment, and procedures for post accident conditions are not required to be described in the SAR because of the wide variety of possible post-accident conditions that may occur.

The design must accommodate the retrieval of spent fuel or solid HLW following design basis accidents. The design and procedures for retrieval must be such that the operations can be conducted in compliance with the requirements of 10 CFR Part 20.

4.4.4 Design Criteria for Other SSCs

Design criteria and bases for other SSCs not important to safety should meet the general regulatory requirements as given in 10 CFR 72.24 (a), (b), (c), (d), (e), (f), (g), (h), (l), and the appropriate requirements as given in 10 CFR 72, Subparts E and F.

The applicant must identify design criteria and bases for SSCs not important to safety. These design criteria and bases for ISFSI and MRS SSCs that are not important to safety may be adequately defined by statements in the SAR identifying the design codes and standards to be met in design and construction. Greater definition is typically appropriate for SSCs that interface with SSCs important to safety.

4.5 Review Procedures

The reviewer should complete the appropriate sections of Table 4.2 at the end of this Chapter. The review includes evaluation of compliance with all regulatory requirements and acceptance criteria given in the FSRP and other NRC documents, as well as accepted codes. NRC may inspect various aspects of the ISFSI or MRS construction process during the SAR review.

4.5.1 Materials To Be Stored

The reviewer should verify that the types of materials to be stored comply with 10 CFR 72.2(a)(1) and (a)(2), and 10 CFR 72.120(b). The reviewer should confirm that the SAR gives spent fuel or high-level radioactive waste acceptance specifications, including upper or lower bound limits of acceptable variability. The reviewer should verify that these acceptance specifications are incorporated in the facility technical specifications. The reviewer should confirm that the SAR gives the criteria for procedures for testing, inspecting, and verifying wastes received for storage at the facility. The reviewer should verify that the SAR defines criteria for procedures for handling, repackaging, and shipping of out-of-specification wastes.

4.5.1.1 Spent Fuel

The reviewer should determine that the spent nuclear fuel is appropriately characterized so that the necessary design and analytical calculations and acceptance tests may be carried out. Analytical calculations include nuclear criticality safety, heat removal, shielding, etc. Fuel characteristics include reactor type, fuel configuration and vendor, enrichment, dimensions, weight, burnup, cooling time, type of cladding, assemblies to be stored per confinement vessel or pool facility, decay heat, fuel pin gas volume and temperature, condition (i.e., intact, undamaged), presence of control components, or other radioactive materials associated with fuel assemblies, and physical form of radionuclides.

In the SAR, the applicant must specify if damaged fuel is to be stored at the ISFSI. Damaged fuel should be canned for storage and transportation. The purpose of canning is to confine gross fuel particles to a known, subcritical volume during off-normal and accident conditions, and to facilitate handling and retrievability. As proof that the fuel is undamaged, the applicant, at a minimum, should review the fuel records and verify that the fuel was undamaged. Also, the applicant should specify that prior to loading, the fuel assemblies will receive an external visual examination for any obvious damage. For fuel assemblies where reactor records are not available, the applicant should provide alternate information which provides reasonable assurance that the fuel is undamaged or that damaged fuel loaded in a storage or transportation cask is canned in addition to the external visual examination for any obvious damage.

4.5.1.2 High-Level Radioactive Waste

The reviewer should determine that the high-level radioactive waste is appropriately characterized so that the necessary design and analytical calculations and acceptance tests may be carried out. For high-level radioactive waste, such characteristics include waste form, decay heat, and inventory of radionuclides.

The reviewer should specifically ensure that the waste form is solid and not liquid. If the waste form contains liquid, as in undried filter residues, the NRC staff must establish waste acceptance specifications and bounding limits of acceptability.

4.5.2 Classification of SSCs

The reviewer should review all SSCs classified as important to safety and the rationale for classification. When reviewing the applicant's rationale for classification, the reviewer should consider the concept of classifying the SSCs into three categories as discussed in Regulatory Guide 7.10, "Establishing Quality Assurance Programs for Packaging Used in the Transport of Radioactive Material, Revision 1," and developed further in NUREG/CR-6407. The reviewer should compare the results of the applicant's classification process with the listings in NUREG/CR-6407 where category A and B items are generally considered important to safety.

The reviewer should determine if the following SSCs and functions that have typically been considered as important to safety are included: (1) components of the confinement vessel and integral components and structures used within the vessel, (2) radiation shielding, (3) SSCs providing capabilities for lifting, handling, and transfer of spent fuel, (4) confinement for pool coolant, (5) instrumentation and controls (I&C) SSCs, if they are used as the primary means for real-time recognition of off-normal conditions, (6) SSCs providing either active or passive decay heat removal, (7) the confinement systems to preclude the release of radioactive liquids, and (8) SSCs for retaining radioactive material within the pool building.

4.5.3 Design Criteria for SSCs That Are Important to Safety

4.5.3.1 General

The reviewer should verify that the SAR identifies the principal design criteria and bases for SSCs important to safety. These design criteria and bases may be presented by reference to a summary discussion or tabular listing in the SAR. Table 4.2 illustrates the headings for such a listing.

The reviewer should check Chapter 4 of the SAR (Design Criteria), as well as sections of the SAR which address confinement, cooling, subcriticality, radiation protection, decommissioning, retrieval capability, and ALARA. Design criteria and bases for the system as a whole must be identified and evaluated.

The reviewer should determine that the criteria derived from the site characteristics and accident analyses (accident and off-normal conditions) are consistent with the analyses used in the qualification of the SSCs. The reviewer should verify that these criteria are equivalent to those proposed in the facility design.

The reviewer should confirm that ALARA goals were considered in development of the applicant's general design criteria. The criteria should reflect any stated applicant ALARA policies.

The reviewer should verify that criteria defining the response of SSCs to normal, off-normal, and accident conditions are satisfactory. The following sections provide general guidance for determining if the proposed criteria are acceptable.

The reviewer should determine the design criteria for normal conditions and operations which do not result in any degradation of the capabilities of the ISFSI or MRS. Routine maintenance, as described in the SAR, should be sufficient to correct any "wear and tear" from normal conditions and operations that would degrade the capabilities of the ISFSI.

The reviewer should determine that the design criteria for off-normal conditions do not permit any degradation of the capabilities of the ISFSI or MRS, assuming contingency operations during and following off-normal conditions. The NRC does not require that radioactive material handling or waste processing functions or capabilities at an ISFSI or MRS continue during an off-normal condition or that such operations resume immediately. The licensee may impose inspections and system checkouts following any event or condition.

The reviewer should determine that design criteria for accident conditions do not permit degradation of SSCs important to safety, including, but not limited to, (1) reduced radioactive material handling and waste processing capability, (2) reduced capability to withstand further accident conditions without excess response, without remedial action, and (3) reduced ability to provide functions for the full system life time without remedial action. The reviewer should determine that design criteria for accidents still prevent (1) criticality, (2) unacceptable releases of radioactive material, (3) unacceptable radiation doses for the public and workers, and (4) loss of retrieval capability.

The NRC staff does not require assumption of multiple failure scenarios of SSCs important to safety unless these multiple failure scenarios are credible consequences of the initiating event.

The NRC requires analysis or testing of SSCs for some events (e.g., cask drop or tipover) even though the events may be determined as non-credible in the accident analysis. Criteria for survival of SSCs important to safety for these "non-mechanistic" events should be the same as the criteria for survival of credible accidents.

4.5.3.2 Structural

For confinement SSCs designed to ASME B&PVC, Section III, the reviewer should verify that the loads, load conditions, and load combinations are defined in accordance with Article 3000 and include design pressure, design temperature, and design mechanical loads for Service Levels A, B, C, and D associated with normal, off-normal, and accident conditions.

The reviewer should ensure that acceptable design codes have been specified for SSCs important to safety that are not confinement casks and internal components, such as critical lifting devices, pool and pool facilities, waste management facilities, and radiation and protective shielding. The reviewer should compare applicant-proposed load combinations with those presented in Table 3-1 of NUREG-1536 which identifies load combinations for SSCs important to safety. The table also categorizes load combinations for normal, off-normal, or accident conditions. The load combination expressions identify which loads should be considered as acting concurrently. The reviewer should ensure that the appropriate loads and load combinations are used and correspond to the appropriate operating conditions for the specific site. The reviewer should verify that the SSCs meet appropriate guidance in Regulatory Guides 1.76 and 1.117 for tornado protection; Regulatory Guides 1.29, 1.60, 1.61, 1.92, and 1.122 for protection against seismic events; Regulatory Guides 1.59, and 1.102 for flood protection; and NUREG-0800 for tornado missile protection.

4.5.3.3 Thermal

The reviewer should verify that the design bases and criteria for thermal conditions are defined and appropriate for the site. The reviewer should ensure that design parameters, such as maximum cladding temperature, pool coolant temperature, reinforced concrete temperature, and other SSCs that are temperature-sensitive are defined. The reviewer should verify that the design criteria meet the appropriate sections of Regulatory Guide 1.120 for fire protection. The following sections provide general guidance for normal, off-normal, and accident conditions.

The reviewer should verify that the thermal design criteria address: (1) extremes of normal ranges of ambient temperature versus storage or operational time durations, (2) maximum site insolation, (3) maximum duration that an active cooling system may be unavailable as a result of normal conditions (e.g., cooling of material in storage, if active cooling used, or of pool water) as the result of a "normal" occurrence, and (4) maximum design basis stored material decay heat load.

The design criteria for off-normal conditions include: (1) extreme off-normal ranges of ambient temperature versus significant time durations, (2) maximum site insolation for high ambient temperature case, and (3) maximum duration that an active cooling system may be unavailable (e.g., cooling of material in storage or in pool water) as the result of an off-normal occurrence.

The design criteria for accident conditions include: (1) accident ranges of ambient temperature versus significant time durations, (2) maximum site insolation for highest ambient temperature case, and (3) maximum duration that an active or passive cooling system may be unavailable (e.g., cooling of material in storage or in pool water) as the result of an accident occurrence.

4.5.3.4 Shielding and Confinement

The reviewer should verify that the design bases and criteria define the shielding and confinement systems. The reviewer should verify that the maximum dose rates for the confinement cask surfaces and exterior of shielding are defined. The reviewer should verify that the dose rate and annual dose to workers are specified. The reviewer should check that the ISFSI controlled area boundary complies with the regulations and that the dose rates and annual dose rates to the public meet the regulations. The reviewer should ensure that the criteria are explicit for normal, off-normal, and accident conditions. For confinement casks, the reviewer should confirm that the method of sealing is defined and meets regulations for redundant seals and that the maximum leak rates are specified and do not result in exceeding dose requirements. The reviewer should verify that monitoring systems are specified and that they meet the regulations for continuous monitoring of SSCs important to safety. Where appropriate, the reviewer should confirm that guidance given in Regulatory Guides 1.143, "Design Guidance for Radioactive Waste Management Systems, Structures, and Components Installed in Light-Water-Cooled Nuclear Power Plants," 8.5, "Criticality and Other Interior Evacuation Signals," 8.25, "Air Sampling in the Workplace," and 8.34, "Monitoring Criteria and Methods to Calculate Occupational Radiation Doses," are considered.

The reviewer should verify that the design criteria or design bases for normal conditions include: (1) locations of on-site personnel with respect to shielding and radiation protection afforded by site characteristics and installation layout, (2) ALARA concepts applied to normal maintenance and operations, and (3) estimation of dose rates or doses for on-site workers and the public based on dispersion characteristics associated with normal weather patterns and bounding radiological source terms along with facility shielding.

The reviewer should verify that the design criteria or design bases for off-normal conditions include (1) ALARA concepts applied to operator action during off-normal events and conditions, and (2) estimation of dose rates or doses for on-site workers and the public based on dispersion characteristics associated with conservative weather patterns.

The reviewer should verify that the design criteria or design bases for accident conditions include maximum accident dose rates to the offsite public based on accident analysis.

4.5.3.5 Criticality

The reviewer should confirm that the method of criticality control, such as geometry, fixed poisons, borated pool water, etc., is specified. The reviewer should confirm that procedures are in place to control minimum boron concentration in fixed poisons in the confinement cask or in the pool. The reviewer should verify that the design criteria require that k_{eff} less than 0.95 (with 95% probability and 95% confidence) for all normal events, abnormal events, and postulated accidents. The reviewer should verify that design criteria require that the calculation of k_{eff} includes the effects of maximum fresh fuel enrichment, optimum moderation, and computer code computational and experimental benchmark bias.

4.5.3.6 Decommissioning Considerations

The reviewer should confirm that the design criteria include requirements for decommissioning as outlined in 10 CFR 72.130. Regulatory Guide 1.86, "Termination of Operating Licenses for Nuclear Reactors," offers guidance on contamination levels on material which can be released.

4.5.3.7 Retrieval Capability

The reviewer should verify that design criteria for retrieval capability of spent fuel or other high-level radioactive waste forms considers normal and off-normal events.

4.5.4 Design Criteria for Other SSCs

The reviewer should verify that the design bases and criteria for other SSCs not important to safety meet the general regulatory requirements as given in 10 CFR 72.24 (a), (b), (c), (d), (e), (f), (g), (h), and (l).

Typical concerns for general design criteria reviews of other SSCs not important to safety include, but are not limited to, adequate functional performance, interfacing with other SSCs, and recognition of appropriate site characteristics.

4.6 Evaluation Findings

Evaluation findings are prepared by the staff upon completion of the SAR review and determination that the regulatory requirements identified in Section 4.3 and staff safety concerns have been properly addressed and factored into the design. If the documentation submitted with the application fully supports positive findings for each of the regulatory requirements, the statements of findings may be as follows (numbering is for convenience in referencing the FSRP section):

F4.1   The SAR and docketed materials adequately identify and characterize the spent fuel to be stored at the site in conformance with the requirements given in 10 CFR 72.2 (a)(1) and (a)(2), and 10 CFR 72.6 (b). The form of the spent fuel is acceptable if the fuel is solid fuel and not in liquid form, and meets the requirements given in 10 CFR 72.120 (b).

F4.2   The SAR and docketed materials adequately identify and characterize the high-level radioactive waste as required by 10 CFR 72.3. The waste form is solid and not liquid as required by 10 CFR 72.120 (b).

F4.3   The structure, systems and components have been classified according to their function as important to safety or not important to safety, and meet the requirements given in 10 CFR 72.3, 10 CFR 72.24 (n), and 10 CFR 72.144 (a) and (c).

F4.4   The SAR and the docketed materials relating to the design bases and criteria meet the general requirements as given in 10 CFR 72.24 (c)(1), (c)(2), (c)(4), and (n); 10 CFR 72.106 (a) and (c); 10 CFR 120 (a) and (b); 10 CFR 122 (a), (b), (c), (d), (e), (f), (g), (h), (i), (j), (k), and (l); 10 CFR 72.144; and 10 CFR 72.182 (a), and (b).

F4.5   The SAR and docketed materials relating to the design bases and criteria for structures categorized as important to safety meet the requirements given in 10 CFR 72.24 (c)(1), (c)(2), (c)(3), and (n); 10 CFR 72.102 (a), (b), (c), (d), (e), and (f); 10 CFR 72.120 (a) and (b); and 10 CFR 72.122 (a), (b)(1), (b)(2) and (b)(3), (c), (d), (f), (g), (h), (i), (j), and (k). For certified confinement casks complying with Subpart L, the regulatory requirements are outlined in 10 CFR 72.236. The SAR meets the guidance given in Regulatory Guides 1.76, 1.117, and NUREG-0800 for tornado and tornado missile protection. The SAR meets the guidance in Regulatory Guides 1.59 and 1.102 for flood protection. The SAR meets Regulatory Guides 1.29, 1.60, 1.61, 1.92, and 1.122 for seismic events.

F4.6   The SAR and docketed materials meet the regulatory requirements for design bases and criteria for thermal consideration as given in 10 CFR 72. 122 (a), (b)(1), (b)(2) and (b)(3), (c), (d), (f), (g), (h), and (i); and 10 CFR 72.128 (a)(4). The SAR meets the regulatory requirements for design criteria of for fire protection given in Regulatory Guide 1.120.

F4.7   The SAR and docketed materials relating to the design bases and criteria for shielding, confinement, radiation protection and ALARA considerations meet the regulatory requirements as given in 10 CFR 72. 24 (c)(1), (c)(2), (c)(4), and (n); 10 CFR 72.104 (a), (b), and (c); 10 CFR 72.106 (a), (b), and (c); 10 CFR 72.122 (a), (b), (c), (d), (e), (f), (g), (h), and (i), 10 CFR 72.126 (a), (b), (c) and (d); and 10

CFR 72.128 (a) and (b). The SAR meets the guidance given in Regulatory Guides 1.143, 8.5, 8.25, and 8.34.

F4.8   The SAR and docketed materials relating to the design bases and criteria for criticality safety meet the regulatory requirements as given in 10 CFR 72.124 (a), (b), and (c).

F4.9   The SAR and docketed materials relating to design criteria for decommission of the facility comply with the regulatory requirements given in 10 CFR 72.130 and the guidance given in Regulatory Guide 1.86.

F4.10   The SAR and docketed materials relating to the design bases and criteria for retrieval capability meet the regulatory requirements as given in 10 CFR 72.122 (a), (b)(1), (b)(2), and (b)(3), (c), (f), (h) and (l).

F4.11   The SAR and docketed materials relating to the design bases and criteria for other SSCs not important to safety, but subject to NRC approval, meet the general regulatory requirements as given in 10 CFR 72.24 (a), (b), (c), (d), (e), (f), (g), (h), (l) and the appropriate requirements as given in Subparts E and F of 10 CFR 72.

4.7 References

NRC documents referenced are identified at Consolidated References, Chapter 17.

ASME Boiler and Pressure Vessel Code, Section III, Division 1, "Rules for Construction of Nuclear Power Plant Components."

Table 4.2 Summary of Design Criteria/Bases for SSCs Important to Safety

Design Criteria (Specify normal/off-normal/accident, if applicable)

Design Bases

> Specifications of radioactive material to be stored
> Bounding normal design event and condition parameters
> Bounding off-normal design event and condition parameters
> Design basis accident design event and condition parameters

Design Life (Initial license restricted to 20 years with potential for renewal)

Structural

> Design codes for:
> > Confinement casks and integral and internal components
> > Other SSCs important to safety
> > Radiation and protective shielding
> > Pool
> > Pool facility SSCs important to safety
> > Waste management facility SSCs important to safety

> Design weights
> > (Account for nominal dimension ranges.)

> Cask design cavity pressures

> Special response and degradation limits

Thermal

> Maximum design temperatures
> > Cladding
> > Reinforced concrete
> > Pool coolant
> > [Other SSCs that are temperature sensitive in range of projected temperatures]
> > Maximum temperature gradients for structures subject to thermal stress

> Insolation

> Fill gas specification

> Maximum stored material decay heat load

Confinement

Method of sealing

Maximum leak rates
>Primary seals
>Redundant seals
>Cask body

Monitoring system specifications

Retrievability

Normal and off-normal
After accident events

Criticality

Maximum fresh unirradiated U^{235} enrichment

Method of Control
>(e.g., geometry, fixed poison, borated pool water)

Minimum boron concentration
>Fixed in confinement cask
>Pool water

Maximum k_{eff}

Burnup credit

Radiation Protection/Shielding

Maximum dose rate
>Confinement cask surface(position)
>Exterior of shielding (transfer/storage mode position)
>Pool surface

Individual workers
>Dose rate
>Annual dose
>Dose per loading operation

ISFSI controlled area boundary
>Normal/off-normal/accident dose rate
>Annual dose

5 INSTALLATION AND STRUCTURAL EVALUATION

5.1 Review Objective

The objective of the installation design review is to ensure compliance with required site features and to support other evaluation areas. The objective of the structural evaluation review is to ensure the structural integrity of structures, systems, and components (SSCs) with emphasis on SSCs important to safety. These SSCs may provide confinement, subcriticality, radiation shielding, and retrievability of the stored materials, and must be appropriately maintained under all credible loads for normal, off-normal, and design basis accident conditions. These conditions also include natural phenomena. Chapter 4, Design Criteria, discusses the categorization of the SSCs into two subsets, "important to safety" and "not important to safety."

Figure 5.1 presents an overview of the structural evaluation review process. The figure shows the information flow from the applicant and from other sections of the review such as thermal analysis, criticality analysis, and accident analysis. The figure also shows the flow of results from the structural evaluation to the Safety Evaluation Report (SER) and to other review areas, such as confinement analysis and limiting conditions for operation.

5.2 Areas of Review

The Safety Analysis Report (SAR) should be reviewed for adequacy of the description and evaluation of the structural integrity for all structures, systems and components which are classified in the SAR and confirmed in Chapter 4 of this Standard Review Plan (SRP) for Spent Fuel Dry Storage Facilities as important to safety or otherwise subject to the NRC approval. The following outline shows the areas of review addressed in Section 5.4, Acceptance Criteria and Section 5.5, Review Procedures:

Confinement Structures, Systems, and Components
> Description of Structural Design
> Design Criteria
> Material Properties
> Structural Analysis

Pool and Pool Confinement Facilities
> Description of Structural Aspects of Pool
> Design Criteria
> Material Properties
> Structural Analysis

Reinforced Concrete Structures
> Description of Structural Design
> Design Criteria
> Material Properties
> Structural Analysis

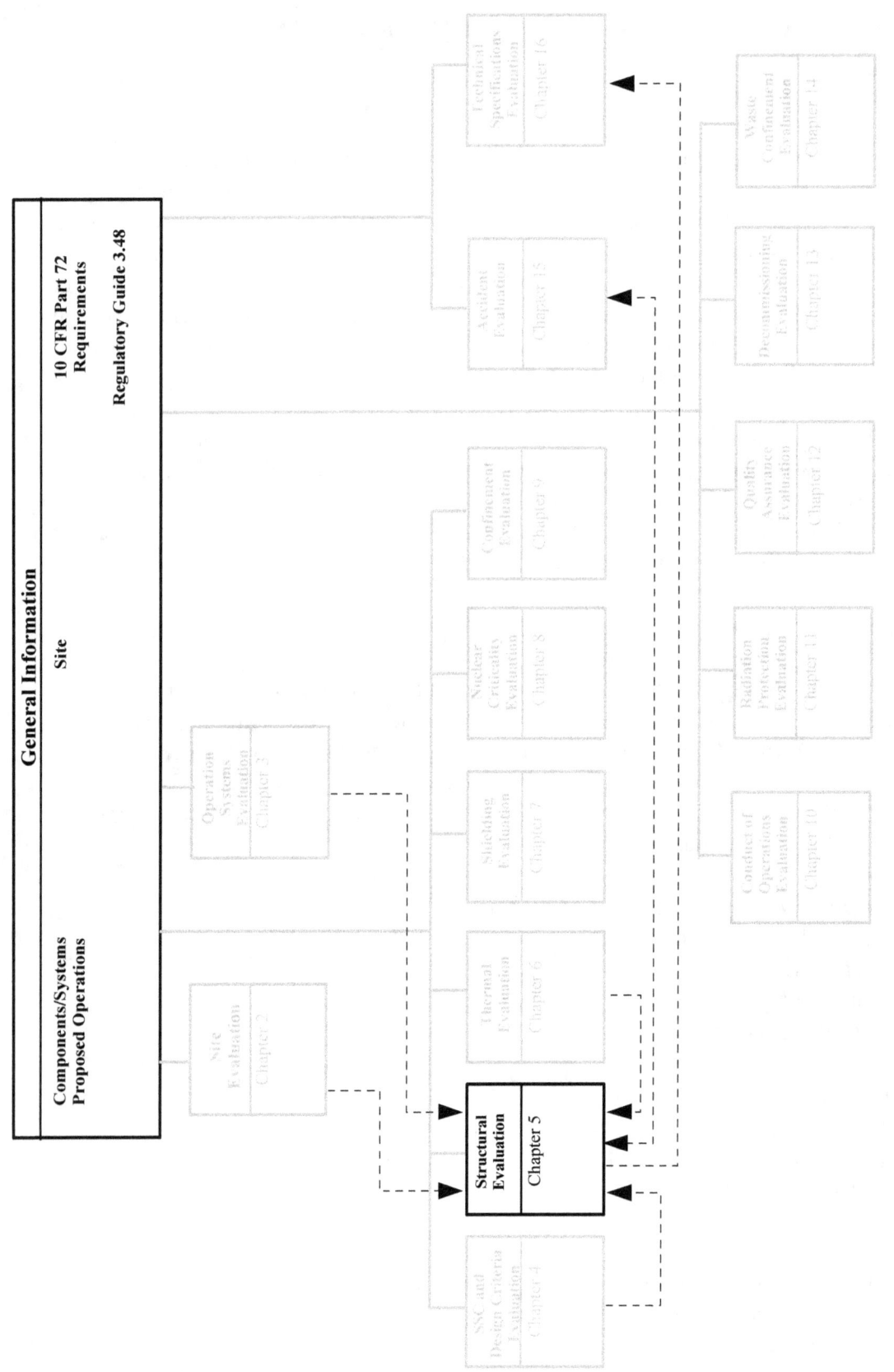

Figure 5.1 Overview of Structural Evaluation

Description of Structural Aspects
Design Criteria
Material Properties
Structural Analysis

Other SSCs

Description of Structural Aspects
Design Criteria
Material Properties
Structural Analysis

5.3 Regulatory Requirements

This section identifies and presents a high-level summary of Title 10 of the Code of Federal Regulations (CFR) Part 72 relevant to the review areas addressed by this chapter. The NRC staff reviewer should read the exact regulatory language. A matrix at the end of this section matches the regulatory requirements identified in this section to the areas of review identified in the previous section.

72.24 Contents of application: Technical information [Contents of SAR]
(a) "A description and safety assessment of the site ... and evaluation of the major SSCs...."
(b) "A description and discussion of the [Independent Spent Fuel Storage Installations] ISFSI or {Monitored Retrievable Storage] MRS structures ...."
(c) "The design of the ISFSI or MRS in sufficient detail ... including:"
        (1) "The design criteria...."
        (2) "The design bases and the relation of the design bases to the design criteria;"
        (3) "Information relative to materials of construction... dimensions ...."
        (4) "Applicable codes and standards."
(d) "An analysis and evaluation of the design and performance of SSCs important to safety...."
(i) "If the proposed ISFSI or MRS incorporates structures... have not been demonstrated...."

72.40 Issuance of license.
(a) "Except as provided in paragraph (c) of this section...."
        (1) "The applicant's proposed ISFSI or MRS design complies with Subpart F;"
        (2) "The proposed site complies with the criteria in Subpart E;"
        (3) "If on the site of a nuclear plant..."

72.82 Inspections and tests.
(c)(2) "For a site with a single storage installation ...."

72.102 Geological and seismological characteristics.

(a)   (1) "East of the Rocky Mountain Front...response spectrum anchored at 0.2 g."
           (2) "For those sites that have been evaluated under paragraph (a)(1)..."
(b) "West of the Rocky Mountain Front...evaluated by the techniques of appendix A of part 100..."
(c) "Sites other than bedrock sites..."
(d) "Site-specific investigations and laboratory analyses...soil conditions..."
(e) "In an evaluation of alternative sites..."
(f) "The design earthquake (DE) for the use in the design of structures..."
           (1) "For sites that have been evaluated under the criteria of appendix A..."
           (2) "Regardless of the results...no less than 0.10 g..."

72.106 Controlled area of an ISFSI or MRS.
(a) "For each ISFSI or MRS site, a controlled area must be established."
(b) "...The minimum distance from the spent fuel....shall be at least 100 meters."
(c) "The controlled area may be traversed by a highway...."

72.120 General considerations.
(a) "Pursuant to...must include the design criteria for the proposed storage installation...."
(b) "The MRS must be designed to store either spent fuel or solid high-level radioactive wastes...."

72.122 Overall requirements.
(a) Quality Standards. "SSCs important to safety must be designed, fabricated...."
(b) Protection against environmental conditions and natural phenomena.
           (1) "SSCs important to safety must be designed to....postulated accidents."
           (2) " SSCs important to safety must be designed to... natural phenomena...."
           "The design bases for these SSCs must reflect:"
                     (i) "Appropriate consideration of the most severe of the natural phenomena...."
                     (ii) "Appropriate combinations of the effects of normal and accident conditions...."
           (3) "Capability must be provided for determining the intensity of natural phenomena...."
(c) Protection against fires and explosions. " SSCs important to safety must be designed...."
(d) Sharing of SSCs. "SSCs important to safety must not be shared...."
(f) Testing and maintenance of system and components. " Systems...permit inspection...."
(g) Emergency capability. "SSCs important to safety must be designed for emergencies...."
(h) Confinement barriers and systems.
(i) Instrumentation and control systems. "Instrumentation and control systems...."
(j) Control room or control area. "A control room or control area, if appropriate ...."
(k) Utility or other services.
(1) Retrievability. "Storage systems must be designed to allow ready retrieval...."

72.128 Criteria for spent fuel, high-level radioactive waste, and other radioactive waste storage and handling.
(a) Spent fuel and high-level radioactive waste storage and handling systems.
(b) Waste treatment.

[The following regulatory requirements apply to ISFSI and MRS confinement casks, if the design of the confinement cask system has been previously certified under 10 CFR 72 Subpart L].

72.236 Specific requirements for spent fuel storage cask approval
(b) "Design bases and design criteria..."
(e) "The cask must be designed to provide redundant sealing of confinement systems."
(f) "The cask must be designed to provide adequate heat removal capacity without active cooling...."
(g) "The cask must be designed to store the spent fuel safely for a minimum of 20 years...."
(k) " The cask must be conspicuously and durably marked with:..."

A matrix showing the primary relationship of these regulations to the specific areas of review in this chapter is given in Table 5.1. The reviewer should independently verify the relationships in this matrix to ensure that no requirements are overlooked because of unique applicant design features.

Table 5.1 Relationship of Regulations and Areas of Review

Areas of Review	10 CFR Part 72 Regulations								
	72.24	72.40	72.82	72.102	72.106	72.120	72.122	72.128	72.236
Confinement SSCs	●	●	●	●	●	●	●	●	●
Pool and Facilities	●	●	●	●	●	●	●	●	
Reinforced Concrete	●	●	●	●	●	●	●	●	
Other SSCs Important To Safety	●	●	●	●	●	●	●	●	
Other SSCs	●	●	●		●	●	●	●	

5.4 Acceptance Criteria

This section identifies the acceptance criteria used for the structural evaluation. Acceptability of the design of the structures, systems, and components as described in the SAR is based on compliance with requirements and Regulatory Guides determined by independent calculations and staff judgments. The design of the SSCs are acceptable if the integrated design meets the general and specific criteria discussed below.

The license approval process for ISFSI and MRS is a one-step licensing process rather than a two-step process as exemplified by 10 CFR Part 50 for a reactor license. Thus, the evaluation of the SAR and the supporting materials for an ISFSI license is the sole occasion in the design and construction sequence that the design and proposed construction are comprehensively reviewed by the NRC staff. The result is that the depth of information required for individual SSCs important to safety is greater for ISFSI and MRS than would be required for similar SSCs in the application for a construction permit under 10 CFR 50.

The confinement systems, including pool facilities, reinforced concrete structures, and other SSCs, which are important to safety or subject to NRC approval, must to have sufficient structural capability to withstand the worst-case loads under accident conditions and natural phenomena events. This may be verified by the reviewer of the SAR, first by verifying acceptable design criteria and then by verifying acceptable analyses, which ensure that the structures preclude:

- unacceptable risk of criticality
- unacceptable release of radioactive materials
- unacceptable radiation levels
- impairment of ready retrievability of stored material

Provided that a certified cask system has not been modified, the use of a certified cask design can be used to satisfy a part of the requirements for the facility license application by reference. Site facilities and infrastructure of concern to the NRC are to have the descriptions, design criteria, and safety analyses as appropriate to safety reviewed. These could include the pool and pool facility SSCs, the waste facilities, space for NRC use, and other elements of the site physical infrastructure.

5.4.1 Confinement Structures, Systems, and Components

5.4.1.1 Description of Confinement Structures

10 CFR 72.24 (a) and (b), 10 CFR 72.82 (c)(2), and 10 CFR 72.106 (a), (b), and (c) outline the contents of the application, which include design descriptions in sufficient detail to support findings in the SER. For confinement SSCs the application must include text descriptions, drawings, figures, tables and specifications that would fully define the structural features of the confinement SSCs.

For a site-specific ISFSI, the application may involve use of a cask certified under 10 CFR 72, Subpart L, including the SAR for the certified cask system by reference. Additional information relating to the cask should also be provided, including the applicant's evaluations that establish that site parameter limits are within the bounds of those established as limiting conditions as set forth in the Certificate of Compliance.

If actual site parameters exceed the bounds of those assumed in the safety analysis submitted for the certified cask system or exceed specified conditions of compliance, then the SAR submitted with the application must fully address those areas affected by the variations. If the design of the proposed cask system is not identical to the certified cask system, the SAR shall include a full description of the cask system (drawings and construction or fabrication specifications), a description of all changes to the certified design, and analyses that show the proposed design satisfies the criteria for the proposed installation.

5.4.1.2 Design Criteria for Confinement Structures

The regulatory requirements given in 10 CFR 72.24 (c)(1), (c)(2), and (c)(4); 10 CFR 72.40 (a)(1); 10 CFR 72.120 (a), and (b); 10 CFR 72.122 (a), (b), (c), (d), (f), (g), (h), (i), (j), (k), and (l); and 10 CFR 72.128 (a) and (b) identify acceptable design criteria. The NRC generally considers the design criteria identified below to be acceptable to meet the structural requirements of 10 CFR 72 for storage confinement casks.

General Structural Requirements

The confinement structures are to have sufficient structural capability so that every cross section of the structure can withstand the worst-case loads and successfully preclude the unacceptable risk of criticality, unacceptable release of radioactive materials to the environment, unacceptable radiation dose to the public or workers, and significant impairment of ready retrievability of the stored nuclear material. Confinement of radioactive material must be maintained under normal, off-normal, and accident conditions.

These criteria do not require that all confinement systems and other structures important to safety survive all design basis accidents and extreme natural phenomena without any permanent deformation or other damage. Some load combination expressions for accident events, for structures important to safety, permit stress levels that exceed yield. These scenarios should be shown to be acceptable by computations, analyses, and/or tests acceptable to the NRC.

Structures important to safety are not required to survive accident events and conditions to the extent that they remain suited for use for the life of the ISFSI or MRS without inspection, repair, or replacement. However, confinement structures are required to maintain confinement integrity under all accident conditions. The NRC does not accept breach of the storage confinement.

If the life of structures important to safety may be degraded by design basis events, requirements and procedures for determination and correction of the degradation, or other acceptable remedial action must be provided.

Spent fuel cladding must be protected against gross rupture caused by degradation resulting from normal, off-normal, or accident conditions.

The cask and any racks for positioning stored fuel or waste material within the cask must not deform under credible loading conditions to the extent that the subcritical condition or the retrievability of the fuel would be jeopardized. The cask must be analyzed to show that it will not slide, tip over, or drop in its storage condition as a result of a credible natural phenomenon event, including tornado winds and tornado missiles, earthquakes, and floods. This criterion is to preclude damage to an entire array. A tip-over or drop is always to be assessed as a bounding condition during handling operations.

Radiation shielding for the cask system, required for protection of the public or workers, must not degrade under normal or off-normal conditions. The shielding function may be acceptably degraded by a design basis event (e.g., loss of liquid neutron shielding resulting from a drop accident). However, the loss of function must be readily apparent.

Applicable Codes and Standards

The applicant must identify the design codes and standards intended for confinement structures. The structural design, fabrication, and testing of the confinement system must comply with an acceptable code or standard. Use of codes and standards that have been accepted by the NRC expedites the evaluation process. The alternative use of other codes and standards may require extensive NRC review and may delay the evaluation process.

An accepted code for design, fabrication, and testing of steel confinement casks is Section III of the ASME Boiler and Pressure Vessel Code (ASME B&PVC). The NRC has accepted use of either Section NB or NC. The NRC has accepted use of Sections NF and NG of the ASME B&PVC, Section III, Division 1 for cask system components used within the confinement cask but not integrated with it. This includes the "basket" which is a structure used inside casks to restrain and position fuel assemblies. Other design codes or standards may be acceptable depending on their application.

The NRC accepts use of Regulatory Guides 7.11, "Fracture Toughness Criteria of Base Material for Ferritic Steel Shipping Cask Containment Vessels with a Maximum Wall Thickness of 4 Inches (0.1 m)," and 7.12, "Fracture Toughness Criteria of Base Material for Ferritic Steel Shipping Cask Containment Vessels with a Wall Thickness Greater than 4 Inches (0.1m) But Not Exceeding 12 Inches (0.3m)," as bases for determining the potential for brittle fracture. These Regulatory Guides also incorporate a portion of NUREG/CR 1815 by reference. The reviewer should be aware of those portions of NUREG/CR 1815 which are excluded by Regulatory Guides 7.11 and 7.12.

The fatigue limits of the cask structural materials may be based on the provisions of the ASME B&PVC, Section III or the guidance provided in Regulatory Guide 7.6, "Design Criteria for the Structural Analysis of Shipping Cask Containment Vessels." Since casks are typically not subjected to cyclic loads, fatigue may not be a significant concern.

Cask Closure Welds After Fuel Loading

The following special considerations are generally accepted by the NRC for the dry storage canister top end closure welds which are made after the canister has been loaded with spent nuclear fuel assemblies. All other dry storage canister bottom end closure welds and shell welds should be designed, fabricated, examined, and tested to the requirements of the appropriate subsections of the ASME Section III Code.

The top end closure welds are to be helium leak tested. No hydrostatic or pressure tests are required if a minimum margin of safety equal to or greater than 1.5 against design pressure was demonstrated by analysis.

The closure weld joint may be either a full thickness penetration weld or a partial penetration groove weld. For a partial penetration groove weld, the maximum clearance between the closure plate and the enclosure shell should be small enough to ensure a good weld and should not exceed the clearance allowed in the weld procedure qualification. The minimum depth of the groove shall be equal to or larger than the enclosure shell thickness. The weld strength of the closure joint is based on the nominal weld area and the design stress intensity values for the weaker of the two materials jointed. However, the minimum ultimate tensile strength of the weld metal should equal or exceed the base metal strength to preclude weld metal failure.

For dry storage canisters made from austenitic stainless steels Type 304, 304L, 304LN, 316, 316L, or 316LN, the top end closure weld may be examined by either the ultrasonic methods (UT) or progressive liquid penetrant (PT) examinations as follows:

> If UT is used, the UT acceptance criteria shall be the same as NB-5332 for pre-service examination.

> If PT is used, the examination shall be performed progressively on the root layer; the lesser of one half of the welded joint thickness, or ½ inch intervals thereafter; and the final surface. In addition, a stress reduction factor of 0.8 shall be applied to the weld strength of the joint.

For dry storage canisters made from austenitic stainless steels other than the Type 304 or 316 materials listed above, the top end closure weld may be examined by PT as described above for Type 304 and 316, except that the thickness and number of intermediate layers to be examined shall be determined by a fracture mechanics assessment of the weld considering the specific geometry, material properties, and loadings. The maximum thickness of each weld pass deposit and PT layer shall not exceed the allowable critical flaw size for a 360 degree circumferential flaw.

For dry storage canisters made from ferritic steels, the top end closure weld should be examined by UT and:

> The critical flaw size and the critical design stress values shall be determined by the linear elastic fracture mechanics methodology specified in ASME Code, Section XI using the applicable service temperature and material properties.

> The UT must be performed in accordance with pre-qualified procedures and methods. The UT examination methodology should be demonstrated to be reasonably accurate and consistently able to detect flaw sizes less than the critical flaw size determined by linear elastic fracture mechanics.

> The UT examination must be performed by tested and certified operators.

> The welding processes, weld inspection criteria, and weld personnel qualifications shall be in conformance with the ASME Code. The welding process and technique used should be evaluated to preclude hydrogen induced cracking.

As an alternative, progressive surface examinations, utilizing PT or magnetic particle examination (MT), are permitted only if unusual design and loading conditions exist. PT or MT must be performed after sufficiently small intervals to ensure that flaws equal to the critical flaw size will be detected. In addition, a stress reduction factor of 0.8 shall be used for the weld strength of the closure joint to account for imperfections or flaws potentially missed by progressive surface examinations. Critical flaw sizes for ferritic steels are generally small. Therefore, PT or MT must be performed on many layers of the weld and this alternative may become unacceptable, due to ALARA concerns. The weld design should provide a sufficient safety margin and should be approved by the NRC on a case-by-case basis.

5.4.1.3 Material Properties

Acceptable criteria for materials used in all structural components and systems are given in 10 CFR 72.24 (c)(3). The applicant must identify standards for materials and properties used in analyses.

The information provided on materials must be consistent with the application of the accepted design criteria, codes, standards, and specifications selected for the storage cask system. For example, if the ASME B&PVC, Section III is used for the design criteria, the materials selected for the cask must be consistent with those allowed by the particular Section of the ASME B&PVC used for design. Acceptable requirements are ASME-adopted specifications given in ASME B&PVC, Section II, Part A "Ferrous Metals," Part B "Nonferrous Metals," Part C "Welding Rods, Electrodes, and Filler Metals," and Part D "Properties." NUREG-1536 provides additional guidance regarding the use of the ASME B&PVC requirements for material properties and specifications.

Compatibility of materials and coatings to be used with the environments to be experienced must be established. This includes compatibility with fluids during loading and unloading operations that may occur on-site. Compatibility verification should specifically include potential reactions in the presence of liquids that may be used in conjunction with loading, unloading, decontamination, wet transfer operations, electrolytes, and water. Reactions may include chemical and galvanic actions, the possibility of production of explosive or toxic gas, and/or degradation.

The SAR should include tables with material properties and allowable stresses and strains associated with temperature, as appropriate. Appropriate corrosion allowances should be established and used in the structural analyses. The potential for brittle fracture must be reviewed. The potential for brittle fracture of some components important to safety has resulted in conditions of use that preclude transfer operations during extremely low temperatures.

5.4.1.4 Structural Analysis

Requirements for acceptable structural analysis are given in 10 CFR 72.24 (d)(1), (d)(2), and (i), as well as 10 CFR 72.122 (b)(1), (b)(2), and (b)(3), (c), (d), (f), (g), (h), (i), (j), (k), and (l). The applicant must provide analyses of load combinations for normal, off-normal, and accident conditions.

The applicant must provide design analyses with adequate detail so that they may be readily audited to permit determination of the sources of expressions used, values of material properties, data from other supporting calculations and assumptions. ANSI N45.2.11 provides guidance for preparation of design analyses which is acceptable to the NRC.

The design analysis for confinement SSCs shall identify all loading conditions and combinations of loadings. The analysis shall establish the design internal and external pressures, the design temperatures, and all the design mechanical loads. The analysis shall identify all combinations of design loads which can occur simultaneously. The specification shall establish service loadings (with appropriate service limits), which are discussed as normal, off-normal, and accident conditions in this SRP. For comparison purposes, normal service corresponds to Service Levels A and B of the ASME B&PVC, Section III; and accident service corresponds to Service Level D.

5.4.1.5 Buckling of Irradiated Fuel Under Bottom End Drop Conditions

Fuel rod buckling analyses under bottom end drop conditions have traditionally been performed to demonstrate integrity of the fuel following a cask drop accident. The analytical method described by Lawrence Livermore National Laboratory (LLNL) in report UCID-21246, is a simplified approach. The analytical method assumes that buckling occurs when a fuel rod segment between the bottom two spacer grids reaches the Euler buckling limit. The analytical method uses material properties for irradiated cladding, considers the weight of the cladding, but neglects the weight of fuel pellets. The NRC considers that, in addition to the weight of the cladding, end drop analyses should include the weight of fuel pellets and irradiated material properties. With the weight of the fuel pellets included, the analytical method of UCID-21246 yields highly conservative results.

The analytical methods in UCID-21246 used to demonstrate fuel integrity following a cask drop accident yield a large margin to the point of actual failure. The calculated onset of buckling does not imply fuel or cladding failure. Where such analyses yield too conservative results, the applicant may use more realistic analyses of dynamic fuel behavior. If the cladding stress remains below yield strength, the fuel integrity is assured.

If the applicant uses the analytical approach described in UCID-21246 for axial buckling to assess fuel integrity for the cask drop accident, the analysis should use the irradiated material properties and should include the weight of fuel pellets.

Alternately, an analysis of fuel integrity which considers the dynamic nature of the drop accident and any restraints on fuel movement resulting from cask design is acceptable if it demonstrates that the cladding stress remains below yield. If a finite element analysis is performed, the analytical model may consider the entire fuel rod length with intermediate supports at each grid support (spacer). Irradiated material properties and weight of fuel pellets should be included in the analysis.

5.4.2 Pool and Pool Confinement Facilities

The pool and pool confinement facilities provide a capability that may be essential to the conduct of ISFSI and MRS loading for storage and unloading functions and that may be needed for retrievability (see guidance in SRP Sections 3.4.8 and 4.4.3.7). The pool and pool confinement facilities are considered to include those systems important to safety that provide for wet transfer, loading, unloading, and temporary holding or long-term storage of spent fuel, high-level waste, and/or other radioactive materials associated with spent fuel or high-level waste storage. Other ISFSI or MRS equipment that may be used within and outside the pool facility, or that are used for lifting or transfer within the facility but are not installed cranes or conveyance systems, are addressed as "other SSCs important to safety" or "other SSCs."

The safety function of the pool and associated equipment is to maintain the spent fuel assemblies in a safe and subcritical array during all credible storage conditions and to provide a safe means of loading the assemblies into shipping casks.

The ISFSI and MRS pools and pool facilities should be designed as though they were to be in constant use for in-pool storage and wet transfer for the life of the ISFSI/MRS license. However, it is anticipated that the actual use of the pool facility may differ from the use of the spent fuel pool at a reactor facility. Therefore, limited or part-time use of the pool should be well-described in the SAR. The use status of the pool facility may have a major impact on the generation of radioactive and other waste. The design may also need to provide for conversion to standby mode or decontamination and decommissioning (D&D) while the rest of the ISFSI or MRS remains in use for dry storage.

5.4.2.1 Description of Pool Facilities

10 CFR 72.24(a) and (b), 10 CFR 72.40(a)(3), 10 CFR 72.82(c)(2), and 10 CFR 72.106(a), (b), and (c) address the descriptive information to be included in a license application. The application must describe pool facilities in sufficient detail to support a detailed review and evaluation. This would include text, descriptions, drawings, flow diagrams, figures, tables, and specifications to fully define the systems and features of the pool facilities.

The NRC accepts use of existing pool and pool confinement facilities that are licensed under 10 CFR 50 for ISFSI or MRS, if concerns for possible sharing of SSCs between separately licensed facilities are satisfied (10 CFR 72.3 (included with definition of ISFSI), 72.24 (a), 72.40 (a)(3), and 72.122 (d)). The existing pool and pool confinement facilities may continue to be licensed under 10 CFR 50, or they may be re-licensed as elements of a wet storage and/or dry storage ISFSI.

5.4.2.2 Design Criteria

The regulatory requirements given in 10 CFR 72.24 (c)(1), (c)(2), and (c)(4); 10 CFR 72.40 (a)(1); 10 CFR 72.120 (a), and (b); 10 CFR 72.122 (a), (b), (c), (d), (f), (g), (h), (i), (j), (k), and (l); 10 CFR 72.128 (a) and (b); and 10 CFR 72.236 (b), (e), (f), (g), and (k) identify acceptable design criteria.

Design criteria for important to safety facilities in 10 CFR 72 are fully applicable to pool and pool confinement facilities. Pool and pool confinement facilities should meet the criteria for structural integrity for similar facilities constructed at a power reactor which must comply with 10 CFR 50. These criteria are principally as stated in 10 CFR 50, Appendix A, General Design Criteria 61, "Fuel Storage and handling and radioactivity control." Some portions of the General Design Criteria 62, "Prevention of criticality in fuel storage and handling," and General Design Criteria 63, "Monitoring fuel and waste storage" apply. Additionally, the General Design Criteria 2, 4, and 5 apply to the design of pool facilities. See NUREG-0800 Sections 9.1.2, Spent Fuel Storage and 9.1.3, Spent Fuel Pool Cooling and Cleanup System for specific acceptance criteria, which derives from 10 CFR Part 50, Appendix A.

The intended usage of the pool and pool facilities may be used in the development of design requirements. Should the intended usage be long-term storage of spent nuclear fuel, the NRC accepts design of elements of the pool facility in accordance with ANSI/ANS 57.2. Should the intended usage be short term or primarily to facilitate wet transfer operations, the NRC accepts design of elements of the pool facility in accordance with ANSI/ANS 57.7. Regardless of whether ANSI/ANS 57.2 or 57.7 is used, it should be noted that 10 CFR 72.2 requires that spent fuel be aged for at least one year after discharge from the core.

The NRC accepts design of the pool liquid containment SSCs as required for Quality Group B (per Regulatory Guide 1.26, "Quality Group Classifications and Standards for Water-, Steam-, and Radioactive Waste-Containing Components of Nuclear Power Plants") that are licensed under 10 CFR 50. This quality group requires design to not less than the requirements of ASME B&PVC, Section III, Class 2 (Division 1, Section NC).

The NRC accepts design of ISFSI and MRS pool facility cooling and make-up water systems (as required) for Quality Group C. This quality group requires design to not less than the requirements of ASME B&PVC, Section III, Class 3 (Division 1, Section ND).

The NRC accepts the guidance for reactor facility pools provided by Regulatory Guide 1.13, "Spent Fuel Storage Facility Design Basis," for ISFSI and MRS pool facilities. The principal criteria for pool facility design included in Regulatory Guide 1.13 are to:

- prevent loss of water from the pool that would uncover the radioactive material

- protect the radioactive material from mechanical damage

- provide capability for limiting the potential offsite exposures in the event of a significant release of radioactivity from the subject materials.

5.4.2.3 Material Properties

Acceptable criteria for materials used in all structural components and systems are given in 10 CFR 72.24 (c)(3). The applicant must identify materials and material properties to be used in the design.

The information describing material properties must be consistent with the application of the accepted design criteria, codes, standards and specifications for the structural components of the pool facility. For example, if pool components forming the primary hydraulic containment or water level control, such as piping, pumps, valves, holding tanks, or filters are designed according to the ASME B&PVC Section III, then the materials selected must be consistent with those allowed by the particular Section of the design code. If the pool is housed in a reinforced concrete building designed according to ACI 349, then material properties should be consistent with the ACI 349 Code. If steel structures are to American Institute of Steel Constructions (AISC) standards, then the steel should have material properties from the Steel Construction Manual.

In addition to the criteria given in 10 CFR 72.24 (c)(3), materials wetted by the pool water should be reviewed for compatibility and chemical stability. The selection of materials should be such that there are no potential mechanisms that will: (1) alter the location of any fixed neutron absorbers used in the design of the storage racks, and/or (2) cause physical distortion of the structures designed to retain the stored fuel assemblies in a fixed location.

5.4.2.4 Structural Analysis

Requirements for acceptable structural analysis are given in 10 CFR 72.24 (d)(1), and (d)(2), (i), as well as 10 CFR 72.122 (b)(1), (b)(2), and (b)(3), (c), (d), (f), (g), (h), (i), (j), (k), and (l).

Design analyses should be prepared such that they may be readily audited to permit determination of the sources of expressions used, values of material properties, data from other supporting calculations, and assumptions. ANSI N45.2.11 provides guidance for preparation of design analyses which is acceptable to the NRC.

The design specification for SSCs comprising the pool and the pool facilities shall identify all loading conditions and combinations of loadings. The specification shall establish the design internal and external pressures, the design temperatures, and all the design mechanical loads. The specification shall identify all combinations of design loads which can occur simultaneously. The specification shall establish service loadings (with appropriate service limits), which are discussed as normal, off-normal, and accident conditions in this SRP. ANSI/ANS 57.2 and ANSI/ANS 57.7 provide guidance for establishing design loads and structural analysis methods. Design codes are discussed.

5.4.3 Reinforced Concrete Structures

5.4.3.1 Description of Concrete Structures

10 CFR 72.24 (a) and (b), 10 CFR 72.82(c)(2), and 10 CFR 72.106(a), (b), and (c) outline the contents of the application, which includes design descriptions in sufficient detail to support a detailed review and evaluation. Concrete structures may have roles in providing radiological shielding, forming ventilation passages, weather enclosures, structural supports, access denial, foundations, earth retention, anchorages, floors, walls, movable shields, and protection against natural phenomena and accidents. The applicant must fully describe any reinforced concrete structures. The description should include text descriptions, drawings, figures, tables, and specifications that would fully define the structural features of the reinforced concrete structures.

Concrete structures may be cast in place, cast at the site, or cast elsewhere. Concrete structures may also be combinations of cast in place and precast sections that are integrated by bolting, welding, fitting, grouting, or placing additional concrete at the site. They may also include concrete that may be cast as part of a composite confinement cask with metallic liner. A metallic liner of a composite confinement cask, its closures, or its internal components should be designed as required for confinement SSCs (5.4.1).

5.4.3.2 Design Criteria

The regulatory requirements given in 10 CFR 72.24 (c)(1), (c)(2), and (c)(4); 10 CFR 72.40 (a)(1); 10 CFR 72.120 (a), and (b); 10 CFR 72.122 (a), (b), (c), (d), (f), (g), (h), (i), (j), (k), and (l); 10 CFR 72.128 (a) and (b); and 10 CFR 72.236(b), (e), (f), (g), and (k) identify acceptable design criteria.

The structural design of the concrete structures shall withstand the effects of credible accident conditions and natural phenomena events without impairment of their capability to perform safety functions. The principal safety functions include maintaining subcriticality, containing radioactive material, providing radiation shielding for the public and workers, and maintaining retrievability of the stored material.

The NRC has accepted special criteria for selection of components of reinforced concrete that may be exposed to elevated temperatures in normal or off-normal conditions. These criteria are given in the SRP Section 6.5.2.3. The acceptability of loads and stresses associated with thermal conditions is analyzed as part of the structural analysis.

Concrete pads that support confinement casks in storage are not "pavements." They should be designed and constructed as foundations under the applicable code (ACI 318 or ACI 349).

Codes and Standards

ANSI/ANS 57.9 is generally applicable to ISFSI design and construction (with exceptions for confinement casks). Table 3-1 of NUREG-1536 includes extracts of ANSI/ANS 57.9 that are especially applicable to concrete structure design and construction. The table also includes corresponding evaluation guidance for review of the SAR documentation.

The NRC has not accepted use of a set of criteria that has been derived by selection of criteria from more than one code. However, the NRC has accepted use of ACI 349 for design and material selection for concrete structures important to safety (but not as confinement cask), but has allowed the optional use of ACI 318 for construction, as described in this Section.

There are codes other than those discussed herein that may be applicable to the design and construction of the concrete elements of ISFSI and MRS. It is acceptable that such codes (e.g., the National Fire Protection Association (NFPA) Electric, Life Safety and Lightning Protection Codes) be included in the design by reference in the SAR documentation. Where designs of structures subject to approval are also covered by such other codes, the review should include evaluation of compliance with those codes.

The NRC accepts use of ACI 349 for design, material selection and specification, and construction of all concrete structures that are not within the scope of ACI 359; except that additional or more stringent requirements given in ANSI/ANS 57.9, as incorporated by reference in NRC Regulatory Guide 3.60, "Design of an Independent Fuel Storage Installation (Dry Storage)," must also be met. Use of ACI 318 for construction of structures designed and with materials selected in accordance with ACI 349 is acceptable.

The following identifies the portions of ACI 349 and ASTM standards that are applicable to design (including material selection and metal embedments) that must be met by those applicants that choose to use ACI 318 for construction. The paragraph references are as in ACI 349-90. Unlisted and excepted sections cover construction requirements, for which the NRC accepts substitution of ACI 318.

Chapter 1,	"General Requirements", Section 1.1 and 1.5 (less references to construction), Section 1.2, Section 1.4
Chapter 2,	"Definitions", All
Chapter 3,	"Materials, All, except Section 3.1, 3.2.3, 3.3.4, 3.5.3.2, 3.6.7, 3.7
Chapter 4,	"Concrete Quality", Section 4.1.4
Chapter 6,	"Form work, Embedded Pipes, and Construction Joints", Section 6.3.6(k), 6.3.8
Chapter 7,	"Details of Reinforcement", All
Chapter 8,	"Analysis and Design" - General Considerations, All
Chapter 9,	All
Chapters 10-19,	All
Appendix A,	All
Appendix B,	"Steel Embedments," All, but note that the load combinations and load variation requirements of ANSI/ANS 57.9 must be met in addition to those of ACI 349 Section 9.2 cited at Section B.3.2 (given in Table 3-1 of NUREG-1536)
Appendix C,	"Special Provisions for Impulsive and Impactive Effects", All, except that the load combinations and load variation requirements of ANSI/ANS 57.9 must be met in addition to those of ACI 349 Section 9.2 (given in Table 3-1 of NUREG-1536).

Concrete Containments

ACI 359, Section CC, is acceptable for prestressed and reinforced concrete that is an integral component of a radioactive material containment vessel that must, in operation or in testing, withstand internal pressure. Application of ACI 359 is based on the containment function, regardless of whether the concrete structure is fixed or portable, or where the concrete structure is fabricated. ACI 359 also applies to structural concrete supports that are constructed as an integral part of the containment.

If ACI 359 is applicable to an ISFSI/MRS structure, it is applicable for the full design, material selection, fabrication, and construction of that structure. The NRC has not accepted the substitution of elements of ACI 349 or ACI 318 for any portion of ACI 359 for an ISFSI/MRS structure. Structures for which ACI 359 is applicable shall also meet the minimum functional requirements of ANSI/ANS 57.9, where specific requirements in the subject area are not included in ACI 359.

5.4.3.3 Material Properties

Acceptable criteria for materials used in all structural components and systems are given in 10 CFR 72.24 (c)(3).

The information describing material properties must be consistent with the application of the accepted design criteria. For concrete structures as referenced in ACI 349-90, this would include ASTM standard specifications applicable to design and material specifications: A 36, A 53, A 82, A 184, A 185, A 242, A 416, A 421, A 496, A 497, A 500, A 501, A 572, A 588, A 615, A 706, A 722, C 33, C 144, C 150, C 595, and C 637.

Fabrication and Construction

Selection and validation of concrete mix to meet design requirements is considered to be a construction function. Specification of cement type, aggregates, and special requirements for durability and elevated temperatures is considered to be a design or material selection function, and therefore, to be governed by ACI 349 (ACI 359 if applicable).

The following identifies sections of ACI 318, Building Code Requirements for Reinforced Concrete (chapters, appendix, and paragraphing per ACI 318-89) that have been accepted by the NRC for construction of ISFSI concrete structures that are not within the scope of ACI 359.

Chapter 1,   "General Requirements", Section 1.1.1, 1.1.2, 1.1.3, and 1.1.5 (less references to design and material properties); Section 1.3
Chapter 2,   "Definitions", use ACI 349 Chapter 2
Chapter 3,   "Materials", Section 3.1, Section 3.8 (except delete A 616 and A 617)
Chapter 4,   "Durability Requirements", All
Chapter 5,   "Concrete Quality, Mixing, and Placing", All
Chapter 6,   "Form work, Embedded Pipes, and Construction Joints", All (less references to design and material properties, these are governed by ACI 349)

ASTM standard specifications acceptable for construction and associated testing are: C 31, C 39, C 42, C 94, C 109, C 172, C 192, C 260, C 494, C 496, C 685, and C 1017.

The following standards relating to construction are identified in ACI 349 and may be used: C 88, C 131, C 289, and C 441.

ASTM standard specifications acceptable for construction and associated testing are: C 31, C 39, C 42, C 94, C 109, C 172, C 192, C 260, C 494, C 496, C 685, and C 1017.

5.4.3.4 Structural Analysis

Requirements for acceptable structural analysis are given in 10 CFR 72.24 (d)(1), and (d)(2), (i), as well as 10 CFR 72.122 (b)(1), (b)(2), and (b)(3), (c), (d), (f), (g), (h), (i), (j), (k), and (l).

Design analyses should be prepared such that they may be readily audited to permit determination of the sources of expressions used, values of material properties, data from other supporting calculations, and assumptions. ANSI N45.2.11 provides guidance for preparation of design analyses which is acceptable to the NRC.

The design specification for concrete structures shall identify all loading conditions and combinations of loadings. The specification shall establish the design internal and external pressures, the design temperatures, and all the design mechanical loads. The specification shall identify all combinations of design loads which can occur simultaneously. The specification shall establish service loadings (with appropriate service limits), which are discussed as normal, off-normal, and accident conditions in this SRP.

The NRC accepts strength design as presented in the current ACI 349 for concrete structures important to safety that are not within the scope of ACI 359. ACI 359 is based on allowable stress design.

Load definitions and load combinations shown in Table 3-1 of NUREG-1536 have been accepted by the NRC for analysis of steel and reinforced concrete ISFSI and MRS structures important to safety. The load combinations are as included or derived from ANSI/ANS 57.9 and ACI 349. Load combinations to be used for concrete structures designed in accordance with ACI 359 should be as given in ACI 359 (Section CC3230)

5.4.4 Other SSCs Important to Safety

5.4.4.1 Description of Other SSCs Important to Safety

10 CFR 72.24 (a) and (b), 10 CFR 72.82(c)(2), and 10 CFR 72.106(a), (b), and (c) outline the contents of the application, which includes design descriptions in sufficient detail to support findings in the SER. For other SSCs important to safety this would include text descriptions, drawings, figures, tables, and specifications that would fully define the structural features of the items identified.

5.4.4.2 Design Criteria

The regulatory requirements given in 10 CFR 72.24(c)(1), (c)(2), and (c)(4); 10 CFR 72.40 (a)(1); 10 CFR 72.120 (a), and (b); 10 CFR 72.122 (a), (b), (c), (d), (f), (g), (h), (i), (j), (k), and (l); 10 CFR 72.128 (a) and (b); and 10 CFR 72.236(b), (e), (f), (g), and (k) identify acceptable design criteria.

Codes and Standards

The NRC accepts use of ANSI/ANS 57.9 and the codes and standards cited therein as the basic references for ISFSI structures important to safety that are not designed in accordance with the ASME B&PVC Section III.

The principal included references applicable to steel structures and components are the following:

- AISC, "Specification for Structural Steel Buildings - Allowable Stress Design and Plastic Design"

- AISC, "Code of Standard Practice for Steel Buildings and Bridges"

- AWS D 1.1, "Structural Welding Code-Steel"

- ASCE 7, "Minimum Design Loads for Buildings and Other Structures," however, note that the load combinations of ANSI/ANS 57.9 are to be used

5.4.4.3 Material Properties

Acceptable criteria for materials used in all structural components and systems are given in 10 CFR 72.24 (c)(3).

5.4.4.4 Structural Analysis

Requirements for acceptable structural analysis are given in 10 CFR 72.24 (d)(1), (d)(2), and (i), as well as 10 CFR 72.122 (b)(1) and (b)(2), (c), (d), (f), (g), (h), (i), (j), (k), and (l).

Design analyses should be prepared such that they may be readily audited to permit determination of the sources of expressions used, values of material properties, data from other supporting calculations, and assumptions. ANSI N45.2.11 provides guidance for preparation of design analyses which is acceptable to the NRC.

The design specification for all other SSCs important to safety shall identify all loading conditions and combinations of loadings. The specification shall establish the design internal and external pressures, the design temperatures, and all the design mechanical loads. The specification shall identify all combinations of design loads which can occur simultaneously. The specification shall establish service loadings (with appropriate service limits), which are discussed as normal, off-normal, and accident conditions in this SRP.

5.4.5 Other SSCs

5.4.5.1 Description of Other SSCs

10 CFR 72.24 (a) and (b), 10 CFR 72.82 (c)(2), and 10 CFR 72.106 (a), (b), and (c) outlines the contents of the application, which includes design descriptions in sufficient detail to support findings in the SER. For other SSCs subject to NRC approval this would include text descriptions, drawings, figures, tables and specifications that would fully define the structural features of the items identified.

5.4.5.2 Design Criteria

The regulatory requirements given in 10 CFR 72.24 (c)(1), (c)(2), and (c)(4); 10 CFR 72.40 (a)(1); 10 CFR 72.120 (a), and (b); 10 CFR 72.122 (a), (b), (c), (d), (f), (g), (h), (i), (j), (k), and (l); 10 CFR 72.128 (a) and (b); and 10 CFR 72.236 (b), (e), (f), (g), and (k) identify acceptable design criteria.

Codes and Standards

The principal structural codes and standards for SSCs which are not important to safety but which are subject to NRC approval include:

*   ASCE 7
*   Uniform Building Code (UBC)
*   AISC, "Specification for Structural Steel Buildings, Allowable Stress Design and Plastic Design"
*   AISC, "Code of Standard Practice"
*   ASME B&PVC, Section VIII

The above include acceptable load definitions and load combinations. Load definitions and load combinations shown in Table 3-1 of NUREG-1536 have been accepted by the NRC for analysis of steel and reinforced concrete ISFSI structures important to safety. These may also be used for structures not important to safety.

5.4.5.3 Material Properties

Acceptable criteria for materials used in all structural components and systems are given in 10 CFR 72.24 (c)(3).

5.4.5.4 Structural Analysis

Requirements for acceptable structural analysis are given in 10 CFR 72.24 (d)(1), (d)(2), and (i), as well as 10 CFR 72.122 (b)(1), (b)(2), and (b)(3), (c), (d), (f), (g), (h), (i), (j), (k), and (l).

Design analyses should be prepared such that they may be readily audited to permit determination of the sources of expressions used, values of material properties, data from other supporting calculations, and assumptions. ANSI N45.2.11 provides guidance for preparation of design analyses which is acceptable to the NRC.

The design specification for all other SSCs subject to NRC approval shall identify all loading conditions and combinations of loadings. The specification shall establish the design internal and external pressures, the design temperatures, and all the design mechanical loads. The specification shall identify all combinations of design loads which can occur simultaneously. The specification shall establish service loadings (with appropriate service limits), which are discussed as normal, off-normal, and accident conditions in this SRP.
Load combinations for analysis of structures not important to safety but subject to NRC approval should be as given in acceptable codes and standards. The load combinations given in ACI 318 or the Uniform Building Code (UBC) are appropriate for SSCs not important to safety.

5.5 Review Procedures

The following procedures are generally applicable to the structural evaluation of all SSCs subject to NRC approval.

Review the entire application, particularly the sections that describe the overall design and operations, as given in Chapters 4 and 5 of the SAR; the design criteria and bases, and structural evaluation information as given in Chapter 3; the accident analysis in Chapter 8 of the SAR; and the operating controls and limits in Chapter 10 of the SAR. If drawings and calculation packages were submitted with the application, review those which are pertinent to the particular structure being evaluated. From Chapter 3, ensure that all the components which are identified as important to safety or otherwise require NRC approval have been included in Chapter 7.

5.5.1 Confinement Structures, Systems, and Components

5.5.1.1 Description of Confinement Structures

Review the descriptive material in the SAR. The text descriptions along with the drawings, figures, tables, and specifications included in the application should fully define the confinement SSCs.

The reviewer should determine if SSCs important to safety are described in sufficient detail in the SAR or its supporting documentation to enable an evaluation of their structural and functional suitability. The configurations are defined by drawings and fabrication specifications. The specifications should include reference to the codes that govern design details not shown on the drawings. The combination of the drawings, specifications, and proper application of the codes and standards cited in the specifications or on the drawings accompanying the license application, should provide a design that is so defined that final fabrication drawings and specifications could be prepared without further information.

The structural components of a storage cask may include: the cask body (including an inner shell, an outer shell, and gamma radiation shielding), any integral structural supports or lifting and handling aids, inner lid (to be welded or bolted), port covers (to be welded or bolted), outer lid (to be welded or bolted), neutron shields and shell, trunnions, fuel basket, exterior components forming elements of the confinement boundary during storage, such as tubes and valves used to monitor the pressure of the storage cavity, and impact limiters.

At a minimum, the SAR documentation should provide the following: (1) the dimensions of all sections of the confinement structure, including locations, sizes, configurations, and weld specifications, (2) structural materials with defining standards or specifications, including test requirements such as brittle fracture testing, (3) fabrication, assembly, and test procedures for assemblies and subassemblies, and (4) weld materials, and weld codes, including pre- and post-heat requirements.

Coordinate with the confinement review Chapter 9 of this SRP to verify that the SAR clearly identifies the confinement boundaries. The confinement boundaries may include the primary confinement vessel, the penetrations, seals, welds, and closure devices. Any redundant sealing joints should be described. Ensure that the applicant has provided proper specifications for all welds and bolted closures.

Review the calculations which quantify the weights and centers of gravity, and verify that the applicant used limiting cases for structural evaluations.

Fabrication and Construction

The NRC has accepted fabrication of confinement casks in accordance with the ASME B&PVC, Section III. Any deviations from use of the particular Section (as identified by the applicant) of the ASME B&PVC, Section III used for design as the code for construction, fabrication, or assembly of the confinement cask, must be explicitly justified in the SAR and accepted by the NRC. The reviewer should especially address any specifications for preparation for welding, materials to be used in welds, performance of welding, and inspection of welds that do not fully comply with the Code.

5.5.1.2 Design Criteria for Confinement Structures

For each of the confinement SSCs being reviewed, the reviewer should review the design criteria, design bases, and design codes proposed by the applicant. In the event that the reviewer does not concur with the SAR, the issue may be resolved to the staff's satisfaction by writing a Request for Additional Information. Acceptable design criteria for codes and standards are discussed in Section 5.4.1.2.

Review the confinement boundary weld designs for compliance with the design code used. Acceptable weld design codes appear in the ASME Code Section III, Sections NB-3352 and NC-3352, "Permissible Types of Welded Joints," and NB-4240 and NC-4240, "Requirements for Weld Joints in Components." Welds must be well characterized on drawings using standard welding symbols and/or notations as discussed in American Welding Standard (AWS) A2.4.

The NRC has previously accepted alternative confinement boundary weld designs that achieve equivalent structural integrity, but do not meet all the provisions of NB-3352 or NC-3352 for full penetration welds or do not meet the non-destructive examination (NDE) requirements for full volumetric examination (NB-5200 or NC-5200, typically for Category C welded joints). The NRC has accepted alternative designs for the welds of the head or flat end plate to the cylindrical portion of the confinement vessel. The NRC has required redundant seals for these alternative designs.

Structural Acceptance Testing

The NRC has accepted use of the codes and standards used for design of the confinement SSCs as the basis for structural acceptance testing. These codes may incorporate other codes, standards, and specifications by reference. The reviewer should verify that for the confinement system, the ASME Section III, Section NB or NC, depending on the Section used for design, is specified for acceptance testing. The reviewer should verify that for the cask internals, (e.g., basket) the ASME Section III is specified for acceptance testing.

Confirm that cask components are fabricated and examined in accordance with an accepted standard used for their design, in overview: Section II ("Materials Specifications and Properties"), Section V ("NDE Specifications and Procedures"), and Section IX ("Qualification Standard for Welding and Brazing Procedures, Welders, Brazers, and Welding and Brazing Operators").

The reviewer should verify that NDE of weldments is well characterized on drawings, using standard NDE symbols and/or notations (as given in AWS A2.4). Check the appropriate documents for a detailed, weld inspection plan in accordance with an approved Quality Assurance program that complies with 10 CFR 72, Subpart G. The inspection plan should:

•   include visual tests (VT), dye penetrant tests (PT), magnetic particle tests (MT), ultrasonic tests (UT), and radiographic tests (RT), as applicable.

•   identify welds that will be examined

•   include the examination sequence

•   identify the type of examination

•   state the appropriate acceptance criteria

- require that inspection personnel be pre-qualified in accordance with the current revision of SNT-TC-1A (as specified by the ASME B&PVC).

The reviewer should verify that confinement boundary welds and welds for components performing redundant sealing meet the requirements of ASME B&PVC Section III, NB-5200 or NC-5200. This generally requires RT or UT for volumetric examination and either PT or MT for surface examination. Redundant seal welds for the confinement boundary which do not meet the configuration for a "pre-qualified," full penetration weld according to the ASME B&PVC, should be avoided in the design process. When a pre-qualified, full penetration weld cannot be used, every effort should be made to permit full volumetric inspection of the weld by means of UT techniques in conformance with NB-5330 or NC-5330.

The NRC has accepted multiple surface examinations of welds combined with helium leak tests for inspecting the final redundant seal welded closures. The reviewer should verify that PT tests are performed in accordance with ASME B&PVC Section V, Article 6. Acceptance criteria for confinement welds should be in accordance with ASME B&PVC Section III NB-5350 or NC-5350. Repair procedures should be in accordance with NB-4450 or NC-4450.

Confirm that RT tests are in accordance with ASME B&PVC Section III, NB-5320 or NC-5320. Confirm that UT tests are in accordance with NB-5330 or NC-5330. Repaired welds should be reexamined in accordance with the original examination method and associated acceptance criteria.

Fabrication controls and specifications should be in place and field verifications performed to prevent post-welding operations (such as grinding) from compromising the design requirements (such as wall thickness). The specifications should be clear that reduction of wall thickness at the weld region is not acceptable.

Structural Pressure Tests and Leak Tests

Confirm that the confinement boundary (including that of the redundant sealing) will be tested at an overpressure, in accordance with ASME B&PVC, Section III, Article NB-6000 or NC-6000. 10 CFR 72.122 requires that the cask system be designed to withstand postulated accidents. The pressure test should be at a pressure level that is not less than the maximum cask cavity pressure with 100 percent failure of the fuel rods. The test pressure should be maintained for a minimum of 10 minutes, after which a visual inspection should be performed to detect any leakage. All accessible welds shall be PT inspected. The test pressure should be clearly specified in the SAR.

Confirm that leak tests will be performed on all confinement boundaries. These include the primary confinement boundary, the boundary of the redundant sealing, and, if applicable, any additional boundaries used in the pressure monitoring system. Leakage criteria in units of std cc/s must be at least as restrictive as those specified in the principal design criteria. The general testing methods (e.g., pressure rise, mass spectrometer) and the required sensitivities should also be indicated. If cask closure depends on more than one seal (e.g., lid, vent port, drain port), the leakage criteria should ensure that the total leakage is within the design requirements. The reviewer should verify that leak testing will be conducted in accordance with ANSI N14.5.

Cask Closure Welds After Fuel Loading

The reviewer should verify that, if the applicant proposes to use the special considerations for loaded dry storage canister top end closure welds described in section 5.4.1.2, that the applicant adequately describes how the requirements of section 5.4.1.2 will be met.

5.5.1.3 Material Properties

Coordinate with the thermal review, Chapter 6 of the SRP, to verify that the material properties used in the structural analysis are appropriate for the load condition (i.e. hot or cold temperature) and that the appropriate temperature at which allowable stress limits are defined is consistent with service temperatures.

For each of the confinement SSCs being reviewed, determine what structural materials are specified, and verify that the information defining the materials is consistent with the accepted design codes and standards. Acceptable material requirements are discussed in Section 5.4.1.3. Chapter 3, Section V of NUREG-1536 provides a comprehensive discussion of review procedures for materials.

In reviewing the structural materials, consider the source of the information. If the applicant has selected the ASME B&PVC Section III for the design code, then the material properties should be taken from Section II of the Code. Confinement vessels may use components which have no structural role except that the mass must be considered. In such cases, sources of material properties need not be taken from Section II of the Code; however, preferred sources include industry and Government standards and specifications.

For ASME B&PVC, Section III, Section NB or NC applications, additional material requirements regarding examination prior to fabrication, testing, analyses, and traceability are applicable. Compliance with the requirements of the following Section III paragraphs, or their equivalent, must be acknowledged in the SAR: NB-2121 or NC-2121 (Permitted Material Specifications), NB-2130 or NC-2130 "Certification of Material," NB-2500 or NC-2500 "Examination and Repair of Pressure Retaining Material," and NB-2400 or NC-2400 "Welding Material."

Review the structural materials that are in direct contact with each other and with other materials, and verify that they will not produce a significant chemical or galvanic reaction and initiate corrosion or generate combustible gas. NRC Bulletin 96-04, "Chemical, Galvanic, or other Reactions in Spent Fuel Storage and Transportation Casks," may be referred to for additional information on this topic. Evaluate the potential for corrosion to ensure that the applicant has provided for appropriate corrosion allowance for materials susceptible to corrosion.

Review the test procedures and performance specifications in the SAR for any material which has the potential for brittle fracture at low operational temperatures. The reviewer should verify that limiting conditions of operation in the technical specifications chapter of this SRP are specified for such materials. Ensure that consistent test procedures are cited in the SAR and that they are applicable. Section III of the ASME B&PVC has consistent test procedures and performance requirements for primary confinement vessels (i.e., Sections NB and NC); however the reviewer may require testing to prevent brittle fracture for internal basket components which exceed the Code requirements (i.e., Subsections NF and NG) for some materials and/or material thicknesses. The basis for this is that two functions of basket components are to prevent criticality and ensure ready retrievability. These functions are outside the scope and intent of Subsections NF and NG. Regulatory Guides 7.11 and 7.12 may be referred to for determining the bases for brittle fracture.

5.5.1.4 Structural Analysis

The reviewer should verify that the design analyses include determination of the sources of expressions used, properties used for structural materials and components, and data derived by other calculations and assumptions.

Load Conditions

Coordinate with the thermal review in Chapter 6 of this SRP to verify that the temperatures and pressures for all confinement structures presented in the SAR correspond to the same temperatures given in the thermal stress analysis.

Coordinate with the operating system review in Chapter 3 of the SRP to verify that the configuration of the confinement structure (i.e., storage cask in a transfer component, or storage cask on the storage pad or in the spent fuel pool, etc.) corresponds to the same configuration used in the various load conditions and load combinations.

Chapter 3, Section V of NUREG-1536 has a detailed discussion of appropriate review procedures for the structural analysis of casks. The following discussion briefly outlines load conditions which are necessary to meet the structural requirements of 10 CFR 72: normal conditions, off-normal conditions, and accident conditions (including natural phenomena).

Normal Conditions

Normal conditions are associated with the normal range of environments for operations and storage. The limits of normal use environments are supported by the Environmental Report, Site Characteristics, and/or the Operating Procedures.

Loads normally applicable to a confinement cask are weight, internal/external pressure, and thermal loads caused by temperature gradients. Normal conditions include handling and transfer operations. The weight is the maximum or design weight of the cask as it is stored and loaded with spent fuel. However, for certain operation and procedures, the weight should include water fill. All orientations of the cask body and closure lids during normal operations and storage conditions should be evaluated.

The reviewer should verify that the stress intensity level is below the stress limits for dead weight, pressure, normal handling and transfer operations, thermal loadings, and all load combinations (i.e., Service Levels A and B of the ASME B&PVC). The reviewer should verify that the maximum weight is used and that all normal temperature conditions are considered. The reviewer should verify that the maximum temperature gradient is considered.

Off-Normal Conditions

Off-normal conditions are considered to include those events that may reasonably be expected to occur during the life of the cask system and that exceed normal conditions. Environmental limits should be stated to support comparison of the cask system design bases with specific site environmental data. Off-normal conditions can involve mishandling, simple negligence of equipment operators, equipment malfunction, loss of power, and severe weather (short of extreme natural phenomena).

The reviewer should verify that the stress intensity level is below the stress limits for off-normal conditions and load combinations (i.e., Service Level C of the ASME B&PVC).

Accident Conditions

Coordinate with the accident analysis review in Chapter 15 of this SRP to verify that all accidents presented in that chapter have been adequately analyzed for structural integrity. Accident conditions are considered to include events that exceed the levels associated with off-normal conditions. Hypothetical accidents may or may not actually occur in the design life of the SSC; however, the reviewer must verify that the structure has been designed to resist the accidents. The reviewer should verify that all accidents have either been analyzed, or alternatively, that the effects of the accident have been shown to be bounded by another credible accident event.

The NRC accepts that the confinement system may experience some permanent deformation but no loss of confinement or other safety function in response to accident conditions. The reviewer should verify that the stress intensity level is below the stress limits for all accident conditions and accident load combinations (i.e., Service Level D of the ASME B&PVC). Other SSCs important to safety may experience some deformation and limited damage in response to accident conditions, if this is readily apparent and remedial actions are identified.

The following accidents should be included as a part of the analysis submitted in the SAR for confinement SSCs. For a more detailed discussion of these accidents, see Chapter 3 of the NUREG-1536.

Cask Drop

The SAR should identify the operating environment experienced by the cask and the drop events (end/side/corner) that could result. The "operating environment" includes the configuration of the confinement SSCs, i.e., a storage cask impacts a storage pad horizontally, or a storage cask inside a transfer component impacts a spent fuel pool floor vertically. The reviewer should verify that the impact surface is characterized sufficiently to quantify the deceleration level.

The maximum height above a receiving (impact) surface to which the cask could be lifted should be used for the design basis accident drops, if the hypothetical drops occur outside of a spent fuel pool building. The analysis should recognize that a drop may involve initial impact with the storage confinement cask at a wide range of orientations. Further, different orientations at the time of initial impact can result in the highest stresses for different elements of the confinement cask and its internal components. The reviewer should verify that the worst drop cases have been examined and that the stress intensity level is below the stress limit (i.e. Service Level D of the ASME B&PVC).

Cask Tip-over

The NRC requires that occurrence of a cask tip-over be assumed and analyzed. For this analysis, the NRC will accept cask tip-over about a lower corner onto a receiving surface from a position of balance with no initial velocity. The NRC has also accepted analysis of cask drops with the longitudinal axis horizontal, which together with a drop with the longitudinal axis vertical, could bound a non-mechanistic tip-over analysis.

Explosive Overpressure

Coordinate the structural review with Chapters 2 and 15 to determine what scenarios were considered in the SAR for explosive overpressure. Explosion-caused overpressure and reflected pressure may result from sources such as: explosion hazards associated with explosives, fuels, and chemicals transported by rail or on public highways; natural gas pipelines; vehicular fires involving equipment used in the transfer of casks; and aircraft crash. With the exception of transfer vehicle accidents, the explosion hazards are typically similar to those for facilities subject to 10 CFR Part 50 reviews.

As an accident condition, the structures are not required to survive an explosion's effects without damage or permanent deformation. The maximum response should be determined and should be shown in the SAR documentation. The reviewer should verify that the component's confinement integrity is maintained by showing that the stress intensity level is below the stress limit (i.e., Service Level D of the ASME B&PVC). Note, the "explosive overpressure" is not meant to be that from a sabotage event. There is currently no design basis sabotage event.

Fire

To check if a hypothetical fire accident was considered, coordinate the structural review with Chapter 2, Site Characteristics, and Chapter 15, Accident Analysis. If a fire was postulated, determine from Chapter 6, the thermal evaluation chapter of the SRP, what the response of the confinement cask was. The structural evaluation for fire should include increased pressures in the confinement cask. Allow for temporary loss of strength at elevated temperatures and permanent loss of strength because of annealing. The reviewer should verify if the response included physical destruction (e.g., surfaces of concrete exposed to intense or prolonged high temperatures).

Flood

Coordinate the structural review with the site characteristics, and identify the severity and frequency of potential flooding. Flood control or mitigation measures should be included in the installation design for the site. Regulatory Guides 1.59, "Design Basis Floods for Nuclear Power Plants," and 1.102, "Flood Protection for Nuclear Power Plants," provide guidance for flood protection.

Confirm that the resistance of the confinement cask to flood hydrostatic pressure is analyzed in accordance with ASME B&PVC, Section III, Section NB or NC (depending on the Section used for design). Table 3-1 in NUREG-1536 includes analyses for tip-over and sliding that are applicable to potential flood forces on an exposed cask and other structures. The reviewer should verify that the confinement cask does not tip over or slide due to the effects of a potential flood.

Tornado Winds

Coordinate with Chapters 2 and 15 of this SRP to determine what wind conditions are applicable to the facility. Regulatory Guide 1.76, "Design Basis Tornado for Nuclear Power Plants," and NUREG-1503 provide applicable tornado parameters. ANSI/ANS 57.9 provides acceptable criteria for resistance to overturning or sliding. ASCE 7 provides an acceptable conversion of wind speed to lateral pressure and coefficients for pressure coefficients.

Confinement casks are generally not vulnerable to damage from overpressure or negative pressure associated with tornadoes or extreme winds. However, they may be vulnerable to secondary effects, such as wind-borne missiles or collapse of a weather enclosure or adjacent stack. Tornado or extreme winds have been a governing load condition in prior reviews for major structures (other than confinement casks) that form part of an ISFSI system.

The NRC has maintained a position that warning of tornadoes should not be assumed. Therefore, the effects of tornadoes during operations such as transfer between the pool facility and a storage site must be evaluated. The reviewer should verify that the confinement cask does not tip over due to the effects of tornado winds.

Tornado Missiles

Tornado winds and missiles are described in Regulatory Guide 1.76, NUREG-1503, and NUREG-0800 (Section 3.5.1.4). The reviewer should verify that the SAR has defined the missile parameters for which the cask system is evaluated. NUREG-0800 (Section 3.3.2) states that the most adverse combined effects of tornado winds, tornado missiles, and tornado differential pressure should be evaluated. The reviewer should verify that the combined effects of tornado loading does not cause a tip-over. Confirm from the calculations that damage to the confinement cask does not result in release of radioactive material, unacceptable radiation dose, or preclude ready retrieval of the fuel.

The NRC has accepted use of the analytical approaches given in ORNL-NSIC-5, Volume 1, Chapter 6 for estimating the potential effects of missile impact on steel sheets, plates, and other structures. Further guidance on acceptable analytical approaches is in NUREG-0800 Section 3.5.3, "Barrier Design Procedures." The NRC has accepted use of Kennedy, R.P., "A Review of Procedures for the Analysis and Design of Concrete Structures to Resist Missile Impact Effects," for analysis and design of reinforced concrete structures to resist missiles.

Earthquake

The reviewer should verify that the confinement SSCs are designed to maintain principal safety functions during the maximum response to an earthquake. The design earthquake is that developed from the analysis of the site and reported in the Environmental Report and SAR Site Characteristics. Confirm that the design earthquake in Chapter 2 and 15 of the SRP corresponds to the value used in the structural evaluation. The design earthquake shall not be less than that required for the site by 10 CFR 72.102. The reviewer should verify that regulatory guidance as provided in Regulatory Guides 1.29, "Seismic Design Classification;" 1.60, "Design Response Spectra for Seismic Design of Nuclear Power Plants;" 1.61, "Damping Values for Seismic Design of Nuclear Power Plants;" 1.92, "Combining Modal Responses and Spatial Components in Seismic Response Analysis;" and NUREG-0800 has been appropriately followed.

Storage confinement casks and SSCs are not required to survive accident-level earthquakes without permanent deformation; however, the reviewer should verify that the stress intensities are less that the stress allowable (i.e., Service Level D of the ASME B&PVC). The reviewer should verify that the confinement cask does not tip over or slide due to the effects of the seismic event.

Structural Analysis Methods for Confinement Structures

NUREG-1536 has a detailed discussion of structural analysis methods and procedures which are appropriate for evaluating structural integrity of confinement SSCs. These procedures include discussion of finite element methods, closed-form calculations, and prototype or scale model testing.

Values for the stress intensity limits, based on the maximum shear stress theory for ductile materials, are defined in the ASME B&PVC. Confirm that the stress intensities are below the stress limits for all load conditions and load combinations.

Compare, when feasible, solutions from finite-element analyses with closed-form calculations. For example, the stress state caused by internal pressure in the cask can be checked with the formulas for the stress in a cylinder with end-caps. A source of closed-form equations for stress analysis which is accepted by the NRC is Young's Roark's Formulas for Stress and Strain.

Prototype or scale model testing may be performed in lieu of impact analysis for cask drop conditions or to support analytical results. Drop tests may be performed to obtain an equivalent static load to be used in analysis. Various methods may be used to obtain key data for the impact, or target surface, and the engineered foundation including the spring constants.

When test results are submitted in the SAR, verify proper scale parameters, including distribution of loadings (weights), geometry (dimensions), and material properties of the cask.

Structural Analysis for Specific Cask Components

A few specific examples of structural analysis for some of the confinement cask components are listed below:

Trunnions

The reviewer should verify the adequacy of the design of the trunnions, their connections with the cask body, and the cask body in the area around the trunnions. The trunnions can be either a single-load path or a dual-load path design. In either case, the design should meet the requirements of ANSI N14.6 or NUREG-0612 for critical loads.

Lifting trunnions should be fabricated and tested in accordance with ANSI N14.6. Since the cask is considered to be a critical load during handling at heights higher than design drop heights (i.e. lifting in the pool building facility), verify that trunnion testing is performed at a minimum of 150 percent of the maximum service load if a dual-load path is employed or at 300 percent of the service load if a single-load path is used. Confirm that any restrictions on cask lifting, resulting from these tests, is included in Chapter 12 of the SAR and in the Technical Specifications listed in the SER.

Fuel Basket

The reviewer should verify that the weight supported by the basket is the maximum or design weight of spent fuel. Consider all credible orientations of the cask and basket during cask drop. End or side drops typically produce the greatest structural demand on various basket components. Compare the stress intensity level of the fuel basket components with stress limits (i.e., Service Level D of the ASME B&PVC).

Evaluate the buckling capacity of the basket. Acceptable guidance for evaluating the buckling capacity of cask basket materials is given in the ASME B&PVC, Section III, Appendix F, and in NUREG/CR-6322.

Closure Lid Bolts

Review the analysis of closure lid bolts (if used in the design). The reviewer should verify that the combined effects of weight, internal pressure(s), thermal stress, O-ring compression force, cask impact forces, and bolt pre-load are used. The weight used in the analysis should be the maximum or design weight of the closure lids and any cask components supported by the lids. Acceptable methods for analysis of closure bolts are given in NUREG/CR-6007.

Buckling of Irradiated Fuel Under Bottom End Drop Conditions

If the applicant uses the analytical method described by Lawrence Livermore National Laboratory (LLNL) in report UCID-21246, for axial buckling to assess fuel integrity for the cask drop accident, the reviewer should verify that the analysis uses the irradiated material properties and includes the weight of fuel pellets.

Alternately, it is acceptable if the applicant uses an analysis of fuel integrity which considers the dynamic nature of the drop accident and any restraints on fuel movement resulting from cask design, if the analysis demonstrates that the cladding stress remains below yield. If a finite element analysis is performed, the analytical model may consider the entire fuel rod length with intermediate supports at each grid support (spacer). Irradiated material properties and weight of fuel pellets should be included in the analysis.

5.5.2 Pool and Pool Confinement Facilities

5.5.2.1 Description of Pool Facilities

Review the descriptive material in Chapter 1 of the SAR and the descriptive information in Chapter 3 of the SAR. The text descriptions along with the drawing figures, tables, flow diagrams, and specifications included in the application should fully define the pool facilities. Review the description of SSCs important to safety, and verify that there is sufficient detail to be able to proceed with the evaluation of the structural integrity and functional suitability. The configurations should be defined by drawings and fabrication specifications. The specifications should include references to the codes which govern the design details. The reviewer should verify that the combination of the drawings, specifications, appropriate codes and standards, and supporting calculations are sufficient.

A pool and pool confinement facilities involve a broader range of components and systems than the confinement structures. However the staff anticipates a diversity of pool facilities ranging from existing conventional pools designed under 10 CFR Part 50 requirements to site-specific designs used for limited, short-duration, wet transfer operations. The facilities may be comprised of some of the following elements which will require verification of structural integrity:

- pool structure, structural supports, and components that form the primary hydraulic confinement, water level control, cooling, and clean-up systems, such as piping, valves, pumps, filters, monitoring stations, and feeders

- pool components that provide for positioning the radioactive materials within the pool to ensure subcriticality (racks), accessibility, and compatibility with lifting interfaces

- pool components that ensure against improper movement of transfer or storage casks during wet loading and unloading operations

- secondary hydraulic containment that precludes releases to the surface or subsurface environment that might result from leaks or rupture of elements of the primary hydraulic containment, including equipment and floor drainage system

- SSCs associated with lifting, loading, unloading, transfer, or other handling of ISFSI/MRS vessels, transfer or transportation casks, other shielding vessels, or radioactive material to be stored

- enclosure(s) of the pool and operations that involve loading, unloading, and handling of the subject radioactive materials and other SSCs forming structural elements of the confinement boundary

- emergency power capability necessary to maintain safe conditions and monitor radioactivity

- internal waste collection and/or confinement, demineralized water make-up system, compressed air system for cask dewatering system (if used)

- SSCs providing compartmentalization and secondary confinement boundaries within (or coincident with) a pool facility's tertiary confinement barrier, such as for control room, electrical and machinery rooms, cask system component holding and inspection, personnel changing and showers, personnel decontamination and monitoring, health physics, and technical and administrative spaces.

Other ISFSI or MRS equipment that may be used within and outside the pool facility or that is used for lifting or transfer within the facility, but is not installed in the facility, such as cranes or conveyance systems, is addressed as "other SSCs important to safety" or "other SSCs."

Coordinate with the confinement review, Chapter 9 of this SRP, to verify that the SAR clearly identifies the confinement boundaries associated with the pool and pool facilities.

5.5.2.2 Design Criteria

For each of the SSCs being reviewed, determine what the design criteria and design bases are from the SAR. Confirm that the design criteria comply with acceptance criteria as outlined in Section 5.4.2.2.

Depending on the type of usage, i.e., long-term storage or short-term wet transfer, verify that the appropriate criteria are applied. ANSI/ANS 57.2 is appropriate for long-term, as well as short-term storage, whereas ANSI/ANS 57.7 may be more appropriate for short-term storage or wet transfer operations.

The reviewer should verify that the following sections of NUREG-0800 (Section 9.1.2) are adequately addressed:

- General Design Criteria 2, as it relates to structures housing the facility and that the facility is capable of withstanding the effects of natural phenomena such as earthquakes, tornadoes, and hurricanes.

- General Design Criteria 4, as it relates to structures housing the facility and that the facility is capable of withstanding the effects of environmental conditions and external missiles such that safety functions are not precluded.

- General Design Criteria 5 as it relates to shared structures, systems and components.

- General Design Criteria 61 as it relates to the facility design for fuel storage and handling of radioactive materials.

- General Design Criteria 62 as it relates to the prevention of criticality of the fuel by means of physical systems.

5.5.2.3 Material Properties

Coordinate with the thermal review, Chapter 6 of the SRP to verify that the material properties used in the structural analysis are appropriate for the load conditions and that the appropriate temperature at which the stress limits are defined is consistent with service temperatures. For each of the SSCs being reviewed, determine what structural materials are specified (e.g., reinforced concrete, steel, etc.), and verify that the material properties conform with the accepted design codes and standards. Section 5.4.2.3 gives references to acceptable codes. Review structural and other materials, and verify that they will produce no significant chemical or galvanic action or cause corrosion degradation that could adversely affect the safety function.

5.5.2.4 Structural Analysis

Design analyses should be prepared such that they may be audited to permit determination of the sources of expressions used, properties used for structural materials and components, and data derived by other calculations and assumptions.

Confirm that the design analysis includes codes and standards, design documentation, and design conditions for: (1) the spent fuel storage and cask handling pools, (2) the spent fuel cask and fuel assembly handling systems, (3) spent fuel storage racks, (4) fuel pool water makeup, cooling, and cleanup systems, (5) heating, ventilating and air conditioning equipment, (6) fuel storage buildings, and (7) electrical power, I&C and communications, as described in ANSI/ANS 57.2 and/or ANSI/ANS 57.7.

If ANSI/ANS 57.2 is used, the review should verify that the SSCs meet the following General Design Criteria (GDC) from 10 CFR Part 50, Appendix A:

- GDC 2: Confirm that regulatory position C.2 of Regulatory Guide 1.13, applicable portions of Regulatory Guides 1.29, 1.117, "Tornado Design Classification," and appropriate paragraphs of ANSI/ANS 57.2 are met.

  The reviewer should verify by review of supporting documentation and appropriate staff confirmatory calculations that position C.2 of Regulatory Guide is met. Position C.2 states that the pool facility should be designed to keep tornado winds and missiles generated by tornado winds from causing significant loss of watertight integrity of the fuel storage pool and to prevent tornado driven missiles from contacting the fuel stored in the pool.

- GDC 4: Confirm that regulatory position C.2 of Regulatory Guides 1.13, 1.115, "Protection Against Low-Trajectory Turbine Missiles," and 1.117, as well as appropriate paragraphs of ANSI/ANS 57.2 are met.

- GDC 5: Confirm that SSCs important to safety are capable of performing the required safety function.

- GDC 61:   Confirm that positions C.1 and C.4 of Regulatory Guide 1.13 and appropriate paragraphs of ANSI/ANS 57.2 are met.

  The reviewer should verify by review of supporting calculations or independent staff confirmatory calculations that positions C.1 and C.4 of Regulatory Guide 1.13 are satisfied. Position C.1 states that the fuel storage facility, including its structures and facilities (with some exceptions in C.6), should be designed to Category I seismic requirements. Position C.4 states that a controlled leakage building should enclose the fuel pool. It should be equipped with an appropriate ventilation and filtration system to limit the potential release of radioactive materials. Although the building need not be designed to withstand extremely high winds, leakage should be suitably controlled during fuel transfer operations. The ventilation and filtration system should be based on the assumption that the cladding of all the fuel rods in one fuel bundle might be breached.

- GDC 62: Confirm that positions C.1 and C.4 of Regulatory Guide 1.13 and appropriate paragraphs of ANSI/ANS 57.2 are met.

- Confirm that the handling of heavy loads (e.g., a spent fuel storage cask or spent fuel shipping cask) conforms with the guidance given in NUREG-0612.

Drop of a confinement cask may include secondary effects with safety implications, such as: deformation of interior structural SSCs that may preclude ready retrievability of the stored materials, structural damage and possible rupture of the pool (without loss of coolant that would uncover the fuel), damage to radioactive materials in the pool, and damage to the transfer cask and/or radiation shielding. These may also involve analyses addressed under the other structural evaluation categories such as the pool and pool facilities, reinforced concrete structures, and other SSCs important to safety.

Regulatory Guide 1.120, "Fire Protection Guidelines for Nuclear Power Plants," provides guidance for fire protection, where applicable, to some confinement systems such as the spent fuel pool area.

5.5.3 Reinforced Concrete Structures

5.5.3.1 Description of Concrete Structures

Review the descriptive material in Chapters 1 and 3 of the SAR. The text descriptions along with the drawings, figures, tables, and specifications included in the application should fully define the reinforced concrete structures. The configurations are defined by drawings and fabrication specifications. The specifications should include reference to the codes that govern the design details. The reviewer should verify that the combinations of drawings, specifications, appropriate codes and standards, and supporting calculations are sufficient.

Confirm that, at a minimum, the SAR documentation provides the following: (1) the dimensions of all sections that have a structural role including locations, sizes, configuration, spacing, enclosure (e.g., spirals, stirrups), and depth of cover or reinforcement for the reinforced concrete SSCs, (2) structural materials with defining standards or specifications, (3) location and specifications for control, contraction, and construction joints, and (4) fabrication codes and standards.

5.5.3.2 Design Criteria

For each of the concrete SSCs being reviewed, determine what the design criteria and design bases are from the SAR. Confirm that the design criteria comply with the acceptance criteria outlined in Section 5.4.3.2.

5.5.3.3 Material Properties

Coordinate with the thermal review, Chapter 6 of the SRP, to verify that the material properties used in the concrete structural analysis are appropriate for the load condition and that the appropriate temperature at which the strength limits are defined is consistent with service temperatures.

For each of the concrete structures being reviewed, determine what structural materials are specified (e.g., concrete composition, reinforcing material, and embedments, etc.), and verify that the material properties conform with the accepted design codes and standards. Section 5.4.3.3 gives complete references for cement type, aggregates, reinforcing, and embedments.

5.5.3.4 Structural Analysis

Design analyses should be prepared such that they may readily be audited to permit determination of sources of expressions used, properties used for structural materials, and data obtained by other calculations and assumptions.

Coordinate with the thermal review in Chapter 6 of this SRP to verify that the temperatures and pressures (where applicable) for all concrete structures presented in the SAR correspond to the same temperatures and pressures given in the thermal loads analysis.

Coordinate with the operation systems review in Chapter 3 of this SRP to verify that the configuration of the concrete structure (i.e., shielding cask or module on the concrete storage pad or shielding cask with confinement cask lift, etc.) corresponds to the same configuration which is used in the various load conditions and load combinations.

Normal Conditions

Normal conditions of concern for concrete structures are: (1) live and dynamic loads associated with transfer of the confinement cask, and installing closures, (2) load or support conditions associated with differential settlement of foundations, and (3) thermal gradients associated with normal operations and ranges of ambient temperatures.

The reviewer should verify that the design strength of the concrete structures exceeds the required strength as outlined in the ACI 349 code. If the ACI 359 code is used for confinement concrete structures, verify that the allowable stresses are not exceeded for normal conditions. The reviewer should verify that the maximum weight is used and that all normal ambient temperatures are considered. The reviewer should verify that the maximum temperature gradient is considered.

Off -Normal Conditions

Off-normal conditions of concern for concrete structures may include: (1) live and dynamic loads associated with equipment or instrument malfunctions, accidental misuse during transfer operations, (2) loads arising from jamming a confinement cask into a concrete structure, (3) impact loads on a concrete structure by a suspended transfer, confinement or storage cask, and (4) off-normal ambient temperature conditions.

The reviewer should verify that the design strength of the concrete structures exceeds the required strength as outlined in the ACI 349 code. If the ACI 359 code is used for confinement concrete structures, verify that the allowable stresses are not exceeded for off-normal conditions. The reviewer should verify that the maximum weight is used and that off-normal ambient temperatures are considered. The reviewer should verify that the maximum temperature gradient for the off-normal temperature is considered.

Accident Conditions

Coordinate with the accident analysis review in Chapter 15 of this SRP to verify that all accidents presented in that chapter have been adequately analyzed for structural integrity. Accident conditions which may be of concern to concrete structures include: (1) loads associated with accidental drops during transfer and handling operations of the confinement cask, (2) conditions arising from extreme thermal gradients in the concrete sections, (3) response to earthquakes, (4) tornadoes and tornado-driven missiles, (5) floods, (6) fires, (7) concrete cask drop, and (8) explosive overpressure.

The reviewer should verify that the design strength of the concrete structures exceeds the required strength as outlined in the ACI 349 code. If the ACI 359 code is used for confinement concrete structures, verify that the allowable stresses are not exceeded for accident conditions. The reviewer should verify that the maximum weight is used and that maximum ambient temperatures are considered. The reviewer should verify that the maximum temperature gradient is considered.

The ACI codes are intended to ensure ductile response beyond initial yield of structural components, regardless of the excess of capacity at yield over the maximum accident-level event. ACI 349 imposes additional conditions on design (over those of ACI 318) that increase the ductility.

Structural Analysis Methods for Concrete Structures

The reviewer should verify that the concrete structures conform with the respective code requirements as given below.

Strength Design

The NRC accepts strength design as presented in the current ACI 349 for concrete structures important to safety. Strength (or "Ultimate Strength") design is the usual approach used in American concrete design. Strength design is the only design approach that has been accepted for ISFSI or MRS concrete structures not within the scope of ACI 359. Strength design is the approach used in the current ACI 318 and ACI 349 codes. Determination that a concrete structure designed by another approach satisfies ACI 349 typically requires clause-by-clause review of the code for compliance.

Allowable Stress Design

The procedures of ACI 359 constitute an allowable stress design approach. The NRC does not accept an allowable stress design approach for SSCs not within the scope of ACI 359.

5.5.4 Other SSCs Important to Safety

5.5.4.1 Description of Other SSCs Important to Safety

Review the descriptive material in Chapters 1 and 3 of the SAR. The text descriptions along with the drawings, figures, tables, and specifications included in the application should fully define the other SSCs important to safety. The configurations are defined by drawings and fabrication specifications. The specifications should include reference to the codes that govern the design details. The reviewer should verify that the combinations of drawings, specifications, appropriate codes and standards, and supporting calculations are sufficient.

Confirm that, at a minimum, the SAR documentation provides the following: (1) the dimensions of all sections that have a structural role including locations, sizes, configuration, and spacing, (2) structural materials with defining standards or specifications, (3) location and specifications for assembly and weld joints, and (4) fabrication codes and standards.

Other SSCs important to safety are considered to be those SSCs not addressed in the other categories. These may include:

- transfer cask used to transfer the confinement cask to and from the storage area

- transfer and shielding vessel used to transfer radioactive material to or from the storage cask within a confinement barrier (such as from a pool to a storage cask to avoid immersion or lifting of the storage cask)

- non-concrete structures that support or shield the confinement cask during storage (excluding pads or hardstands for placement of moveable cask systems)

- lifting slings, spreaders, hooks, eyes, shackles, etc. used in lifts in which failure of a SSCs could jeopardize the basic safety requirements

- non-concrete foundations for structures important to safety (excluding pads or hardstands for placement of moveable cask systems), such as piles

- emergency power facilities and equipment and other electric equipment, if required to maintain nuclear materials in a safe condition in a general power outage. [Note: There is no safety requirement that handling or testing operations be continued during a power outage. Certified cask systems must be safe in storage without active systems.]

- SSCs associated with contaminated waste handling, treatment, reduction, packaging, and on-site storage

- SSCs associated with on-site inter-modal transfer of nuclear material containers, such as cranes used at truck, rail and barge/ship docks [Note: The cranes may not be important to safety depending on acceptable safety analysis.]

• SSCs whose response to accident conditions could have unacceptable consequences for maintenance of the basic safety requirements for the ISFSI/MRS (subcriticality, containment, radiation level limits, and retrievability)

5.5.4.2 Design Criteria

For each of the SSCs being reviewed, determine what the design criteria and design bases are from the SAR. Confirm that the design criteria comply with the acceptance criteria outlined in Section 5.4.4.2. The general structural requirements preclude unacceptable risk of criticality, unacceptable release of radioactive materials to the environment, unacceptable radiation dose, and impairment of ready retrievability of stored materials.

5.5.4.3 Material Properties

Coordinate with the thermal review, Chapter 6 of the SRP, to verify that the material properties used in the structural analysis are appropriate for the load condition and that the appropriate temperature at which the strength limits are defined are consistent with service temperatures.

For each of the structures being reviewed, determine what structural materials are specified (e.g., structural steel, etc.), and verify that the material properties conform with the accepted design codes and standards. Ensure that material properties which vary as a function of temperature, radiation, or other environments are adequately defined. The reviewer should verify that materials subject to corrosion and other degradation mechanisms are adequately protected or otherwise accounted for.

5.5.4.4 Structural Analysis

Design analyses should be prepared such that they may be readily audited to permit determination of sources of expressions used, properties used for structural materials, and data obtained by other calculations and assumptions.

Coordinate with the thermal review in Chapter 6 of this SRP to verify that the temperatures and pressures (where applicable) for all other SSCs important to safety, presented in the SAR, correspond to the same temperatures and pressures given in the thermal loads analysis.

Coordinate with the operation systems review in Chapter 3 of the SRP to verify that the configuration of the other SSCs (i.e. transfer device lifting the confinement cask, etc.) correspond to the same configuration which is used in the various load combinations.

Coordinate with the accident review in Chapter 15 of the SRP to verify that the accidents identified there correspond to the accident conditions evaluated in this chapter. Ensure that all load combinations, as outlined in Table 3-1 of NUREG-1536 have been appropriately evaluated.

5.5.5 Other SSCs

5.5.5.1 Description of Other SSCs

Review the descriptive material in Chapters 1 and 3 of the SAR. The text descriptions along with the drawings, figures, tables, and specifications included in the application should fully define the other SSCs not important to safety, but subject to NRC approval. The configurations are defined by drawings and fabrication specifications. The specifications should include reference to the codes that govern the design details. The reviewer should verify that the combinations of drawings, specifications, appropriate codes and standards, and supporting calculations are sufficient.

Confirm that, at a minimum, the SAR documentation provides the following: (1) the dimensions of all sections that have a structural role including locations, sizes, configuration, and spacing, (2) structural materials with defining standards or specifications, (3) location and specifications for assembly and weld joints, and (4) fabrication codes and standards.

SSCs not important to safety but subject to NRC approval must be described sufficiently to provide an adequate basis for that approval. Typically this would include descriptive information about the function, applicable codes, and standards for design and manufacture or procurement.

Other SSCs subject to NRC approval may include, in illustration, SSCs as listed below:

- pads and hardstands for storage of confinement casks

- demineralized water makeup system (see NUREG-0800, Section 9.2.3)

- SSCs on site associated with facilities other than for the ISFSI or MRS but which are shared by the ISFSI/MRS facilities, or which are physically connected to SSCs supporting the ISFSI/MRS and that have safety or safeguards and security related functions

- SSCs associated with a standby power capability

- SSCs associated with transfer of confinement and transfer casks on site, including cask loading and extraction equipment, trailers, prime movers, crane, and equipment unique to the cask system whose failure would not jeopardize the basic safety requirements of the confinement system

- on site radioactive material transfer route structures, such as bridges, roads, and rail crossings

- fixed or mobile structures that provide space for NRC use

- SSCs including cranes and other equipment for inter-modal transfer of containers holding nuclear materials, such as truck, rail, and barge/ship docks whose failure would not jeopardize the basic safety criteria

- structures and earthworks to prevent on site facility flooding

- SSCs, including equipment, that provide fire protection or that may be required to mitigate the effects of accident events

- other SSCs required for compliance with code safety requirements, such as for lightning protection

- training facilities and associated equipment for health physics, procedural, and other training

5.5.5.2 Design Criteria

For each of the SSCs being reviewed, determine what the design criteria and design bases are from the SAR. Confirm that the design criteria comply with the acceptance criteria outlined in Section 5.4.5.2.

5.5.5.3 Material Properties

Coordinate with the thermal review, Chapter 6 of the SRP to verify that the material properties used in the structural analysis are appropriate for the load condition and that the appropriate temperature at which the strength limits are defined is consistent with service temperatures.

For each of the structures being reviewed, determine what structural materials are specified (e.g., concrete composition, reinforcing material, and embedments, structural steel, etc.), and verify that the material properties conform with the accepted design codes and standards.

Ensure that material properties which vary as a function of temperature, radiation, or other environments are adequately defined. The reviewer should verify that materials subject to corrosion and other degradation mechanisms are adequately protected or otherwise accounted for.

5.5.5.4 Structural Analysis

Design analyses should be prepared such that they may be readily audited to permit determination of sources of expressions used, properties used for structural materials, and data obtained by other calculations and assumptions.

Coordinate with the thermal review in Chapter 6 of this SRP to verify that the temperatures and pressures (where applicable) for other SSCs presented in the SAR, and subject to NRC approval, correspond to the same temperatures and pressures given in the thermal loads analysis.

Coordinate with the operation systems review in Chapter 3 of the SRP to verify that the configuration of the other SSCs subject to NRC approval corresponds to the same configuration which is used in the various load combinations.

Coordinate with the accident review in Chapter 15 of the SRP to verify that the accidents identified there correspond to the accident conditions evaluated in this chapter. Ensure that all load combinations, as outlined in Table 3-1 of NUREG-1536, have been appropriately evaluated.

The information and evaluation required for these SSCs is typically to lesser levels than that required for SSCs important to safety as described in the respective part of this Section. For example, the structural capacities or design and construction codes may be stated and evaluated, but there typically is no review of structural analyses or other analyses supporting selection or assessment of projected performance.

5.6 Evaluation Findings

The evaluation findings are prepared by the reviewer on satisfaction of the regulatory requirements relating to the installation design and the structural evaluation, as identified in Section 5.3. Based on the review of the applicant's description, proposed design criteria, appropriate use of material properties, and adequate structural analysis of the five categories of structures, systems and components, the staff concludes that the SSCs are in conformance with NRC regulations. The five categories of SSCs, or areas of review are: (1) confinement structures, systems and components, (2) pool and pool confinement facilities, (3) reinforced concrete structures, (4) other SSCs important to safety, and (5) other SSCs subject to NRC approval. The SER should address each acceptance criteria provided in Section 5.4 of this SRP similar to the following (finding numbering is for convenience in referencing within the SRP and SER):

F5.1   The SAR and docketed materials relating to the description of confinement structures, systems and components meet the requirements of 10 CFR 72.24 (a) and (b), 10 CFR 72.82 (c)(2), and 10 CFR 72.106 (a), (b), and (c).

F5.2   The SAR and docketed materials relating to design criteria, including applicable codes and standards meet the requirements of 10 CFR 72.24 (c)(1), (c)(2), and (c)(4); 10 CFR 72.40 (a)(1); 10 CFR 72.120 (a) and (b); 10 CFR 72.122 (a), (b), (c), (d), (f), (g), (h), (i), (j), (k), and (l); 10 CFR 72.128 (a) and (b); and 10 CFR 72.236 (b), (e), (f), (g), and (k). Additionally, the potential for brittle fracture has been considered by meeting the guidance provided in Regulatory Guides 7.11 and 7.12. The confinement structures meet the guidance provided in applicable parts of Regulatory Guides 1.29, 1.60, 1.61, and 1.92 for protection against seismic events. The confinement structures meet the guidance provided in applicable parts of Regulatory Guides 1.76 and 1.117 and NUREG-1503 for tornado protection.

F5.3   The SAR and docketed materials relating to suitable material properties for use in the design and construction of the SSCs meet the requirements of 10 CFR 72.24 (c)(3).

F5.4   The SAR and docketed materials provide adequate analytical and/or test reports to ensure that structural integrity of the SSCs and meet the requirements of 10 CFR 72.24 (d)(1), (d)(2), and (i), and 10 CFR 72.122 (b)(1), (b)(2), and (b)(3), (c), (d), (f), (g), (h), (i), (j), (k), and (l).

F5.5   The SAR and docketed materials relating to the description of pool and pool facilities meet the requirements of 10 CFR 72.24 (a) and (b), 10 CFR 72.40 (a)(3), 10 CFR 72.82 (c)(2), and 10 CFR 72.106 (a), (b), and (c).

F5.6   The SAR and docketed materials relating to design criteria, including applicable codes and standards meet the requirements of 10 CFR 72.24 (c)(1), (c)(2), and (c)(4); 10 CFR 72.40 (a)(1); 10 CFR 72.120 (a) and (b); 10 CFR 72.122 (a), (b), (c), (d), (f), (g), (h), (i), (j), (k), and (l); 10 CFR 72.128 (a) and (b); and 10 CFR 72.236 (b), (e), (f), (g), and (k). Additionally the pool and pool facilities meet the General Design Criteria 2, 4, 5, 61, and portions of 62 and 63 of 10 CFR Part 50, Appendix A. The pool meets the guidance provided in applicable parts of Regulatory Guides 1.13, and 1.26, and ANSI/ANS 57.9 as well as 57.7 and/or 57.2. The pool and pool facilities meet the guidance provided in applicable parts of Regulatory Guides 1.29, 1.60, 1.61, 1.92, and 1.122, "Development of Floor Design Response Spectra for Seismic Design of Floor-Supported Equipment or Components," for protection against seismic events. The pool and pool facility meet the guidance provided in applicable parts of Regulatory Guides 1.76 and 1.117 and NUREG-1503 for tornado protection.

F5.7   The SAR and docketed materials relating to suitable material properties for use in the design and construction of the pool and pool facilities meet the requirements of 10 CFR 72.24 (c)(3).

F5.8   The SAR and docketed materials provide adequate analytical and/or test reports to ensure that structural integrity of the pool and pool facilities and meet the requirements of 10 CFR 72.24 (d)(1), (d)(2), and (i), and 10 CFR 72.122 (b)(1), (b)(2), and (b)(3), (c), (d), (f), (g), (h), (i), (j), (k), and (l).

F5.9   The SAR and docketed materials relating to the description of reinforced concrete meet the requirements of 10 CFR 72.24 (a) and (b), 10 CFR 72.82 (c)(2), and 10 CFR 72.106 (a), (b), and (c).

F5.10   The SAR and docketed materials relating to design criteria, including applicable codes and standards meet the requirements of 10 CFR 72.24 (c)(1), (c)(2), and (c)(4); 10 CFR 72.40 (a)(1); 10 CFR 72.120 (a) and (b); 10 CFR 72.122 (a), (b), (c), (d), (f), (g), (h), (i), (j), (k), and (l); 10 CFR 72.128 (a) and (b); and 10 CFR 72.236 (b), (e), (f), (g), and (k). The concrete structures meet the guidance provided in applicable parts of Regulatory Guides 1.29, 1.60, 1.61, 1.92, and 1.122 for protection against seismic events. The concrete structures meet the guidance provided in applicable parts of Regulatory Guides 1.76 and 1.117 and NUREG-1503 for tornado protection.

F5.11   The SAR and docketed materials relating to suitable material properties for use in the design and construction of the SSCs meet the requirements of 10 CFR 72.24 (c)(3).

F5.12   The SAR and docketed materials provide adequate analytical and/or test reports to ensure that structural integrity of the SSCs and meet the requirements of 10 CFR 72.24 (d)(1), (d)(2), and (i), and 10 CFR 72.122 (b)(1), (b)(2), and (b)(3), (c), (d), (f), (g), (h), (i), (j), (k), and (l).

F5.13   The SAR and docketed materials relating to the description of other SSCs important to safety meet the requirements of 10 CFR 72.24 (a) and (b), 10 CFR 72.82 (c)(2), and 10 CFR 72.106 (a), (b), and (c).

F5.14   The SAR and docketed materials relating to design criteria, including applicable codes and standards meet the requirements of 10 CFR 72.24 (c)(1), (c)(2), and (c)(4); 10 CFR 72.40 (a)(1); 10 CFR 72.120 (a) and (b); 10 CFR 72.122 (a), (b), (c), (d), (f), (g), (h), (i), (j), (k), and (l); 10 CFR 72.128 (a) and (b); and 10 CFR 72.236 (b), (e), (f), (g), and (k). The other SSCs important to safety meet the guidance provided in applicable parts of Regulatory Guides 1.29, 1.60, 1.61, and 1.92 for protection against seismic events. The other SSCs important to safety meet the guidance provided in applicable parts of Regulatory Guides 1.76 and 1.117 and NUREG-1503 for tornado protection. The other SSCs important to safety meet guidance provided in Regulatory Guides 1.59 and 1.102 for flood protection.

F5.15   The SAR and docketed materials relating to suitable material properties for use in the design and construction of the SSCs meet the requirements of 10 CFR 72.24 (c)(3).

F5.16   The SAR and docketed materials provide adequate analytical and/or test reports to ensure that structural integrity of the SSCs and meet the requirements of 10 CFR 72.24 (d)(1), (d)(2), and (i), and 10 CFR 72.122 (b)(1), (b)(2), and (b)(3), (c), (d), (f), (g), (h), (i), (j), (k), and (l).

F5.17   The SAR and docketed materials relating to the description of other SSCs subject to NRC approval meet the requirements of 10 CFR 72.24 (a) and (b), 10 CFR 72.82 (c)(2), and 10 CFR 72.106 (a), (b), and (c).

F5.18   The SAR and docketed materials relating to design criteria, including applicable codes and standards meet the requirements of 10 CFR 72.24 (c)(1), (c)(2), and (c)(4); 10 CFR 72.40 (a)(1); 10 CFR 72.120 (a) and (b); 10 CFR 72.122 (a), (b), (c), (d), (f), (g), (h), (i), (j), (k), and (l); 10 CFR 72.128 (a) and (b); and 10 CFR 72.236 (b), (e), (f), (g), and (k).

F5.19 The SAR and docketed materials relating to suitable material properties for use in the design and construction of the SSCs meet the requirements of 10 CFR 72.24 (c)(3).

F5.20 The SAR and docketed materials provide adequate analytical and/or test reports to ensure that structural integrity of the SSCs and meet the requirements of 10 CFR 72.24 (d)(1), (d)(2), and (i), and 10 CFR 72.122 (b)(1), (b)(2), and (b)(3), (c), (d), (f), (g), (h), (i), (j), (k), and (l).

5.7 References

NRC documents referenced are identified at Consolidated References, Chapter 17.

Codes, Standards, and Specifications

American Concrete Institute (ACI)
ACI 318, "Building Code Requirements for Reinforced Concrete."

ACI 349, "Code Requirements for Nuclear Safety Related Concrete Structures," and ACI 349R, "Commentary."

ACI 359, "Code for Concrete Reactor Vessels and Containments" (also designated as ASME Boiler and Pressure Vessel Code, Section III, "Rules for Construction of Nuclear Power Plant Components," Division 2), American Concrete Institute and American Society of Mechanical Engineers (Joint Committee).

American Institute of Steel Construction (AISC)
AISC, "Specification for Structural Steel Buildings, Allowable Stress Design and Plastic Design," published in the AISC "Manual of Steel Construction."

American National Standards Institute (ANSI)
ANSI N14.6-1993, "American National Standard for Radioactive Materials - Special Lifting Devices for Shipping Containers Weighing 10,000 Pounds (4500 kg) or More," Institute of Nuclear Materials Management, 1993.

ANSI N45.2.11-1974, "Quality Assurance Requirements for the Design of Nuclear Power Plants," 1974.

ANSI N210-1976/ANS-57.2, "Design Objectives for Light Water Reactor Spent Fuel Pool Storage Facilities at Nuclear Power Stations" [Referenced in NUREG-0800, Draft Revision 4, 1996].

ANSI/ANS-57.2-1983, "Design Requirements for Light Water Reactor Spent Fuel Storage Facilities at Nuclear Power Plants."

ANSI/ANS-57.7-1988, "Design Criteria for an Independent Spent Fuel Storage Installation (Water Pool Type)."

ANSI/ANS-57.9-1984, "Design Criteria for an Independent Spent Fuel Storage Installation (Dry Storage Type)," American Nuclear Society (ANS) [Referenced to the extent that ANSI/ANS 57.9-1984 is stated as suitable in Regulatory Guide 3.60].

American Society for Testing and Materials (ASTM)
A 36, "Standard Specification for Structural Steel."

A 53, "Standard Specification for Welded and Seamless Steel Pipe."

A 82, "Standard Specification for Cold-Drawn Steel Wire for Concrete Reinforcement."

A 184, "Standard Specification for Fabricated Deformed Steel Bar Mats for Concrete Reinforcement."

A 185, "Standard Specification for Welded Steel Wire Fabric for Concrete Reinforcement."

A 242, "Standard Specification for High-Strength Low-Alloy Structural Steel."

A 416, "Standard Specification for Uncoated Seven-Wire Stress-Relieved Steel Strand for Prestressed Concrete."

A 421, "Standard Specification for Uncoated Stress-Relieved Steel Wire for Prestressed Concrete."

A 441, "Standard Specification for High-Strength Low-Alloy Structural Manganese Vanadium Steel."

A 496, "Standard Specification for Deformed Steel Wire for Concrete Reinforcement."

A 497, "Standard Specification for Welded Deformed Steel Wire Fabric for Concrete Reinforcement."

A 500, "Standard Specification for Cold-Formed Welded and Seamless Carbon Steel Structural Tubing in Rounds and Shapes."

A 501, "Standard Specification for Hot-Formed Welded and Seamless Carbon Steel Structural Tubing."

A 572, "Standard Specification for High-Strength Low-Alloy Columbium-Vanadium Steels of Structural Quality."

A 588, "Standard Specification for High-Strength Low-Alloy Structural Steel with 50,000 psi Minimum Yield Point to 4 in. Thick."

A 615, "Standard Specification for Deformed and Plain Billet-Steel Bars for Concrete Reinforcement."

A 706, "Standard Specification for Low-Alloy Steel Deformed Bars for Concrete Reinforcement."

A 772, "Standard Specification for Uncoated High-Strength Steel Bar for Prestressing Concrete."

C 31, "Standard Method of Making and Curing Concrete Test Specimens in the Field."

C 33, "Standard Specification for Concrete Aggregates."

C 39, "Standard Method of Test for Compressive Strength of Cylindrical Concrete Specimens."

C 42, "Standard Method of Obtaining and Testing Drilled Cores and Sawed Beams of Concrete."

C 88, "Standard Method of Test for Soundness of Aggregates by Use of Sodium Sulfate or Magnesium Sulfate."

C 94, "Standard Specification for Ready-Mixed Concrete."

C 109, "Standard Method of Test for Compressive Strength of Hydraulic Cement Mortars (Using 2-inch or 50-m-m Cube Specimens)."

C 131, "Standard Test Method for Resistance to Degradation of Small-Size Coarse Aggregate by Abrasion and Impact in the Los Angeles Machine."

C 144, "Standard Specification for Aggregate for Masonry Mortar."

C 150, "Standard Specification for Portland Cement."

C 172, "Standard Method of Sampling Fresh Concrete."

C 192, "Standard Method of Making and Curing Concrete Test Specimens in the Laboratory."

C 260, "Standard Specification for Air-Entraining Admixtures for Concrete."

C 289, "Standard Method of Test for Potential Reactivity of Aggregates (Chemical Method)."

C 441, "Standard Method of Test for Effectiveness of Mineral Admixtures in Preventing Excessive Expansion of Concrete, Due to the Alkali-Aggregate Reaction."

C 494, "Standard Specification for Chemical Admixtures for Concrete."

C 496, "Standard Method of Test for Splitting Tensile Strength of Cylindrical Concrete Specimens."

C 595, "Standard Specification for Blended Hydraulic Cements."

C 618, "Standard Specification for Fly Ash and Raw or Calcined Natural Pozzolan for Use as a Mineral Admixture in Portland Cement Concrete."

C 637, "Standard Specification for Aggregates for Radiation-Shielding Concrete."

C 685, "Standard Specification for Concrete Made by Volumetric Batching and Continuous Mixing."

C 1017, "Standard Specification for Chemical Admixtures for Use Producing Flowing Concrete."

American Society of Civil Engineers (ASCE)
ASCE 4, "Seismic Analysis of Safety-Related Nuclear Structures."

ASCE 7 (formerly ANSI A58.1), "Minimum Design Loads for Buildings and Other Structures."

American Society of Mechanical Engineers (ASME)
ASME Boiler and Pressure Vessel Code, Section III, "Rules for Construction of Nuclear Power Plant Components."

ASME Boiler and Pressure Vessel Code, Section IX, "Welding and Brazing Qualifications."
American Welding Society (AWS)
AWS A2.4, "Standard Symbols for Welding, Brazing and Nondestructive Examination."

AWS D1.4, "Structural Welding Code - Reinforcing Steel."

International Conference of Building Officials (ICBO)
"Uniform Building Code" (UBC).

National Fire Protection Association (NFPA)
NFPA, "National Electric Code."

NFPA, "Code for Safety to Life from Fire in Buildings and Structures."

NFPA, "Lightning Protection Code."

Manuals and Texts

Hoerner, S.F., "Fluid Dynamic Drag," 1965, Hoerner Fluid Dynamics, P.O. Box 342, Brick Town, NJ 08723.

Marker, B.R., et al, "NIKE3D-A Nonlinear, Implicit, Three-Dimensional Finite Element Code for Solid and Structural Mechanics-User's Manual," UCRL-MA-105268, Lawrence Livermore National Laboratory, January 1991.

Young, W.C., "Roark's Formulas for Stress and Strain," McGraw-Hill.

Technical Reports

Cottrell, W.B., and Savolainen, A.W., "U.S. Reactor Containment Technology," ORNL-NSIC-5, Vol. 1, Chapter 6, Oak Ridge National Laboratory.

Kennedy, R.P., "A Review of Procedures for the Analysis and Design of Concrete Structures to Resist Missile Impact Effects," Holmes and Narver, Inc., September 1975.

Levy, I.S., et al., "Recommended Temperature Limits for Dry Storage of Spent Light Water Zircalloy Clad Fuel Rods in Inert Gas," PNL-6189, Pacific Northwest Laboratory.

6 THERMAL EVALUATION

6.1 Review Objective

The objective of the thermal review is to ensure that the decay heat removal (DHR) system is capable of reliable operation so that the temperatures of materials used for systems, structures, and components (SSCs) important to safety, fuel assembly cladding material, and solidified high-level waste packages remain within the allowable limits under normal, off-normal, and accident conditions. Wet and dry fuel assembly transfer systems are evaluated for adequate decay heat removal under normal, off-normal, and accident conditions. Fire hazards analysis and fire protection measures for the MRS or ISFSI are evaluated. The review also confirms that the thermal design of the ISFSI or MRS has been analyzed with acceptable analytical and/or test methods.

The approach to thermal review and evaluation presented in this chapter builds upon the guidance provided for the certification review of casks in Chapter 4 of NUREG-1536, "Standard Review Plan for Dry Storage Systems." The additional guidance of this chapter is necessary because the site specific applications reviewed under this guidance will contain site specific features (e.g., temperature limits) and other systems (e.g., pools, structures using reinforced concrete). If the ISFSI or MRS uses a cask which has received a certificate of compliance, key assumptions, bounding site characteristics and environmental conditions, and cask/ISFSI interface requirements identified in the cask Safety Analysis Report (SAR) and certificate of compliance are also examined and compared to the ISFSI or MRS design and environmental conditions.

Figure 6.1 presents an overview of the thermal evaluation process. The figure shows that the thermal review draws information from different sections of the application including supporting calculations. The figure also shows that the results of the thermal review are both used by other technical review areas and are documented in the NRC staff-prepared Safety Evaluation Report (SER).

6.2 Areas of Review

The following outline shows the areas of review addressed in Section 6.4, Acceptance Criteria, and Section 6.5, Review Procedures:

Decay Heat Removal Systems

Material Temperature Limits

Thermal Loads and Environmental Conditions

Analytical Methods, Models and Calculations

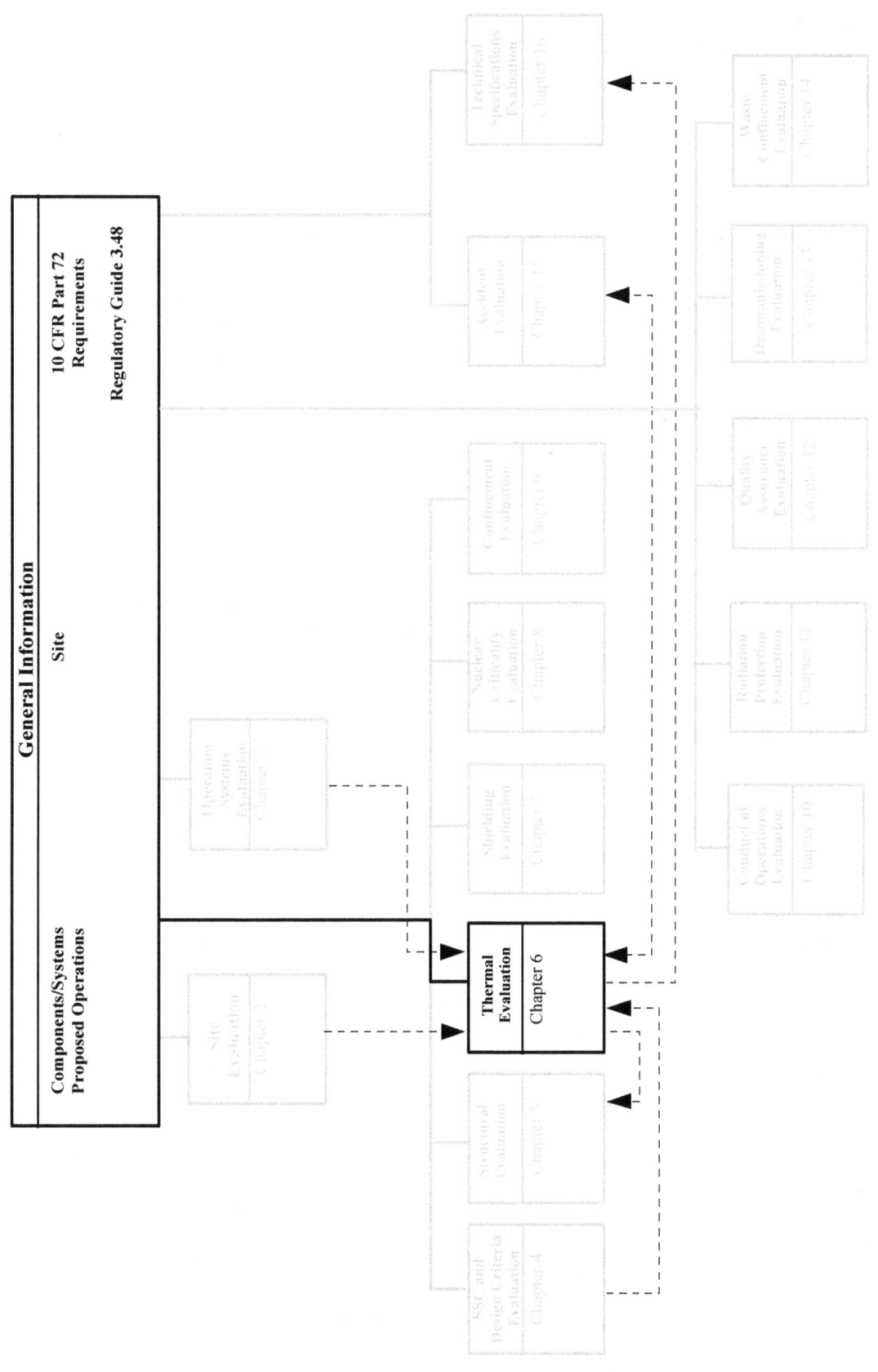

Figure 6.1 Overview of Thermal Evaluation

Fire and Explosion Protection
> General Considerations
> Spent Fuel Casks
> SSCs Important to Safety Guidance for Fire Protection Program

6.3 Regulatory Requirements

This section identifies and presents a high-level summary of Title 10 of the Code of Federal Regulations (CFR) Part 72 relevant to the review areas addressed by this chapter. The NRC staff reviewer should read the exact regulatory language. A matrix at the end of this section matches the regulatory requirements identified in this section to the areas of review identified in the previous section.

72.92 Design basis external natural events
(a) "...natural phenomena that affect the ISFSI design must be identified and assessed according to their potential effects on the safe operation."

72.122 Overall requirements
(c) Protection against fires and explosions
"...must be designed and located so that they continue to perform their safety functions effectively under credible fire and explosion exposure conditions...Explosion and fire detection, alarm, and suppression systems shall be designed and provided."
(d) Sharing of structures, systems and components
"...must not impair the capability of either facility."
(h) Confinement barriers and systems
> (1) "The spent fuel cladding must be protected during storage against degradation that leads to gross ruptures or confined such that degradation of the fuel during storage will not pose operational safety problems."
> (2) "For underwater storage of spent fuel or high-level radioactive waste... systems for maintaining water purity and pool water level must be designed so that any abnormal operations or failure in those systems from any cause will not cause the water level to fall below safe limits."
> (4) "Storage confinement systems must have the capability for continuous monitoring."

(j) Control room or control area (note: as applied to fire and explosion protection)
"A control room or control area...must be designed to monitor the ISFSI or MRS safely under normal conditions and to provide safe control ...under off-normal or accident conditions."

72.128 Criteria for spent fuel, high-level radioactive waste, and other radioactive waste storage and handling.
(a) Spent fuel and high-level radioactive waste storage and handling systems. "... must be designed to ensure adequate safety under normal and accident conditions...These systems must be designed with"
> (4) "A heat-removal capability having testability and reliability."

A matrix which shows the primary relationship of these regulations to the specific areas of review associated with this Standard Review Plan (SRP) chapter is given in Table 6.1. The NRC staff reviewer should verify the association of regulatory requirements with the areas of review presented in the matrix to ensure that no requirements are overlooked as a result of unique applicant design features.

Table 6.1 Relationship of Regulations and Areas of Review

Area of Review	10 CFR Part 72 Regulations		
	72.92	72.122	72.128
Decay Heat Removal Systems		●	●
Material Temperature Limits			●
Thermal Loads and Environmental Conditions	●	●	
Methods, Models, and Calculations		●	●
Fire and Explosion Protection		●	

6.4 Acceptance Criteria

This section identifies the acceptance criteria used for the thermal evaluation review. Specific acceptance criteria are delineated in this section.

6.4.1 Decay Heat Removal Systems

The spent fuel cladding must be protected during storage against degradation that leads to gross fuel rupture (10 CFR 72.122(h)). Decay heat removal systems shall have testability and reliability consistent with their importance to safety (10 CFR 72.128(a)).

The applicant must provide a description of the proposed heat removal system. The description must describe the mechanisms for removing decay heat including any active components or operator actions necessary for operation during normal, off-normal, and accident conditions. If the decay heat removal system is for a pool, the description must address the layout of piping and equipment, control systems for managing flow, and instrumentation systems for monitoring water conditions.

The applicant must provide evidence that the decay heat removal system will operate reliably under normal, off-normal, and accident conditions.

6.4.2 Material Temperature Limits

SSCs important to safety shall be maintained within their minimum and maximum temperature criteria for normal, off-normal, and accident conditions so as to support the performance of the intended safety function (10 CFR 72.128(a).

The applicant must identify the temperature limits for fuel cladding, solidified waste packages, and materials used for SSCs that are important to safety. The applicant shall also provide a basis for the temperature limits. The temperature limits for fuel cladding should include consideration of mechanisms that can lead to gross cladding rupture.

Fuel cladding temperature during dry storage shall be maintained below the expected damage-threshold temperatures for normal conditions and a minimum of 20 years dry storage for ISFSI or MRS design and environmental conditions. The fuel cladding temperature should also generally be maintained below 570 °C (1058 °F) for short-term off-normal, short-term accident, and fuel transfer operations (e.g., vacuum drying of the cask or dry transfer) (PNL-4835).

6.4.3 Thermal Loads and Environmental Conditions

The applicant must identify the design basis thermal loads from the spent fuel or high-level waste, as well as the thermal loads associated with insolation and the site parameters that determine the rate at which heat can be removed from the ISFSI or MRS (10 CFR 72.92).

The heat removal system must accommodate the decay heat of the spent fuel or high-level waste and the site normal, off-normal, and accident thermal conditions (10 CFR 72.122(b)).

6.4.4 Analytical Methods, Models, and Calculations

The applicant shall present a thermal analysis that demonstrates the ability to manage design basis heat loads and have the various materials remain within temperature limits. The analysis shall be conducted for normal, off-normal, and accident conditions. The analysis shall also present temperature and temperature gradient information that is necessary to support the structural analysis. The applicant shall identify the codes or analytical methods used for thermal analysis and discuss the basis for the parameters selected for the analysis.

For each fuel assembly-type proposed for storage, the dry storage system shall ensure a very low probability (e.g., 0.5 percent), per fuel rod, of cladding breach during long-term (e.g., 40 year) storage (10 CFR 72.122(h), PNL-6189).

The maximum internal pressure of the cask shall remain within its design pressure for normal, off-normal, and accident conditions assuming 1 percent, 10 percent, and 100 percent ruptured fuel rods respectively. Assumptions for pressure calculations include release of 100 percent of the fill gas and 30 percent of the significant radioactive gases in the fuel rods (10 CFR 72.128(3), 10 CFR 72.122(h)).

Under the conditions where any of the cask component or fuel cladding temperatures are close (within 5%) to their limiting values during an accident or the maximum normal operating pressure is within 10% of its design basis pressure, or any other special considerations affected by fission gas concentrations, the applicant should analyze the potential impact of the fission gas in the cask on the cask component and fuel cladding temperature limits and the internal cask pressure.

The pool system shall be designed so that, for all postulated events, the pool water level is maintained at a level above the top of the active fuel to ensure adequate decay heat removal and shielding (ANS/ANS 57.7).

6.4.5 Fire and Explosion Protection

Spent fuel assemblies, other radioactive materials, and SSCs important to safety shall have adequate protection against fires and explosions to minimize and control the release of radioactive material to the environment (10 CFR 72.122(c)).

Measures for fire prevention, fire detection, fire suppression, and fire containment for the protection of the spent fuel assemblies and SSCs important to safety shall be provided. 10 CFR 72.122(c) requires that:

•   SSCs important to safety must be designed and located so that they can continue to perform their safety functions effectively under credible fire and explosion exposure conditions.

•   Non-combustible and heat resistant materials must be used wherever practical throughout the ISFSI or MRS, particularly in locations vital to the maintenance of safety control functions.

•   Explosion and fire detection, alarm, and suppression systems shall be designed and provided with sufficient capacity and capability to minimize the adverse effects of fires and explosions on SSCs important to safety.

•   The design of the ISFSI or MRS must include provisions to protect against adverse effects that might result from the operation or failure of the fire suppression system.

In addition, 10 CFR 72.122(j) requires that a control room or control area, if appropriate for the ISFSI or MRS design, must be designed to permit occupancy and actions to be taken to monitor the safety of the ISFSI or MRS under normal conditions and to provide safe control of the ISFSI or MRS under off-normal or accident conditions.

6.5 Review Procedures

The following provides review guidance relevant to the thermal evaluation. This guidance is based on the required products of the review and lessons learned from prior reviews. Additional review guidance is available in Chapter 4, Section V of NUREG-1536.

6.5.1 Decay Heat Removal Systems

6.5.1.1 General Considerations

ISFSI or MRS decay heat removal systems may be passive (natural convection and thermal radiation) for dry storage or may include active cooling systems (motors, pumps, heat exchangers, valve actuators, and switchgear) for wet or dry storage. The reviewer should verify that the application for the ISFSI or MRS clearly establishes that the storage system will function within the original design basis thermal limits under normal, off-normal, and accident conditions. The reviewer should examine the thermal analysis, material temperature limits, and key assumptions of the analysis to ensure that the ISFSI or MRS design and environmental conditions are within the envelope of the storage system original analysis and associated technical specifications.

The reviewer should confirm that the design criteria include maximum heat output of the radioactive materials; temperature levels for the ambient air under normal, off-normal, and accident-level conditions; and associated insolation. The maximum times that the stored material will be subject to ambient elevated temperatures should be identified. Definition of daily cycles of temperature may be important for heat removal for some storage systems. The accompanying insolation cycle may be important for storage systems with direct solar exposures.

The reviewer should evaluate the temperature distributions and temperature criteria that are used in determination of thermal stresses for all DHR system components exposed to heat generated by the fuel assembly. These DHR components include the cask, transfer equipment, and any shielding components. Evaluation of stresses or loads caused by temperature gradients and interacting materials at different temperatures or with different coefficients of thermal expansion is performed under Section 5, "Installation Design and Structural Evaluation."

The reviewer should verify that technical specifications relating to heat removal capability have been included in the technical specification chapter of the SAR. These may have been proposed by the applicant in compliance with 10 CFR 72.26 or may result from the review and evaluation of submittals relating to those areas. The following two paragraphs illustrate technical specifications related to thermal evaluations which have been accepted in previous applications:

- Surveillance requirement: Performance of the heat removal system will be verified by tests conducted upon placing the first full storage container in its storage position. These tests determine heat removal by measurement of air flow and temperatures and will be used to confirm the adequacy of the thermal analysis by comparison of the actual conditions of heat generation by the stored fuel assembly and ambient conditions.

- Surveillance requirement: Periodic surveillance will be performed to ensure that there is no blockage of cooling air flow in the heat removal system. This surveillance [typically based on the minimum time for stored material cladding or other material important to safety (e.g., shielding) to reach a threshold temperature in the event of a complete blockage occurring immediately following the prior surveillance and the minimum time to repair or correct the blockage condition] shall be no less frequent than _____ [insert time interval]. [Alternatives may link the surveillance interval to ambient temperature. Procedures for performing the surveillance may be as proposed by the applicant or may be left unstated in the SER, with the procedures to be developed by the applicant subsequent to license approval.]

Required Thermal Analysis Scenarios

The reviewer should confirm that the following thermal scenarios are considered, to determine that temperature limits are met. The list is not necessarily exhaustive.

- For storage conditions at maximum normal, off-normal, and accident-level ambient temperatures and insolation

- Temperature in storage with partial or full blockage of ventilation passages (if applicable to ISFSI design such as concrete type storage modules or vaults)

- In wet temporary storage with partial or full failure of an active heat removal system or loss of electrical power

- In transfer configuration within a transfer cask at maximum normal, off-normal, and accident-level ambient temperatures and insolation

- During cask dewatering and/or cask purging operations with the interior at a near vacuum

- In a configuration associated with stored material retrieval (retrieval may be required at any time following loading of a cask)

- During cask sealing and opening operations when there is a liquid (water or borated water) in the cask cavity which has been lowered to permit welding or cutting operations on the closure

- Cask reflood for unloading operations

- Facility fire or explosion (internal and external event to the facility)

6.5.1.2 Dry Storage Systems

A dry storage system may consist of a cask used on an ISFSI pad or stored in a dry vaulted system. The boundary conditions on the cask surface or dry storage system depend on the environment surrounding the cask. The reviewer should confirm that the temperature of the environment for normal and off-normal conditions is specified in the SAR. The reviewer should verify the appropriateness of specified incident and absorbed insolance. The reviewer should evaluate the mechanisms and models for the dissipation of the absorbed insolance and decay heat from the surface of the cask to the environment which are identified and described. The reviewer should review the cask SAR to ensure the site conditions are enveloped by the cask thermal analysis. The reviewer should evaluate the thermal performance of the cask in accordance with Chapter 4 of NUREG-1536.

The reviewer should verify that if any of the cask component or fuel cladding temperatures are close (within 5%) to their limiting values during an accident or the maximum normal operating pressure is within 10% of its design basis pressure, or any other special considerations affected by fission gas concentrations, the applicant has analyzed the potential impact of the fission gas in the cask on the cask component and fuel cladding temperature limits and the internal cask pressure.

The reviewer should confirm that the liquid in the cask does not boil during fuel assembly transfer operations to avoid uncontrolled pressures on the cask and the connected dewatering, purging, and recharging system(s) and/or further discharge of liquid providing radiation shielding over the top of the contained radioactive materials. The reviewer should confirm that an adequate subcooling margin has been identified in the application and corresponding operating procedure to prevent boiling which may result in an inadvertent criticality due to optimum moderator conditions. Boiling is also not acceptable because of its impact on doses due to reduced water shielding and potential hydrodynamic loads on cask internal components. This may be cask specific depending on the design of the fuel assembly basket and key assumptions of the criticality analysis. The reviewer should ensure that the ISFSI or MRS maximum temperature (under normal conditions) of the pool water and/or other water used in the cask cavity during loading and unloading operations is below the temperature assumed in the cask criticality safety analysis if a time restriction exists in the corresponding technical specifications.

The reviewer should verify that limiting conditions for the operations have been imposed in the technical specifications which ensure that the temperature will remain acceptable during the process and that normal cooling will begin before the temperature criterion is exceeded if the fuel cladding temperature calculation is based on heatup over a limited time period.

For unloading operations, evaluate fuel cladding temperature and cask pressure calculations supporting procedural steps for cool down of the casks (both transportation and storage casks) and reflood of the casks internals presented in Chapter 5 of the SAR. The applicant's analysis should specify and justify the appropriate temperature and flow rate of the quench fluid, assuming maximum fuel cladding temperatures. Engineering judgement, combined with relevant operational experience of unloading of spent fuel assembly from transportation and storage casks, may support limits on flow and quench fluid temperature. The reviewer should coordinate this review with Section 15 (Accident Analysis review), Section 5 (Structural review), and Section 3 (Operation Systems review). Further technical guidance for reviewing cask unloading analyses is provided in NUREG-1536, Chapter 5, Section V.1.

6.5.1.3 Pool Systems

The ISFSI or MRS facility may employ a wet transfer system or fuel storage pool. The reviewer should confirm that the pool system satisfies the requirements of 72.122(h)(2). In addition, the NRC accepts ISFSI and MRS pool and pool confinement facilities that comply with the criteria for such facilities in 10 CFR 50 and implementing NRC guidance. 10 CFR 50, Appendix A, "General Design Criteria for Nuclear Power Plants," criteria especially applicable to an ISFSI or MRS pool cooling system include, in part, General Design Criteria (GDC) 2, 4, 5, 61 and 63 for a DHR system to transfer heat from SSCs important to safety to an ultimate heat sink, GDC 61 on fuel assembly storage, handling, radioactivity control, and GDC 63 on monitoring of fuel assembly and stored waste.

These GDC provide criteria so that pool systems have the capability to transfer heat loads from safety-related SSCs to a heat sink under normal, off-normal, and accident conditions. The GDC also provide criteria for suitable redundancy of components so that safety functions can be performed assuming a single active failure of a component coincident with the loss of all offsite power, and the capability to isolate components, systems, or piping, if required, so that the system safety function will not be compromised.

The reviewer should identify the pool water temperatures used as limits for normal, off-normal, and accident-level conditions. The possible range of boiling temperatures for the pool coolant solution should be stated with consideration of ranges of the solution, elevation above mean sea level, barometric pressure, and air pressure differential maintained between the pool facility and the outside. The reviewer should confirm that limiting conditions for operation in the Technical Specifications contain water level and temperature limits.

The NRC accepts criteria for pool cooling systems as included and/or identified in NUREG-0800, Sections 9.1.2 and 9.1.3, for application to ISFSI and MRS pools. The review procedures given below are for a typical system. Evaluate the spent fuel assembly pool cooling and cleanup system and its makeup system with respect to their capability to perform the necessary safety functions during all conditions, including normal, off-normal, and accident conditions.

The reviewer should verify the capability of the system to transfer heat loads from safety-related SSCs to a heat sink under both normal, off-normal, and accident conditions.

The reviewer should verify that spent fuel cooling systems have sufficient redundancy of components so that safety functions can be performed assuming a single active failure of a component coincident with the loss of all offsite power.

The reviewer should confirm that, for the maximum normal heat load with normal cooling systems in operation and assuming a single active failure with a loss of all offsite power, the bulk temperature of the pool will be kept at or below 60°C (140°F) with maximum heat generation, at or below 32°C (90°F) for an average annual temperature, below 43°C (110°F) for at least 95% of the time and the liquid level in the pool should be maintained (10 CFR 72.122(4)(h)(2), ANSI/ANS 57.2, ANSI/ANS 57.7). The associated parameters for the decay heat load of the fuel assemblies, the temperature of the pool water, and the heatup time or rate of pool temperature rise for the stated storage conditions, are reviewed on the basis of independent analyses or comparative analyses of pool conditions that have been previously found acceptable.

The reviewer should confirm that the spent fuel assembly pool and cooling systems have been designed so that in the event of failure of inlets, outlets, piping, or drains, the pool level will not be inadvertently drained below a point above the top of the active fuel assemblies which maintains the design dose rates due to water shielding. Pipes or external lines extending into the pool that are equipped with siphon breakers, check valves, or other devices to prevent drainage are acceptable as a means of implementing this requirement.

The reviewer should review the information provided in the SAR pertaining to the design bases and criteria and the safety evaluation section to confirm that the safety function of the system for normal operations is identified. The SAR section on the system functional performance requirements should also be reviewed to determine that it describes the minimum system heat transfer and system flow requirements for normal facility operation, component operational degradation requirements (i.e., pump leakage, etc.), and describes the procedures that will be followed to detect and correct these conditions. The reviewer, using failure modes and effects analyses, should determine if the system is capable of sustaining the loss of any active component and evaluate, on the basis of previously approved systems or independent calculations, if the minimum system requirements (cooling load and flow) are met for these failure conditions. The system piping and instrumentation diagrams (P&IDs), layout drawings, and component descriptions are then reviewed for the following points:

* Confirm that essential portions of the system are correctly identified and are isolable from the nonessential portions of the system.

* Review the P&IDs to verify that they clearly indicate the physical division between each portion and indicate required classification change.

* Review system drawings to ensure that they show the means for accomplishing isolation.

* Review the system description to identify minimum performance requirements for the isolation valves.

* Review the drawings and description to verify that adequate isolation valves separate non-essential portions and components from the essential portions.

6.5.1.4 Dry Transfer Systems

The reviewer should confirm that the dry transfer system ensures that under normal, off-normal, and accident conditions that the fuel cladding temperature will not exceed 570°C (1058°F)(See Section 6.5.2.2).

If the fuel cladding temperature calculation is based on heatup over a limited time period, the reviewer should verify that limiting conditions for the operations have been imposed in the technical specifications which ensure that the temperature will remain acceptable during the process and that normal cooling will begin before the temperature criterion is exceeded.

6.5.2 Material Temperature Limits

6.5.2.1 General Considerations

One of the most important results of the thermal evaluation is confirmation that the fuel cladding temperature is sufficiently low to prevent cladding failure during storage. Identify the allowable temperature levels for stored materials in the SAR for long term storage and for short term and abnormal conditions (guidance on data required is provided in SRP Section 4.4.1). Material temperature restrictions should factor in uncertainties in fabrication of the material and thermal modeling of the DHR system.

The reviewer should review design features and design criteria, typically presented in Sections 1 and 3 of the SAR, for additional detail. The reviewer should examine heat loads from both stored contents and external sources, evaluate temperature limits for each fuel assembly type, and assess models used by the applicant for thermal analyses.

The reviewer should verify that temperature restrictions on other SSCs important to safety are identified and justified in the application. The acceptable temperature limits for other materials that may provide integral confinement (e.g., cask mechanical seals) of the radioactive material, shielding, subcriticality assurance, or heat removal are dependent on the material and its importance. The reviewer should verify that the temperature limit criteria and the basis for that selection are proposed by the applicant. Considerations for determining temperature limits for the material can include: (1) temperature at which the structural strength of the material is affected and time at temperature required to cause the effect, (2) temperature at which chemical reactions may take place (at a significant rate) that affect shielding, subcriticality assurance, or structural integrity, (3) temperature at which the black body characteristics of the material used for modeling may be affected, (4) allowance to provide for uncertainties in the temperatures that may occur, (5) temperatures that may be reached in normal, off-normal, and accident-level conditions and events, and (6) potential combinations of temperature and environment (such as may produce significant reaction with borated water).

The acceptable temperature for the stored radioactive material may provide temperature limits for the thermal performance of the casks or the SSCs. Other temperature limit considerations can include temperatures where: (1) retrievability of the original material is potentially degraded, (2) significant outgassing may occur, (3) outgassing of radioactive gases may occur, (4) chemical reactions may occur at a significant rate, and (5) state changes may occur for at least some of the materials.

Elevated temperatures may be of concern due to effects on strength, heat treatment, durability, other properties, or change of state. Reinforced concrete is addressed separately, below. Thermal properties may be needed for materials that are analyzed for loads on SSC. Confirm that the source of thermal property data is an acceptable reference, such as ASME B&PVC, Section II, Material Specifications and Section III appendices. Use of other sources may be necessary for non-standard materials such as neutron absorbers and cask seals.

6.5.2.2 Fuel Cladding

The reviewer should verify that cladding temperatures for each fuel assembly type proposed for storage at the facility will be below their expected damage thresholds for normal conditions of storage. Zircalloy fuel cladding temperature limits at the beginning of dry storage are typically below 380°C (716°F) for a 5-year cooled fuel assembly and 340°C (612°F) for a 10-year cooled fuel assembly for normal conditions and a minimum of 20 years cask storage (PNL-4835, PNL-6189, and PNL-6364). Other temperature limit values for fuel cooled less than 5 years or more than 10 years can be calculated using the same methodology. Temperature limits will be lower with increased fuel assembly cooling time (or increased burnup) mainly due to lower decay heat rates of older fuel. The previously discussed specific values of zircalloy fuel cladding temperature limit for 5-year and 10-year cooled fuel are representative but should not be construed as the exact acceptable values.

The temperature limits may be calculated using methodologies that are based on expected cladding behavior during storage. The NUREG-1536 endorsement of the diffusion-controlled cavity growth (DCCG) methodology to calculate the maximum cladding temperature limits during dry storage is restrictive and relatively inflexible. The use of other methodologies that account for the full range of materials behavior under the expected storage conditions, such as the Commercial Spent Fuel Management Program (CSFM) methodology as described in PNL-6189 and PNL-6364, are acceptable to the staff for calculation of cladding temperature limits. Alternative methodologies may be approved by the staff if they are sufficiently justified. However, these alternative methodologies must be validated with experimental data and associated modeling uncertainties must be addressed.

For short term off-normal and accident conditions, the staff accepts zircalloy fuel cladding temperatures maintained typically below 570°C (1058°F). The short term off-normal and accident temperature of 570°C (1058°F) for zircalloy-clad fuel assemblies is currently accepted as a suitable criterion for fuel assembly transfer operations. This limit may be lowered for high burnup fuel assembly (e.g., greater than ~28,000 MWD/MTU) due to increased internal rod pressure from fission gas buildup. The applicant should verify that these cladding temperature limits are below the limit for facility specific operations (e.g., fuel assembly transfer) and the worst case credible accident.

The staff may approve the storage of fuel assemblies having burnups greater than 45,000 MWd/MTU (also designated as high burnup fuel) provided that the applicant can demonstrate that the cladding will be protected from degradation which could lead to gross rupture (10 CFR 72.122 (h)(1)) and that the storage system is designed to allow ready retrieval of the spent fuel from the storage system (10 CFR 72.122(l)). If such a demonstration cannot be performed, high burnup fuel assemblies could be enclosed by approved baskets to confine the fuel so that degradation of the fuel during storage will not pose problems with respect to its transportation or removal from storage. Such an enclosure would also maintain subcriticality based on optimum moderation conditions and no potential for buckling and failure of fuel rods, grid spacers, and end fittings under the hypothetical accident conditions.

The Standard Review Plan for Dry Cask Storage Systems (NUREG-1536) does not presently address storage of high burnup fuel. For spent fuel having burnups less than 45,000 MWd/MTU, there is sufficient experimental data to support the long-term and short-term temperature limits identified above. Thus, the staff has generally accepted storage of spent fuel with burnup up to 45,000 MWd/MTU. However, there is limited data to show that the cladding of spent fuel with burnups greater than 45,000 MWd/MTU will remain undamaged during the licensing period. Limited information suggests increased cladding oxidation, increased hoop stresses and changes to fuel pellet integrity with increasing burnup up to and beyond 60,000 MWd/MTU. These burnup dependent effects could potentially lead to failure of the cladding and dispersal of the fuel during transfer and handling operations.

The reviewer should confirm that the applicant has provided the following information to show that high burnup fuels will remain intact for the licensing period:

- Experimentally derived creep data (e.g., time to creep rupture, strain rate under storage temperature and pressure conditions, etc.) and descriptions of the anticipated degradation mechanisms. This information should ensure that creep strains are well below those that would result in cladding damage or excessive deformation. Verify that the tests were performed using high burnup fuel, or comparable cladding material specimens, under conditions (i.e., temperature, stress and strain rate) that approximate those expected for dry storage. Accelerated tests are acceptable in the event that long duration tests are impractical. However, the effects of creep resulting from different creep and/or deformation mechanisms, which are likely to occur over different temperature and stress regimes, should be considered and evaluated for its effect on cladding.

- Calculations, or measurements, of the cladding hoop stress. This information will aid in establishing both the parameters of the accelerated creep tests outlined above, and the accuracy of the cladding life prediction. Verify that the stress calculation includes the effects of: (1) a reduction of thickness due to cladding oxidation, (2) the initial fuel rod backfill gas pressure, (3) the buildup of fission products in the fuel rod, and (4) the generation of other gases (e.g., helium, etc.) due to effects caused by the irradiation of any internal cladding coatings. Experimental data should be used and described, as necessary, to verify any assumed values for the oxide thickness or the increase in pressure caused by the buildup of gases.

- Estimates of the amount of hydrogen absorbed by the cladding during reactor operation and the extent of hydride formation in the cladding. This information should ensure that the concentration levels associated with hydride embrittled zirconium alloys are well below those that could significantly reduce the ductility, or overall integrity, of the cladding.

- Information about the integrity of the fuel pellets (i.e., post-reactor operation pellet size, estimated size and quantity of pellet fragments, etc.). This information should support criticality analyses of potentially reconfigured fuel.

Additional guidance on fuel and fuel cladding is provided in Chapter 4 of NUREG-1536.

6.5.2.3 Special Thermal Criteria for Reinforced Concrete

The reviewer should confirm that the maximum calculated concrete temperature meets the criteria for elevated concrete temperatures stated in ACI 349 Section A.4 . The NRC also accepts the following temperature requirements as an alternative to those given in ACI 349 Section A.4, but only for the temperature range between 66°C (150°F) and 149°C (300°F) occurring in normal and off-normal conditions:

- If concrete temperatures of general or local areas do not exceed 93°C (200°F) in normal or off-normal conditions, tests to prove capability of the concrete for elevated temperatures or reduction of concrete strength used for design are not required.

- If concrete temperatures of general or local areas exceed 93°C (200°F) but would not exceed 149°C (300°F), no tests to prove capability for elevated temperatures and no reduction of concrete strength are required if Type II cement is used and aggregates, fine and coarse, meet the following criteria:

    - Satisfy ASTM C33 requirements and other requirements referenced in ACI 349 for aggregates.

    - Have demonstrated a coefficient of thermal expansion (tangent in temperature range of 21°C (70°F) to 38°C (100°F)) no greater than 3.3×10^{-6} cm/cm/°C (6×10^{-6} in./in./°F) or be one or a mixture of the following minerals: limestone, dolomite, marble, basalt, granite, gabbro, or rhyolite.

- For a case in which off-normal temperatures exceed 93°C(200°F) but are less than 107°C(225°F), the list of acceptable aggregates cited in the above paragraph may be amended to include two additional minerals, quartz and sandstone; however, their use is limited to fine aggregates only.

The NRC has not accepted alternative criteria to the temperature limitations expressed in ACI 359 for SSCs designed according to that code.

6.5.2.4 Extreme Low Temperatures

The reviewer should verify that the site characteristics and environmental conditions for low temperature are enveloped by the cask SAR. Extreme low temperatures may be of concern due to the potential for embrittlement of ferritic steel and other materials that could be used for SSCs important to safety. Thermal analysis is not required for determination of possible minimum temperatures. The minimum temperatures are determined from site conditions.

The reviewer should confirm that the structural analysis assumes that material that will be exposed to the outside environment may be at the ambient temperature. Extreme low temperatures may result in the largest temperature gradients and loads in interconnected structures due to SSCs at different temperatures and/or with different coefficients of thermal expansion.

6.5.3 Thermal Loads and Environmental Conditions

The reviewer should examine the specification for the design basis fuel assembly decay heat presented in Section 2 of the SAR and the corresponding sections of the cask(s) SAR(s) if the cask has received previous NRC approval. The reviewer should ensure that this decay heat is consistent with the specified burnup and cooling times, if included. Typically, decay heat is calculated using the same computer codes as those used to determine radiation source terms. The reviewer should coordinate the review of fuel assembly source terms for consistency with the shielding review in Section 7, as appropriate. Alternatively, the decay heat from the design basis fuel assembly may also be derived from Regulatory Guide 3.54, "Spent Fuel Heat Generation in an Independent Spent Fuel Storage Installation." Except for neutrino energy, all decay heat should be considered to be deposited in the fuel.

The reviewer should confirm that, if control components or other assembly hardware (e.g., shrouds) are included with the fuel assemblies, their heat loads are specified and justified.

The reviewer should review the insolation assumptions and ambient environmental temperature in the SAR(s) for the cask(s) proposed for use at the ISFSI or MRS. Verify that the ISFSI or MRS site characteristics and environmental conditions are bounded by the cask(s) analysis. The ISFSI or MRS applicant should confirm this in the SAR. In general, the staff accepts insolation considerations presented in 10 CFR Part 71 for 10 CFR Part 72 applications. Because of the large thermal inertia of a storage cask, the insolation values listed in 10 CFR 71.71 may be averaged over a 24-hour day assuming steady-state conditions. If a less conservative approach is presented, the SAR must thoroughly describe and justify its use.

The reviewer should compare the MRS or ISFSI environmental data with statements in the cask(s) SAR about assumed bounding temperatures ranges, ambient temperature conditions, and variations of external heat sources over time. When calculating maximum thermal gradients and temperature differences within individual components or between locations, changes in temperature over time may need to be determined. These changes over time should consider the thermal properties, including emissivity, solar absorption coefficients, thermal conductivity, heat capacity, and density of specific components. The reviewer should confirm that the assumed temperatures and temperature variations with time are stated in the SAR for normal, off-normal, and accident-level conditions. Evaluate the elevated ambient temperatures and enhanced heat transfer due to off-normal and accident-level situations (e.g., vehicular, building, or forest fire) to ensure that they are quantified and supported by analysis.

The reviewer should confirm that the conditions that may result in high temperature gradients or pressures are identified in the SAR. The conditions may be transitory and may be controllable or subject to limits. For cask unloading operation (see Section 6.5.1.2 on dry storage systems), ensure that limits are provided for reflood rate and fluid temperature. For concrete, spalling due to temperature gradients is typically considered to have minor (at most) structural significance, but it could partially block ventilation passages, depending on the design.

6.5.4 Analytical Methods, Models, and Calculations

The reviewer should evaluate models used for thermal evaluations to ensure that they are compatible with the analytical approach. The models should be conservative for the analysis in which used. The models should permit analysis and quantification of the heat transfer mechanisms. Guidance on computational methods and computer codes to model dry cask storage systems is provided in Chapter 4 of NUREG-1536. Regulatory Guide 3.54 provides guidance on the calculation of spent nuclear fuel decay heat.

The reviewer should ensure that models of the pool cooling system piping and heat exchange system are based on process flow analytical models. The flow and temperatures within the pool do not need to be modeled if the temperature limits of the stored material are significantly higher than the boiling temperature at the depth of the material for the site. However, if the pool includes significant restrictions to flow adjacent to the stored material, there should be an analysis to demonstrate that boiling in an area of the pool will not occur.

The reviewer should confirm that calculations determine the highest temperatures that would be reached by coolant in the pool under normal, off-normal, and accident conditions. Calculations should be for steady-state and for transient conditions. The calculations should be sufficient to demonstrate balance between heat removal and heat generation and that the most critical situations have been analyzed.

The reviewer should ensure that the calculations provided with the SAR permit full review of the assumptions, input, calculations, and results. The calculations should include temperatures at sufficient points to ensure that the hottest fuel cladding and points on other SSCs important to safety are included. The calculations should provide the most severe thermal gradients for material subject to significant thermal stresses (typically the reinforced concrete in confinement vessels).

The NRC has accepted thermal calculations of cask heat removal and associated temperatures by use of the ANSYS™ (ANSYS, Inc.) and the HEATING (NUREG/CR-0200) codes with appropriate models. Both of these are capable of general steady state and transient calculations. The NRC does not accept two simplified, more approximate codes: SCANS (NUREG/CR-4554) and CASKS (NUREG/CR-6242). Chapter 4 of NUREG-1536 provides additional discussion on the use of computer codes for thermal analysis.

The reviewer should perform confirmatory evaluations of the thermal performance of SSCs important to safety. This should specifically include steady-state temperature distributions, local heat balances, temperatures reached, and temperature distributions within any reinforced concrete SSCs for the bounding ambient temperatures. The reviewer should verify that the maximum temperatures have been calculated for all SSCs important to safety with temperature limits that may be approached. Evaluation by the reviewer should include:

- Heat balance at the outer surface of the cask to verify that the heat from the spent fuel assembly and insolation equal that removed by convection and radiation

- Assessment of the heat transfer coefficients used to confirm appropriateness for the storage conditions

- Estimation of temperature of the cask inner surface (as by calculating the temperature distribution across the cask body with simple heat balance approximations)

- Comparison of the difference between the cask inner surface temperature and the maximum cladding temperature with that of similar confinement casks/baskets reviewed in previous SARs

The reviewer should model and evaluate a portion of the cask or basket to ensure that the SAR results are conservative if a more detailed confirmatory review is considered to be appropriate. The staff may perform an extensive confirmatory evaluation if major errors are suspected or marginal conservatism exists in the applicant's modeling approach.

The confirmatory evaluation by the staff may result in a requirement that the applicant perform design-verification testing of an as-built cask system to validate the thermal analysis presented in the SAR. The test conditions, configuration, and type and location of instrumentation used, if any, should be adequately described.

The NRC accepts simplifying assumptions for the effects of reinforcing steel in determining the thermal performance and temperature distributions of reinforced concrete. Use of a homogeneous material, instead of modeling the concrete and reinforcing steel as separate elements, is acceptable if the substitute hypothetical material has appropriately adjusted thermal properties and the reinforcing steel is covered with concrete in accordance with the applicable structural code. Thermal performance and/or temperature distributions for reinforced concrete designs which have features that provide for significant thermal transfer below the concrete surface (as by internal studs welded to an exposed steel plate) may require more detailed analysis.

6.5.5 Protection from Fire and Explosions

6.5.5.1 General Considerations

The reviewer should ensure that the applicant performed a fire and explosives hazards analysis of the facility and, if warranted, instituted a fire protection program (FPP). The reviewer should verify that the following SAR specific criteria provide information and describe a basis acceptable to the staff that may be used to meet the requirements of 10 CFR 72.122(c) and 72.122(j):

- NUREG-0800 Branch Technical Position (BTP) SPLB 9.5-1 as it relates to the design provisions given to implement the FPP

- Regulatory Guide 1.78, "Assumptions for Evaluating the Habitability of a Nuclear Power Plant Control Room During a Postulated Hazardous Chemical Release," as it relates to habitable areas, such as the control room and to the use of specific fire extinguishing agents

- NRC technical position on fire protection for fuel cycle facilities

  Depending on the design, magnitude, scope and fire hazards of a proposed ISFSI, MRS, or centralized interim storage facility, the applicant may have to institute a fire protection program to satisfy the requirements of 72.122(c). Guidelines for a fire protection program are provided in Section 6.6.4.

6.5.5.2 Spent Fuel Casks

The ISFSI or MRS may use NRC-approved dry storage casks approved under Subpart L of 10 CFR 72 provided, in part, that the applicant satisfies the fire requirements identified in the Certificate of Compliance and 72.122(c).

The reviewer should verify that the fire conditions of the worst case, credible site fire do not exceed the fire assumptions made in the fire analysis of the cask. Using the accident condition temperatures at the MRS or ISFSI, verify that the post accident pressure of the gas in the cask cavity is within the cask design pressure. The pressure should be determined based on the assumption that 100% of the fuel rods have failed.

Under the conditions where any of the cask component or fuel cladding temperatures are close (within 5%) to their limiting values during an accident or the maximum normal operating pressure is within 10% of its design basis pressure, or any other special considerations affected by fission gas concentrations, the applicant should analyze the potential impact of the fission gas in the cask on the cask component and fuel cladding temperature limits and the internal cask pressure.

The reviewer should evaluate the site-specific analysis for explosions and verify that the cask analysis envelopes the site conditions. Impact on SSCs and the cask should be performed as part of the structural review. As noted in NUREG-1536, explosion-caused overpressure and reflected pressure may be associated with explosives and chemicals transported by rail or on public highways, natural gas pipelines, and vehicular fires of equipment used in the transfer of casks. Explosions may result from detonation of an air-gaseous fuel mixture. With the exception of a transfer vehicle accident, the explosion hazards are typically similar to those for facilities subject to 10 CFR Part 50 reviews. Note, this explosive overpressure is not meant to be that from a radiological sabotage event.

The reviewer should verify that the cask materials, such as protective coatings, are compatible with pool or other water used in the cask cavity so as to preclude or minimize the potential for combustible gas generation (see NRC bulletin 96-04 for background).

6.5.5.3 SSCs Important to Safety

A small amount of exterior concrete spalling may result from a fire or other high temperature condition and/or application of fire, water or rain on heated surfaces. The small amount is not expected to affect heat transfer or reduce shielding significantly, and therefore, does not need to be estimated or evaluated in the SAR. Any significant spalling damage is readily detectable, and appropriate recovery or corrective measures may be presumed. NRC accepts that concrete temperatures may exceed the temperature criteria of ACI 349 for accidents if the temperatures result from a fire. In that case, corrective action may be required for continued safe storage.

The reviewer should verify that fire protection for spent fuel pool cooling and waste confinements systems important to safety, has adequate fire and explosive protection (see FPP guidelines below).

6.5.5.4 Guidance for a Fire Protection Program

The reviewer should verify that a FPP provides assurance that a fire will not significantly increase the risk of radioactive releases to the environment in accordance with the general design criteria of 72.122(c). A defense-in-depth approach should achieve balance among prevention, detection, containment, and suppression of fires. Confirm that there is a fire protection policy for the protection of SSCs important to safety at each facility and for the procedures, equipment, and personnel required to implement the program at the site. The FPP consists of fire detection and extinguishing systems and equipment, administrative controls and procedures, and trained personnel.

Portions of the review procedures of NUREG-0800 Section 9.5-1 and the guidelines of the NRC technical position on fire protection for fuel cycle facilities may be applicable to the MRS or ISFSI contingent on the design of the installation and associated fire hazards. Many of the national codes and standards cited in these NRC guidance documents, in particular the codes and standards of the National Fire Protection Association (NFPA), could be applicable to the ISFSI or MRS facility.

The reviewer should review the SAR to determine that the appropriate levels of management and trained, experienced personnel are responsible for the design and implementation of the fire protection program in accordance with NUREG-0800 Branch Technical Position (BTP) 9.5-1.

The reviewer should review the analysis in the SAR of the fire potential in important to safety facility areas and the hazard of fires to these areas to determine that the proposed fire protection program is able to minimize radioactive releases to the environment.

The reviewer should evaluate the FPP P&IDs and facility layout drawings to verify that facility arrangement, buildings, and structural and compartment features which affect the methods used for fire protection, fire control, and control of hazards are acceptable for the protection of safety-related equipment.

The reviewer should determine that design criteria and bases for the detection and suppression systems for smoke, heat and flame control are in accordance with the BTP guidelines and provide adequate protection for SSCs important to safety. The reviewer determines that fire protection support systems, such as emergency lighting and communication systems, floor drain systems, and ventilation and exhaust systems are designed to operate consistent with this objective. The reviewer should review the results of an FPP failure modes and effect analysis to assure that the entire fire protection system for one safety-related area cannot be impaired by a single failure.

The reviewer should review the technical specifications proposed by the applicant for fire protection. The reviewer will determine that the limiting conditions for operation and surveillance requirements of the technical specifications are in agreement with the requirements developed as a result of the staff review.

Guidance for fire detection and suppression along with the fire protection water system is provided in NUREG-0800 BTP 9.5-1.

The reviewer should confirm that the control room or control area ventilation system P&IDs show monitors located in the system intakes that are capable of detecting radiation, smoke, and toxic chemicals. The monitors should actuate alarms in the control room. The reviewer should confirm that provisions for isolation of the control room upon smoke detection at the air intakes are shown on the P&IDs. The isolation may be actuated manually for most cases. Automatic isolation may be required in special cases, such as for fires resulting from aircraft crashes. The reviewer should consult NUREG-0800 BTP 9.5-1 for additional guidance.

The reviewer should confirm that protection for the spent fuel pool area should be provided by local hose stations and portable extinguishers. Automatic fire detection should be provided to alarm and annunciate in the control room and to alarm locally. Verify that fire barriers, automatic fire suppression and detection, and ventilation controls are provided. The reviewer should verify that records storage areas are located and protected so that a fire in these areas does not expose systems or equipment important to safety (see Regulatory Guide 1.28, "Quality Assurance Program Requirements ").

The reviewer should verify that miscellaneous areas, such as shops, warehouses, auxiliary boiler rooms, fuel oil tanks, and flammable and combustible liquid storage tanks are located and protected so that a fire or effects of a fire, including smoke, will not adversely affect any SSCs important to safety.

The reviewer should confirm that acetylene-oxygen gas cylinder storage locations are not in areas that contain or expose safety-related equipment or the fire protection systems that serve those safety-related areas. A permit system should be required to use this equipment in safety-related areas of the facility (also see Position C.2 of NUREG-0800 of BTP 9.5-1). The reviewer should verify that unused ion exchange resins are not stored in areas that contain or expose safety-related equipment. The reviewer should verify that hazardous chemicals are not to be stored in areas that contain or expose safety-related equipment.

The reviewer should verify that materials that collect and contain radioactivity, such as spent ion exchange resins, charcoal filters, and HEPA filters are stored in closed metal tanks or containers that are located in areas free from ignition sources or combustibles. These materials should be protected from exposure to fires in adjacent areas as well. Consideration should be given to requirements for removal of decay heat from entrained radioactive materials.

6.6 Evaluation Findings

The reviewer prepares a safety evaluation report on satisfaction of the regulatory requirements as identified in Section 6.3. The SER should address each acceptance criteria provided in Section 6.4 of this SRP similar to the following (finding numbering is for convenience in referencing within the SRP and SER):

F6.1   SSCs important to safety are described in sufficient detail in Sections _____ of the SAR to enable an evaluation of their heat removal effectiveness. Cask structures, systems and components important to safety remain within their operating temperature ranges in accordance with 10 CFR 72.122.

F6.2   [If applicable] The [dry storage system designation] is designed with a heat-removal capability having testability and reliability consistent with its importance to safety as required by 10 CFR 72.128.

F6.3   [If applicable] The spent fuel cladding is protected against degradation that leads to gross ruptures by maintaining the cladding temperature for _____ -year cooled fuel assembly (fuel assembly type) below _____ °C in an [applicable gas] environment. Protection of the cladding against degradation will allow ready retrieval of spent fuel assembly for further processing or disposal as required by 10 CFR 72.122.

F6.4   The staff concludes that the site specific fire and explosions hazards is acceptable and that the fire protection program meets the requirements of 10 CFR 72.122(c). This conclusion is based on the applicant meeting the guidelines of NUREG-0800

BTP 9.5-1, as well as applicable industry standards. In meeting these guidelines the applicant has provided an acceptable basis for the [ISFSI or MRS] design and location of safety-related structures and systems to minimize the probability and effect of fires and explosions; has used noncombustible and heat resistant materials whenever practical; has provided of fire detection and fire fighting systems of appropriate capacity and capability to minimize adverse effects of fire on systems important to safety.

6.7 References

NRC documents referenced are identified at Consolidated References, Section 17.

Codes, Standards, and Specifications

ACI 349, "Code Requirements for Nuclear Safety Related Concrete Structures," American Concrete Institute.

ACI 359, "Code for Concrete Reactor Vessels and Containments" (also designated as ASME Boiler and Pressure Vessel Code, Section III, "Rules for Construction of Nuclear Power Plant Components," Division 2), American Concrete Institute and American Society of Mechanical Engineers (Joint Committee).

ANSI/ANS-57.2-1983, "Design Requirements for Light Water Reactor Spent Fuel Storage Facilities at Nuclear Power Plants," American Nuclear Society, 1983.

ANSI/ANS-57.7-1988, "Design Criteria for an Independent Spent Fuel Storage Installation (water pool type)," American Nuclear Society, 1988.

ASME Boiler and Pressure Vessel Code, Section III, "Rules for Construction of Nuclear Power Plant Components," Division 1, American Society of Mechanical Engineers.

National Fire Protection Association (NFPA) codes and standards.

Manuals and Texts

"ANSYS Basic Analysis Procedures Guide," Fourth Edition, ANSYS Release 5.6, ANSYS, Inc., Canonsburg, Pennsylvania, November 1999.

Technical Reports

Cunningham, M.E., et al., "Control of Degradation of Spent LWR Fuel During Dry Storage in n Inert Atmosphere," PNL-6364, PNL, October 1987.

Levy, I.S., et al., "Recommended Temperature Limits for Dry Storage of Spent Light Water Zircalloy Clad Fuel Rods in Inert Gas," PNL-6189, PNL, May 1987.

Manteufel, R.D. and N.E. Todreas, "Effective Thermal Conductivity and Edge Configuration Model for Spent Fuel Assembly," <u>Nuclear</u> <u>Technology</u>, Vol. 105, pp. 421–440, March 1994.

PNL-4835, "Technical Basis for Storage of Zircalloy-Clad Spent Fuel in Inert Gases," PNL, September 1983.

Thomas, G. R., and R. W. Carlson, "Evaluation of Use of Homogenized Fuel Assemblies in the Thermal Analysis of Spent Fuel Storage Casks," Lawrence Livermore National Laboratory, Publication No. UCRL-ID-134567, July 1999.

7 SHIELDING EVALUATION

7.1 Review Objective

The primary objective of this review is to determine whether the shielding design features of the independent spent fuel storage installations (ISFSI) or monitored retrievable storage (MRS) meet NRC criteria for protection against direct radiation from the material to be stored. In particular, this evaluation should establish the validity of dose rate estimates made in the applicant's Safety Analysis Report (SAR). These estimates are in turn used in the radiation protection review (described in Chapter 11) to determine (a) compliance with regulatory limits for allowable dose rates, and (b) conformance with criteria for maintaining radiation exposures as low as reasonably achievable (ALARA).

The scope of this chapter is limited to evaluating the shielding for the spent fuel or high-level waste to be stored. Other radiation sources at the ISFSI or MRS for which shielding may be required are addressed in Chapters 11 and 14.

The interrelationship between the site-generated waste review and other areas of review is illustrated in Figure 7.1. The figure shows that the evaluation draws on information in the application, as well as results of other technical reviews. The figure also shows that the results of this review are used in other technical reviews.

7.2 Areas of Review

The following outline shows the areas of review addressed in Section 7.4, Acceptance Criteria, and Section 7.5, Review Procedures:

Contained Radiation Sources
>   Gamma Sources
>   Neutron Sources

Storage and Transfer Systems
>   Design Criteria
>   Design Features

Shielding Composition and Details
>   Composition and Material Properties
>   Shielding Details

Analysis of Shielding Effectiveness
>   Computational Methods and Data
>   Dose Rate Estimates
>   Confirmatory Calculations

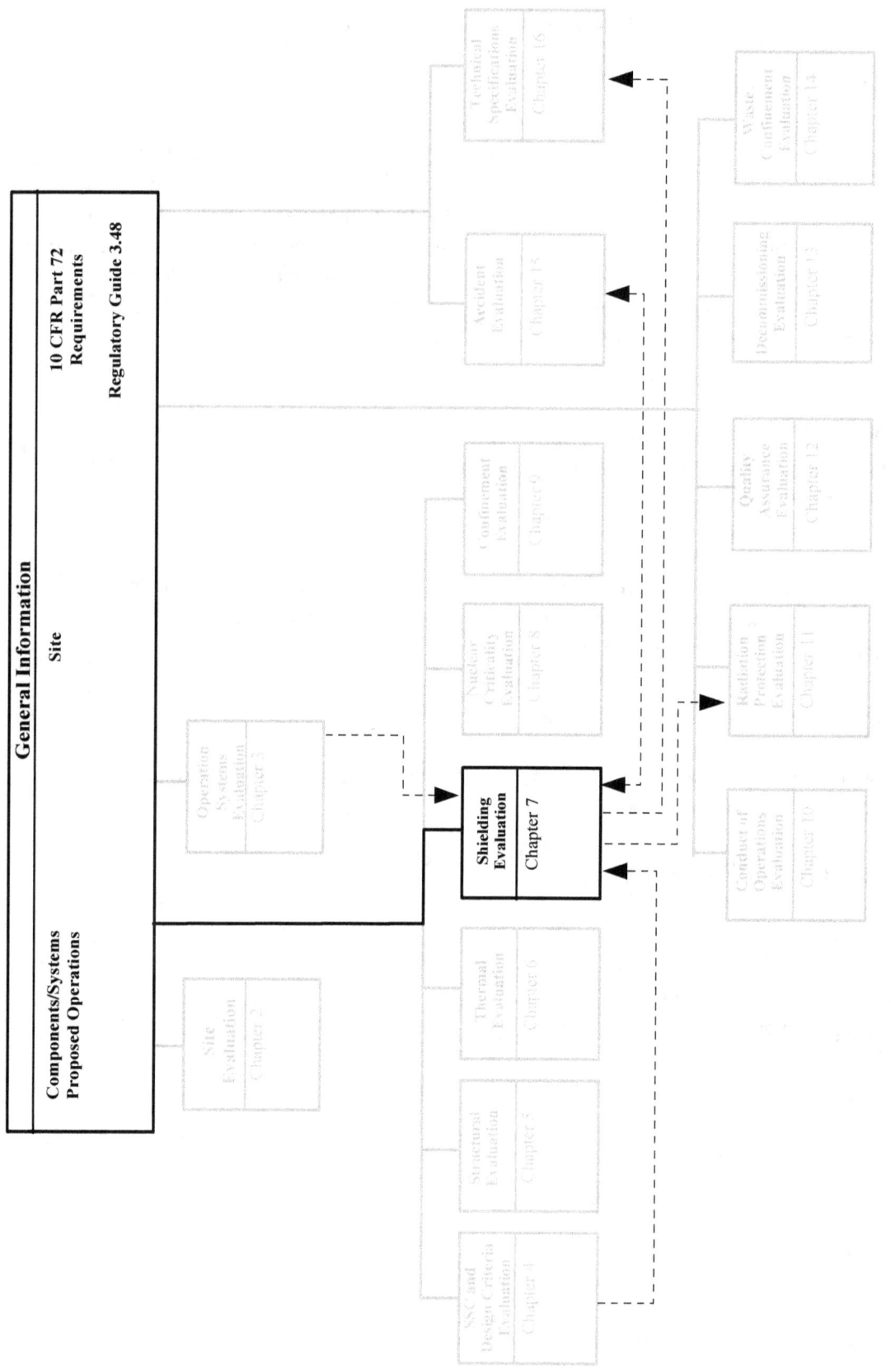

Figure 7.1 Overview of Shielding Evaluation

7.3 Regulatory Requirements

This section identifies and presents a high-level summary of Title 10 of the Code of Federal Regulations (CFR) Part 72 relevant to the review areas addressed by this chapter. The NRC staff reviewer should read the exact regulatory language. A matrix at the end of this section matches the regulatory requirements identified in this section to the areas of review identified in the previous section.

20.1201 Occupational dose limits for adults.
(a) "The licensee shall control the occupational dose to...
      (1) An annual limit, which is the more limiting of-
            (i) The total effective dose equivalent being equal to 5 rem (0.05 Sv); or
            (ii) The sum of the deep-dose equivalent and the committed dose equivalent to any individual organ or tissue other than the lens of the eye being equal to 50 rem (0.5 Sv)."

20.1301 Dose limits for individual members of the public.
(a) "Each licensee shall conduct operations so that-
    (1) The total effective dose equivalent to individual members of the public from the licensed operation does not exceed 0.1 rem (1 millisievert) in a year. . .
    (2) The dose in any unrestricted area from external sources does not exceed 0.002 rem (0.02 mSv) in any one hour."

20.1302 Compliance with dose limits for individual members of the public.
(b) "A licensee shall show compliance with the annual dose limit in Section 20.1301 by-
    (1) Demonstrating ... that the total effective dose equivalent...does not exceed the annual dose limit."

72.24 Contents of application: Technical information
"Each application for a license under this part must include a Safety Analysis Report describing . . ."
(c)   (3) "Information relative to ... all structures, systems, and components important to safety."
(e) The means for controlling and limiting occupational radiation exposures within the limits given in Part 20 of this chapter [to] ... as low as is reasonably achievable."

72.104 Criteria for radioactive materials in effluents and direct radiation from an ISFSI or MRS.
(a) "During normal operations and anticipated occurrences, the annual dose equivalent to any real individual who is located beyond the controlled area must not exceed 25 mrem to the whole body, 75 mrem to the thyroid and 25 mrem to any other organ as a result of exposure to:
    (1) Planned discharges ...,
    (2) Direct radiation..., and
    (3) Any other radiation from uranium fuel cycle operations within the region."

72.126 Criteria for radiological protection.
(a) "Exposure control. Radiation protection systems must be provided for all... onsite personnel ... exposed to radiation ...The design must include means to:
      (6) Shield personnel from radiation exposure."

72.128 Criteria for spent fuel, high-level radioactive waste, and other radioactive waste storage and handling.
(a) "...Spent fuel storage, high-level radioactive waste storage, and other systems ...must be designed to ensure adequate safety.... These systems must be designed with
      (2) Suitable shielding..."

A matrix which shows the primary relationship of these regulations to the specific areas of review associated with this SRP chapter is given in Table 7.1. The NRC staff reviewer should verify the association of regulatory requirements with the areas of review presented in the matrix to ensure that no requirements are overlooked as a result of unique applicant design features.

Table 7.1 Relationship of Regulations and Areas of Review

Areas of Review	10 CFR Parts 20 and 72 Regulations						
	20.1201	20.1301	20.1302	72.24	72.104	72.126	72.128
Contained Radiation Sources						•	
Storage and Transfer Systems				•	•	•	•
Shielding Composition and Details				•		•	
Analysis of Shielding Effectiveness	•	•	•	•	•	•	•

7.4 Acceptance Criteria

The information submitted in the SAR must be of sufficient scope and detail to allow for a thorough evaluation of proposed shielding, including the performance of independent dose rate estimates. All applicable regulatory requirements must be satisfied, and the methods for determining compliance must be acceptable to NRC. The following sections provide criteria for acceptability of SAR informational content and the details and method of evaluation of the proposed shielding features.

Primary guidance related to the information to be included in the SAR is provided by Regulatory Guide 3.48, "Standard Format and Content for the Safety Analysis Report for an Independent Spent Fuel Storage Installation, (Dry Storage)" and NUREG-1536, Chapter 5. The guidance in this section summarizes and supplements the guidance provide by those sources.

7.4.1 Contained Radiation Sources

10 CFR 72.24 describes the required contents of the application. To meet those requirements, the SAR must describe each type of contained radiation source used as a basis for shield design calculations. The physical and chemical form, source geometry, radionuclide content, and estimated curie value and bases for estimation must be described in a manner suitable for use as input for shielding calculations.

7.4.1.1 Gamma Sources

A tabulation of radiological characteristics for each gamma-ray source type must be provided, including isotopic composition and photon yields by X- and gamma-ray energy group. The SAR must specify gamma source terms for both spent fuel and activated materials. The energy group structure from the source term calculation must correspond to that of the cross-section set of the shielding calculation. The computer methodology or database application used to compute source term strength must be specifically identified.

The SAR must describe the extent to which radioactivity may be induced by interactions involving neutrons originating in the stored materials. The SAR must provide source term descriptions for induced radioactivity and the bases (assumptions and analytical methods) used for their estimation. Alternatively, the SAR may describe the bases for excluding induced radioactivity source terms.

7.4.1.2 Neutron Sources

The SAR must describe the neutron source terms and tabulate the neutron yield by energy group. The SAR must describe the bases used to determine the source terms.

7.4.2 Storage and Transfer Systems

10 CFR 72.126 and 72.128 require that the storage and handling systems requiring shielding be described. The SAR must provide design criteria and descriptions of design features for shielded containers.

7.4.2.1 Design Criteria

10 CFR 20.1201, 20.1301, 20.1302, and 10 CFR 72.104 provide dose rate criteria for occupational exposure and for members of the public. The principal design criteria (presented in SAR Section 3) must specify the criteria that have been used as a basis for protection against direct radiation. Design criteria must include the identification of maximum dose rates for each type of shielded container (transfer cask, storage cask, etc.). Design dose rates must also be specified for occupancy areas and correlated with occupancy times and distance to sources. An estimate of collective doses (person-rem per year) must be provided for each occupancy area and for various operations (see Chapter 11, Radiation Protection).

The design dose rates must consider ALARA objectives. The SAR must identify choices between otherwise comparable alternatives affected by ALARA considerations and show that further reduction of collective doses from direct radiation is not practicable.

7.4.2.2 Design Features

The SAR must describe the transfer and storage systems, including the use of shielding, to reduce direct radiation dose rates. The SAR must describe various uses of shielding features at the proposed ISFSI or MRS, including any of the following that apply:

- Shielding provided by the radioactive material being stored

- Neutron capture provided by borated water in casks and storage pools, and by borated materials incorporated into casks

- Gamma and neutron shielding provided by the structural and nonstructural materials (e.g., lead) forming the walls and ends of the storage or transfer casks

- Temporarily positioned shielding used during operations for preparing the storage confinement cask for storage or retrieval, and/or during transfer into the storage position at the storage location

- Shielding provided by pool facility interior and exterior walls

- Shielding provided by natural or man-made earth barriers between the radioactive material and the area beyond the controlled area boundary.

7.4.3 Shielding Composition and Details

10 CFR 72.24 requires that the application include information relative to materials and arrangements of all structures, systems, and components important to safety.

7.4.3.1 Composition and Material Properties

The SAR must describe the composition of shielding materials and geometries. The SAR must give material compositions, densities and references for these data, for all materials used. The SAR must give references to the source of the data and the validation for the data for nonstandard materials (e.g., proprietary neutron shield material).

The SAR must describe the potential for shielding material to experience changes in material properties at temperature extremes. The SAR should give and reference temperature sensitivities of shielding materials.

7.4.3.2 Shielding Details

The SAR must describe the geometric arrangement of shielding. The SAR must use illustrations to identify the spatial relationships among sources, shielding, and design dose rate locations. The SAR must clearly indicate the physical dimensions of sources and shielding materials.

The SAR must identify penetrations, voids, or irregular geometries that provide potential paths for gamma or neutron streaming. These potential streaming paths must be clearly identifiable on submitted drawings. The SAR must describe design features used to minimize streaming through these penetrations.

The SAR must clearly state any differences between shielding features during normal or off-normal conditions and accident level conditions.

7.4.4 Analysis of Shielding Effectiveness

The SAR must describe the computational models, data, and assumptions used in evaluating shielding effectiveness, and must provide dose rate estimates for areas of concern.

7.4.4.1 Computational Methods and Data

The SAR must identify the computer models used in evaluating shielding for each significant radiation source identified in Section 7.4.1 and reference the appropriate documentation. For each computer program used, the SAR must provide test problem solutions that demonstrate substantial similarity to solutions from other sources (hand calculations, published literature results, etc.). The SAR must provide a summary that compares the test problem solutions in either graphical or numeric form. These solutions may be referenced and need not be submitted in the SAR if the references are widely available or have been previously submitted to the NRC for the same model and version.

The SAR must clearly present the data used as input for computational purposes. The SAR must identify any differences between actual material properties or physical dimensions and those used in the analytical method (e.g., for simplifying the computational process). The SAR must defend any simplifying assumptions by showing that the approach used will result in conservative (bounding) estimates.

The SAR must state the basis for the flux-to-dose-rate conversion in its shielding analysis, including conversions that are done by a code using its own data library. The NRC accepts flux-to-dose rate conversion factors in ANSI/ANS 6.1.1.

The SAR must include a representative computer code input file used in type of shielding computation performed for the installation.

7.4.4.2 Dose Rate Estimates

The SAR evaluation of shielding effectiveness must include estimates of dose rates in representative areas around the storage and transfer systems. The SAR estimates must account for such factors as distance to occupied areas, duration of operations, expected occupancy rates, contributions from radionuclide releases and other factors . These criteria are identified and evaluated in the radiation protection evaluation described in Chapter 11. The criteria below relate primarily to the completeness of information provided in the SAR.

The SAR must clearly indicate the physical locations on and around storage or transfer casks for which dose rate calculations have been performed. These locations must include points on or in the immediate vicinity of cask surfaces where workers will perform operations during loading, retrieval, handling operations, and any projected maintenance and surveillance. For storage casks with labyrinthine air flow passages, the SAR must include dose rate estimates for the air inlets and air outlets. The SAR must identify points that have the highest calculated dose rates.

The SAR must include dose rate estimates for all onsite areas at which workers will be exposed to elevated dose rates. The SAR must compute dose rates within restricted areas in enough detail to estimate doses received by workers performing ISFSI or MRS functions.

The SAR must present dose rate estimates for representative points on the perimeter of the controlled area and at locations beyond the controlled area boundary. The SAR must specify these estimates with respect to distance and direction in a manner that will allow for estimation of population dose within a 5-mile (8-km) radius of the site. These dose rates must include contributions from both direct line-of-sight and air-scattered radiation emerging from casks or other shielded sources.

For storage confinement casks, the SAR must calculate the dose rate at 1 meter from the cask surface for off-normal events and conditions that result in a significant dose rate increase. The model used for these calculations must be consistent with the expected condition of the cask after the event.

7.5 Review Procedures

7.5.1 Radiation Sources

The reviewer should verify that all potential radiation sources have been identified, even if analysis shows that they produce negligible contributions to dose. In addition to the intact spent fuel to be stored, the reviewer should consider whether other sources at the ISFSI or MRS may require shielding. These sources might include:

•   High-level waste in a form ready to be stored

•   Other radioactive material to be stored (e.g., with failed cladding, awaiting encapsulation, in wet storage, etc.)

- Radioactive material to be retrieved from storage

- Radioactive material elsewhere within the controlled area or on the site, in transportation or transfer casks, in transit or holding.

The reviewer should verify that the physical and chemical form, source geometry, radionuclide content, and estimated curie value and bases for estimation are provided for each source type. For spent fuel, the reviewer should ensure that the SAR has assumed burn-up and decay properties that will provide bounding results.

The reviewer should review the use of computer codes for estimation of each of the source terms. The reviewer should determine whether the codes are appropriate for the cases in which they are applied. The reviewer should consider whether the required level of precision (which can vary depending on the relative source contributions) is achieved. Spent fuel source terms are usually determined by using ORIGEN-S (e.g., as an SAS2 sequence of SCALE)(Petrie 1995), ORIGEN2 (ORNL 1991), or the U.S. Department of Energy - Office of Civilian Radioactive Waste Management Characteristics Database (TRW Environmental Safety Systems). Although the latter two are easy to use, both have energy group structure limitations (discussed below for gamma sources). If ORIGEN2 is used, the reviewer should determine whether the SAR includes verification that the chosen cross-section library is appropriate for the fuel being considered. Many libraries are not appropriate for a burnup that exceeds 33,000 MWd/MTU.

The reviewer should compare use of codes to determine source terms with prior uses of the same code from prior reviews. The reviewer should note the restriction on use of proprietary data of a possible competitor to the SAR to support the SAR's preparation or modification of application documents.

If the SAR has used a computer code that is not an industry standard, the reviewer should ensure that appropriate descriptive information, including validation and verification status, and reference documentation have been provided. The reviewer should determine whether the code is suitable for the source term estimation cases in which it is used and whether it has been appropriately applied.

7.5.1.1 Gamma Sources

The reviewer should verify that the SAR has specified gamma source terms as a function of energy for each type of gamma-ray source. The reviewer should ensure that activated materials are included in addition to the spent fuel or high-level waste to be stored. The reviewer should ensure that the source terms specified for spent fuel are clearly stated as per assembly, per total assemblies, or per metric ton.

The reviewer should determine whether the energy group structure from the source term calculation corresponds to that of the cross-section set used in the shielding calculation. If they do not correspond, the reviewer should determine whether the SAR must regroup the photons, by using the nuclide activities from the source term calculation as input to a simple decay code with a variable group structure or by interpolating from one structure to the other. In general, regrouping is necessary only for gamma energies from approximately 0.8 to 2.5 MeV since these are the main contributors to the dose rate through typical types of shielding; the SAR must include the full range of energies in shielding calculations.

The reviewer should assess the SAR treatment of secondary radiation contributions and determine whether these have been appropriately addressed in the source term description. The reviewer should pay particular attention to neutron interactions that may result in the production of energetic gammas near the storage confinement cask surface. The SAR must provide source term descriptions for any materials that would be activated by the radioactivity of the stored materials.

7.5.1.2 Neutron Sources

The reviewer should verify that neutron source terms are expressed as a function of energy. Neutron source terms generally result from both spontaneous fission and alpha-n reactions in the fuel. Depending on the method used to determine these source terms, the SAR may need to determine the energy group structure independently. Since the contribution from alpha-n reactions is usually small, this determination can be made by selecting the nuclide with the largest contribution to spontaneous fission (e.g., ^{244}Cm) and using that spectrum for all neutrons.

A specified source term is difficult for most cask users to determine and for inspectors to verify. The specification of a minimum initial enrichment is a more straightforward basis for defining the allowed contents. The reviewer should verify that the specification bounds all assemblies proposed for the casks in the application. Specific limits are needed for inclusion in the Certificate of Compliance. Lower enriched fuel, irradiated to the same burnup as higher enriched fuel, produces a higher neutron source. Consequently, the reviewer should verify that the SAR specifies the minimum initial enrichment as an operating control and limit for cask use. Alternately, the SAR must specifically justify the use of a neutron source term, in the shielding analysis, that specifically bounds the neutron sources for fuel assemblies to be placed in the cask. The SAR should not attempt to establish specific source terms as the operating controls and limits for cask use without adequate justification acceptable to the staff.

7.5.2 Storage and Transfer Systems

The reviewer should determine that the SAR descriptions of storage and transfer systems requiring shielding provide the information necessary to evaluate the shielding in the context of its proposed use.

7.5.2.1 Design Criteria

The reviewer should verify that specific design criteria have been used as a basis for protection against direct radiation and that the criteria include the specification of maximum dose rates for each type of shielded container. The reviewer should consult with the radiation protection reviewer and determine whether design dose rates consistent with ALARA objectives have been specified for occupancy areas.

7.5.2.2 Design Features

The reviewer should evaluate the SAR descriptions regarding the proposed uses of shielding features on storage and transfer systems. The reviewer should review the items for consideration identified in Section 7.4.2.2 and evaluate their applicability to the proposed installation.

7.5.3 Shielding Composition and Details

7.5.3.1 Composition and Material Properties

The reviewer should review the SAR descriptions of shielding material composition. The descriptions should identify and describe all materials taken into consideration in determining shielding requirements. These include:

- Materials that have other functions but their mass also provides shielding (especially gamma shielding by structural materials, and gamma and neutron shielding by concrete and pool water)

- Materials especially selected and positioned for gamma shielding, such as lead

- Materials especially selected and positioned for neutron shielding, such as water, concrete, and proprietary shielding materials.

The reviewer should ensure that material specifications for nonstandard materials (e.g., proprietary neutron shield material) include appropriate references for the parameter values used.

The reviewer should assess the temperature sensitivities of shielding materials identified in the SAR and determine whether the shielding effectiveness can be compromised by exposure to temperatures that can be reached under normal, off-normal, and accident level conditions. For example, elevated temperatures can reduce neutron shielding provided by hydrogen content through loss of bound or free water in concrete or other hydrogenous shielding materials. The reviewer should ensure that the elemental composition and density of shielding materials are conservatively adjusted to account for any degradation from aging, high temperature, accumulated radiation exposure, and manufacturing tolerances.

The reviewer should review the illustrations and descriptions of the geometric arrangement of shielding and ensure that all physical dimensions are clearly indicated. The reviewer should assess whether the spatial relationship between sources, shielding, and design dose rate area is adequately described. The reviewer should consider that design of shielding can be oriented either on the radiation sources or a point to be protected. Because an ISFSI or MRS typically has extensive access areas for workers, the potential exists for direct radiation exposure of the offsite population in all directions. As a result, shielding is typically oriented on the sources. As the attenuation effectiveness of a given thickness of shielding material is independent of the distance from the source, the most effective positioning of shielding is as close to the source as feasible.

The reviewer should ensure that penetrations, voids, or irregular geometries that provide potential paths for gamma or neutron streaming are specifically accounted for or otherwise treated in a conservative manner.

Finally, the reviewer should verify that differences, if any, between shielding that exists under normal, off-normal, or accident-level conditions have been clearly stated.

7.5.4 Analysis of Shielding Effectiveness

7.5.4.1 Computational Methods

The reviewer should evaluate the computer programs used for the shielding analysis. There are several recognized programs widely used for shielding analysis. These include codes that use Monte Carlo, deterministic transport, and point-kernel techniques for problem solution. The point-kernel technique is generally appropriate only for gammas since casks typically do not contain sufficient hydrogenous material to apply removal cross-sections for neutrons.

The NRC has several models previously accepted for ISFSI source and shielding analyses; however, since their previous use does not constitute generic NRC approval, the reviewer is cautioned that these codes can produce errors when used incorrectly. The reviewer should determine that the SAR has design control measures that will ensure the quality of computer programs used for shield analysis. The programs previously accepted by NRC for ISFSI source and shielding analyses include:

- ANISN (one-dimensional neutron attenuation code, RSIC CCC-J14 Micro)

- MicroSkyshine (air-scattering code)

- MORSE (Monte Carlo multigroup three-dimensional neutron and gamma transport computer code)

- MCBEND (Monte Carlo multigroup three-dimensional neutron and gamma transport computer code similar to MORSE developed by the United Kingdom (UK) National Radiation Protection Board (NRPB))

- QAD-CGGP (three-dimensional point kernel gamma transport shielding computer code)

- RANKERN (three-dimensional point kernel gamma transport shielding computer code similar to QAD-CGGP)

- MARC-1 (a suite of linked computer codes used for calculating the radiological effects of releases of radionuclides to the environment developed by the UK NRPB)

- LINGAP and HMARC (modules of MARC-1 used to calculate the effects of an atmospheric release)

- SKYSHINE-II (air-scattering code, NUREG/CR-0781)

- STREAMING (code for calculation of attenuation of a gamma flux incident on a variety of shielding penetrations, such as ducts and voids).

Some other shielding codes which have potential application to ISFSI or MRS sources include:

- TORT\DORT (three- and two-dimensional discrete-ordinate neutron/photon transport codes) (ORNL-6268)

- ONEDANT/TWODANT (one- and two-dimensional multigroup discrete-ordinate transport codes)(LANL LA-9184-M, Rev)(LANL LA-10049-M Rev)

- MCNP (Monte Carlo n-particle transport code)(LANL, MCNP 4A)

- SCALE (a modular code system for performing standardized computer analyses for licensing evaluation).

The reviewer should verify that the SAR describes each of the models used in the shielding evaluation. For each model used, the reviewer should verify that the following information has been provided:

- The author, source, dated version

- A description of the model, and the extent and limitation of its application

- The computer program solutions to a series of test problems, demonstrating substantial similarity to solutions obtained from hand calculations, analytical results published in the literature, acceptable experimental tests, a similar program, or benchmark problems.

The reviewer should review the solution comparisons provided by the SAR and determine whether satisfactory agreement of computer and test solutions (or resolution of deviations) is evident. The reviewer should identify any deviations that have not previously been justified to staff satisfaction and transmit the finding to the SAR with a request for additional technical justification regarding application of the code.

Ideally (though not a requirement), the program used for evaluation of shielded storage containers has been validated with actual dose rate measurements from similar or prototypical spent fuel or high-level waste storage systems.

The reviewer should assess whether the number of dimensions of the code is appropriate for the dose rates being calculated. Generally, at least a two-dimensional calculation is necessary. One-dimensional codes provide little information about off-axis locations and streaming paths that may be significant to determining occupational exposure. This also applies to computation of dose rates at the end of storage confinement casks.

For storage confinement casks, source term locations for the contained radioactive materials and the internal positioning, spacers, and structural support (i.e., the "basket") may need to be modeled by region to minimize avoidable error. A uniform source distribution would be conservative for the top and bottom but not for the center region, because of the typical burnup profile for spent fuel rods (greater at the top and bottom). Regions used for analysis typically are the top (to the height of the stored radioactive materials), middle, and bottom thirds. Within the cask, the fuel and basket materials may be homogenized by region to simplify shielding calculations. In addition, the other materials within the cask may not be uniformly distributed over its height, to the extent that the amount homogenized with the fuel, within regions, may be different. This other material typically includes the basket components: positioning sleeves, spacers for the sleeves, and structural supports.

For storage confinement casks with a water or borated water charge, during the loading and retrieval operations the liquid may be homogenized with the fuel and basket materials to the height of the stored radioactive materials. Water and basket parts above the height of the fuel may be homogenized for the shielding calculation. Different heights of the liquid charge may need to be modeled for different operations, depending on the type of cover seal. The liquid height is typically reduced to permit welding and cutting but should be left at a height that retains cover over the stored material. The lowest liquid height that may result from the specified procedures for cask sealing and retrieval operations should be used for the calculations.

The reviewer should consult with the thermal analysis reviewer to determine whether shielding can be compromised by boiling of the liquid charge. The reviewer should verify that a bounding technical specification has been established to prevent boiling.

SAR calculations must include a representative computer code input file and the cross-section library used by the code. The reviewer should review the input file to ensure that data for the source-shield configuration being modeled are properly entered into the code. The reviewer should verify that the cross-section library used by the code is appropriate for analyzing cask shielding problems. If the SAR has not independently determined a source term for neutron-induced gamma radiation or subcritical multiplication of neutrons, the reviewer should ensure that a coupled cross-section set was used and that the SAR executed the code in a manner that accounts for these secondary source terms.

The reviewer should review the basis used for flux-to-dose-rate conversion. The NRC accepts flux-to-dose rate conversion factors in ANSI/ANS 6.1.1. The shielding code may also perform the conversion by using its own data library.

7.5.4.2 Dose Rate Estimates

On the basis of experience, comparison to similar systems, or scoping calculations, the reviewer should make an initial assessment of whether the dose rates appear reasonable and whether their variation with location is consistent with the geometry and shielding characteristics of the cask system.

The reviewer should verify that dose rates have been estimated for all locations that are accessible to occupational personnel during cask loading, transport to the ISFSI or MRS, and maintenance and surveillance operations. Generally these locations include points at or near various cask components and in the immediate vicinity of the cask. The reviewer should verify that the points with the highest calculated dose rates are identified.

The reviewer should verify that the dose rate estimates have appropriately considered the following areas that have been of special concern in past reviews:

- Conservatism of simplifying assumptions used for the analysis, and assertions that a nonconservative assumption is more than compensated for by other conservative assumptions

- Dose rate estimates at points where radiation streaming may occur from design details, such as failure to offset penetrations of cask lids for venting, dewatering, and recharging

- Dose rate estimates at the top of a cask during preparation for storage operations that assume a water height above contained fuel rods, while procedure descriptions do not ensure that there will be a minimum height of water cover when the water level is lowered for seal welding

- Inclusion in analyses of the potential negative effect of scattering that increases the dose rates to accessible areas at the side of stored material because of mass placed over the source, a scattering that may more than offset the reduction in skyshine contribution because of the overhead shielding

The reviewer should check all solid angles from the shielded source for gaps or significantly reduced shielding that could result in local "hot spots." Similarly, if a work station is shielded from multiple sources of radiation, the reviewer should check the solid angles about that station for potential gaps or other sources of elevated dose rates.

The reviewer should verify that shielding analyses include consideration of various radioactive material situations for normal, off-normal, and accident-level conditions. For example, the external neutron shield of a cask may be damaged by a tip-over accident or degraded in a fire; drop in the level of pool water can uncover parts of fuel rods (if this is a feasible accident-level situation).

The reviewer should consult with the radiation protection reviewer who will use dose rate estimates that result from the shielding review (in addition to other information) to determine whether appropriately detailed SAR calculations (dose rates and collective dose estimates) show that the radiation shielding features are sufficient to meet the requirements of 10 CFR 72.104, 72.106, and ALARA objectives.

Confirmatory Calculations

The reviewer should verify, by independent calculations, the SAR shielding analyses performed for normal operations that result in the greatest collective dose estimates. These operations include cask sealing, dewatering, and purging the cask during preparation of the cask for storage; monitoring and repressurization of the space between redundant seals of cask in storage; or surveillance of the ventilation ports over the life of the ISFSI or MRS. The computations should be sufficient to verify the probable accuracy of the SAR estimates.

In determining the level of effort appropriate for the confirmatory calculations, the reviewer should consider:

- The degree of sophistication and the margin of safety in the SAR analysis

- Comparison of SAR dose rates with those of similar casks that have been previously reviewed, if applicable

- The fact that actual doses will be monitored and limited by 10 CFR 20 requirements and site-specific license conditions

- Applicant's experience using the methods and codes, and in validating the methods and codes with actual measured dose rates

- Use of methods or codes not previously reviewed by the staff

- Any significant departures from prior cask system designs or procedures (e.g., unusual shield geometry, new types of materials, different source terms, increased manual operations near the storage confinement cask).

The reviewer should examine the SAR's input to the computer program used for the shielding analysis, use of that input in the program, reasonableness of results, and use of the results in developing projected doses. The reviewer should verify use of proper dimensions, material properties, and an appropriate cross-section set. The reviewer should independently evaluate gamma and neutron source terms.

Depending on the accident analysis and the magnitude of the potential collective dose (person-rem) from direct radiation from in-place radioactive material, the reviewer should determine whether a separate computation should be performed to verify the projected greatest dose that may be received from a design basis accident level event. If the direct radiation dose from in-place material is small relative to the collective dose from routine operations, this analysis may be omitted.

7.6 Evaluation Findings

Evaluation findings are prepared by the staff on satisfaction of the regulatory requirements related to design features for protection against direct radiation, as identified in Section 7.3.

These findings are determined in conjunction with the radiation protection and accident analysis reviews, described in Chapters 11 and 15. If the documentation submitted with the application fully supports positive findings for each of the regulatory requirements, the statements of findings should be as follows (finding numbering is for convenience in referencing within the FSRP and SER):

F7.1   The design of the shielding system(s) of the [ISFSI/MRS] satisfies the criteria for radiological protection of 10 CFR 72.126(a)(6).

F7.2   The design of the [ISFSI/MRS] provides acceptable means for limiting occupational radiation exposures within the limits given in 10 CFR 20.1201 and for meeting the objective of maintaining exposures as low as is reasonably achievable, in compliance with 10 CFR 72.24(e).

F7.3   The design of the [ISFSI/MRS] provides acceptable means for limiting exposure of the public to direct and scattered radiation within the limits given in 10 CFR 72.104.

F7.4   The design of the [ISFSI/MRS] provides suitable shielding for radioactive protection under normal and accident conditions, in compliance with 10 CFR 72.128(a)(2).

7.7 References

NRC documents referenced are identified at Consolidated References, Section 17.

Codes, Standards, and Specifications

ANSI/ANS 6.1.1, "American National Standard for Neutron and Gamma-Ray Fluence to Dose Factors," Institute for Nuclear Materials Management/American Nuclear Society.

"ANISN/PC-Multigroup One-Dimensional Discrete Ordinates Transport Code System with Anisotropic Scattering," CCC-J14 Micro, Radiation Shielding Information Center, Oak Ridge, TN, 1990.

"QAD-CGGP, A Combinatorial Geometry Version of QAD-P5A, A Point Kernel Code System for Neutron and Gamma-Ray Shielding Calculations Using the GP Buildup Factor," CCC-493, Radiation Shielding Information Center, Oak Ridge, TN, 1994.

Manuals and Texts

Alcouffe, R.E., etal., "User's Guide for TWODANT: A Code Package for Two-Dimensional, Diffusion Accelerated, Neutral Particle Transport," LA-10049-M Rev., LANL, April 1992.

"MCBEND - A Monte Carlo Program for Shielding Calculations; User Guide for Version 6," Issue 4, Answers (MCBEND) 2: UKAEA Winfrith, UK, March 1990.

O'Dell, R.D., et al., "Revised User's Manual for ONEDANT: A Code for One-Dimensional, Diffusion Accelerated, Neutral Particle Transport," LA-9184-M, Rev., LANL, December 1989.

"RANKERN, A Point Kernel Integration Code for Complicated Geometry Probelems; User Guide to Version 12," Issue 2, ANSWERS (RANK)2, UKAEE, UK, September 1987.

TRW Environmental Safety Systems, Inc., "DOE Characteristics Database, User Manual for the CDB-R," November 16, 1992.

Worku, G., et al., "MicroShield Version 4.2 User's Manual," Grove Engineering, Inc., Rockville, Maryland, 1995.

Technical Reports

Hill, M.D., J.R. Simmonds, and J.A. Jones, "NRPB Methodology for Assessing the Radiological Consequences of Accidental Releases of Radionuclide to Atmosphere - MARC-1," NRPB-R224, September 1989.

Los Alamos National Laboratory (LANL), "MCNP 4A, Monte Carlo N-Particle Transport Code System," LANL, December 1993.

Oak Ridge National Laboratory (ORNL), "ORIGEN2: Isotope Generation and Depletion Code-Matrix Exponential Method," 1991.

Rhoads, W.A., "The TORT Three-Dimensional Discrete Ordinates Neutron/Photon Transport Code," ORNL-6268, ORNL, November 1987.

8 CRITICALITY EVALUATION

8.1 Review Objective

The objective of the review and evaluation is to ensure that the stored materials remain subcritical under normal, off-normal, and accident conditions during all operations, transfers, and storage at the site.

Figure 8.1 presents an overview of the criticality evaluation process. The figure shows that the criticality review draws information from different sections of the application including supporting calculations.

8.2 Areas of Review

The following outline shows the areas of review addressed in Section 8.4, Acceptance Criteria, and Section 8.5, Review Procedures:

Criticality Design Criteria and Features
     Criteria
     Features

Stored Material Specifications

Analytical Means
     Model configuration
     Material Properties

Applicant Criticality Analysis
     Computer Program
     Multiplication Factor
     Benchmark Comparisons
     Independent Criticality Analysis

8.3 Regulatory Requirements

This section identifies and presents a high-level summary of Title 10 of the Code of Federal Regulations (CFR) Part 72 relevant to the review areas addressed by this chapter. The NRC staff reviewer should read the exact regulatory language. A matrix at the end of this section matches the regulatory requirements identified in this section to the areas of review identified in the previous section.

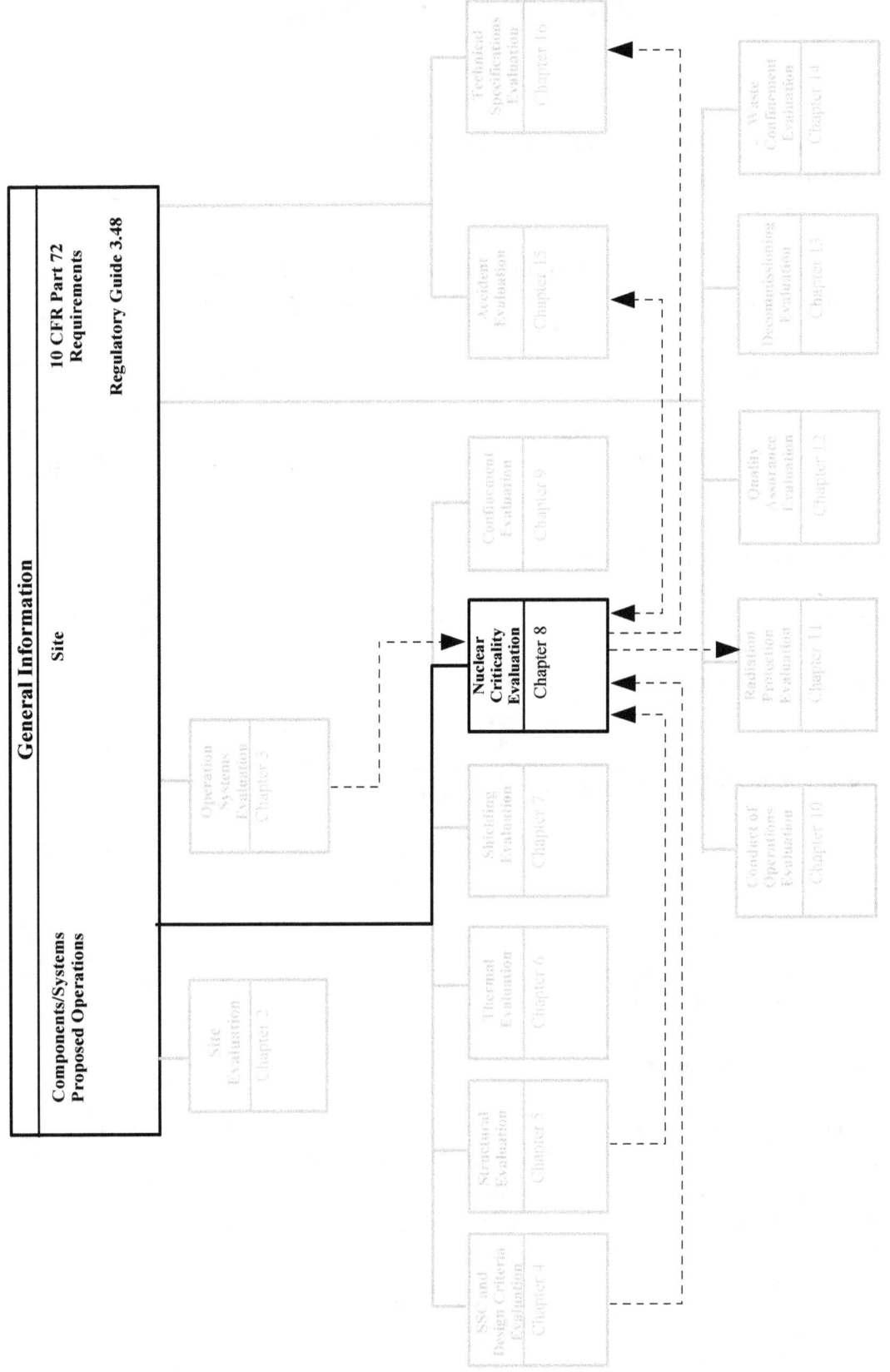

Figure 8.1 Overview of Nuclear Criticality Evaluation

72.40 Issuance of license.

(a) "... license meets the standards and requirements of the Act and the regulations of the Commission."

(13) "...without endangering the health and safety of the public."

72.124 Criteria for nuclear criticality safety.

(a) "Design for criticality safety... maintained subcritical... before a nuclear criticality accident is possible, at least two unlikely, independent, and concurrent or sequential changes have occurred in the conditions essential to nuclear criticality safety...margins of safety for the nuclear criticality parameters that are commensurate with the uncertainties in the data and methods used in calculations... demonstrate safety... under accident conditions."

(b) "Methods of criticality control... favorable geometry, permanently fixed neutron absorbing materials (poisons), or both... solid neutron absorbing materials... positive means to verify their continued efficacy."

A matrix which shows the primary relationship of these regulations to the specific areas of review associated with this Standard Review Plan (SRP) chapter is given in Table 8.1. The NRC staff reviewer should verify the association of regulatory requirements with the areas of review presented in the matrix to ensure that no requirements are overlooked as a result of unique applicant design features.

Table 8.1 Relationship of Regulations and Areas of Review

Areas of Review	10 CFR Part 72 Regulations	
	72.40	72.124
Design Criteria	●	●
Stored Material Specifications		●
Analytical Means		●
Applicant Criticality Analysis	●	●

8.4 Acceptance Criteria

This section identifies the acceptance criteria used for the criticality review. Four types of criteria are described. The first describes criticality design criteria and features including required conditions, assumptions, and scenarios. The second identifies the requirements for the specifications regarding stored nuclear material that are acceptable to the NRC. The third describes the features of criticality analysis models which are acceptable to the NRC. The fourth identifies the features of applicant criticality analyses including the specific computer program, benchmarks, and multiplication factor determination which constitute an acceptable submittal for criticality safety.

Under the right conditions, certain isotopes of selected heavy elements, especially uranium and plutonium, have the ability to split or fission after absorbing a neutron and release energy along with several new neutrons from this fission process. The fission process can be self-sustaining or even grow by a chain reaction, which can produce as many or more neutrons than are absorbed. In criticality terminology, the term, k-effective or k_{eff} is the net ratio of neutrons produced per neutron absorbed in a mass of fissionable material. A k_{eff} of 1.0 indicates a critical mass whereas a k_{eff} of less than 1.0 is an indication of a subcritical condition.

8.4.1 Criticality Design Criteria and Features

8.4.1.1 Criteria

The regulatory requirements given in 10 CFR 72.40 and 10 CFR 72.124 identify acceptable design criteria. The NRC generally considers the design criteria identified below to be acceptable to meet the criticality requirements of 10 CFR 72 for storage confinement casks:

- The multiplication factor, k_{eff}, including all biases and uncertainties at a 95 percent confidence level, must not exceed 0.95 under all credible normal, off-normal, and accident conditions and events.

- Conditions for criticality safety (satisfaction of the limit on multiplication factor, k_{eff}) of subject radioactive material while at the Independent Spent Fuel Storage Installations (ISFSI) or Monitored Retrievable Storage (MRS) must include:

    - no burnup credit. (The conservative assumption of fresh unburned fuel provides a worst case criticality analysis; however, 10 CFR 72.3 requires that spent fuel have been irradiated and cooled at least one year as a condition for storage.) Alternately, burnup credit may be taken using the guidelines described in section 8.4.5 of this SRP.

    - no credit taken for flammable neutron absorbers or for any solid poisons that may melt or lose any significant mass from the original solid form by melting or vaporization at any of the temperature and pressure conditions that may be experienced while in use

    - no credit taken for liquid neutron shielding material (except that k_{eff} for the situation of a loaded confinement cask with liquid that serves as both shielding and absorber and is used in the confinement cask during loading operations or in the pool shall be based on presence of the water and bounding level(s) of poison)

    - no more than 75 percent credit for fixed neutron absorbers, unless comprehensive fabrication acceptance tests capable of verifying the presence and uniformity of the neutron absorber are implemented

    - determination and use of optimum (i.e., most reactive) moderator density

- The multiplication factor limit on k_{eff}, must be met for all conditions and events while at the ISFSI and MRS. This does not require determination of k_{eff} for every situation. However, it must be demonstrated that the situations that have the highest k_{eff} have been analyzed and that thereby the normal, off-normal, and accident and conditions with the lowest margins of safety have been analyzed; or are enveloped by the analyses conducted and included in the SAR and its supporting documentation (ANSI/ANS 8.17-1984). Conditions and events to be considered include, but are not limited to, the following:

  - in dry storage

  - in temporary or prolonged storage in the pool

  - during cask loading and unloading operations, including:

    -- transferring subject radioactive material to or from the storage cask or a transfer container (and possible drops, and including possible selection of the wrong material for loading);

    -- lift and translation movements (and possible drops);

    -- dewatering and charging operations, including situations in which a cask could be full of (unborated) steam; or with a partial fill of liquid water (borated if applicable to the procedures) and the remainder steam.

  - during all on-site transfer, and transportation operations

  - during and following possible drops and other accident events including credible flood from natural or man-made causes

  - during earthquakes

  - during and following a non-mechanistic cask tip-over (if a more severe event is not identified by accident analysis)

  - with the stored material in the densest configuration permitted by the basket or other separators and with the most conservative assumptions for tolerance stack-ups

- At least two unlikely, independent, and concurrent or sequential changes to the conditions essential to criticality safety, under normal, off-normal, and accident conditions, must occur before an accidental criticality is possible (ANSI/ANS 8.1-1983). For analysis, "accidental criticality" is defined as exceeding k_{eff} of 0.95 with a confidence level less than 95% and a 95% probability that k_{eff} =0.95 will not be exceeded.

- Criticality safety of the design must be based on favorable geometry (preferred), permanent fixed neutron absorbing materials (poisons), or both.

- Where solid neutron-absorbing materials are used, the design must provide a means to verify their initial efficacy, such as manufacturer's data or in-situ measurements (ANSI/ANS 8.21). Chapter 6 of NUREG-1536 provides a basis for accepting the 20-year continued efficacy of fixed neutron poisons.

- Unless it is shown that all spent fuel to be stored will be contained within completely intact cladding, the occurrence of pinholes and cracks in the cladding (and water fill of the voids within the cladding) must be assumed for the criticality analysis if it results in a higher k_{eff}. The water fill in the fuel-to-cladding gap should be assumed to be unborated since this is conservative from a criticality safety viewpoint.

The aforementioned criteria should be specifically incorporated into the SAR and applicable supporting criticality calculations.

8.4.1.2 Features

The regulatory requirements given in 10 CFR 72.124(b) identify acceptable design criteria for criticality control. The NRC generally considers the design criteria identified below to be acceptable to meet the criticality control requirements of 10 CFR 72 for storage confinement casks.

The NRC accepts use of borated water in the pool and during cask loading and unloading (of subject radioactive material to or from the storage confinement cask) operations as a means of criticality control if the minimum boron content is a technical specification. If borated water is used for criticality control, then administrative controls and/or design features must be used to ensure: (1) that pool boron concentration is maintained throughout the pool; and, (2) that accidental flooding of a cask with unborated water cannot occur (as in retrieval operations or in an interrupted loading operation requiring cask reflooding in anticipation of off-loading or additional operations on or about the cask). The alternative is that the criticality analysis assume accidental flooding with unborated water.

If borated water is not to be used, the SAR should specify if any dummies (for fuel rods or other fissionable items to be stored) are to be used in storage positions within the confinement cask or pool to displace water. Credit for this displacement of water in the criticality analysis requires acceptable evidence that the storage positions not occupied by the subject radioactive materials will always be occupied by the dummies.

Borated water and any other liquids are not acceptable as a means of criticality control for a cask in dry storage. This includes use of any credit in criticality analysis for presence (outside the cask confinement barrier) of a liquid that may provide neutron shielding. Presence and optimum (most reactive) moderator density of the liquid shall be assumed if it increases k_{eff}.

If more than one certified or licensed basket design of the same supplier could fit in the cask, the type basket to be used should be among the data stamped on the plate on the exterior of the storage confinement cask.

The NRC accepts comparative neutron flux measurement made external to a confinement cask as a positive means for verifying continued efficacy of solid neutron absorbing materials incorporated in the storage cask system, if the following are acceptable:

- testing procedures,
- instrumentation,
- accuracies, and
- determination of baseline measurements for subsequent use in comparisons.

The NRC has accepted a requirement for acceptance testing of the poisons during fabrication as a positive means for verifying continued efficacy of solid neutron absorbing materials incorporated in the storage cask system. This testing should show that: (1) the material is not subject to degradation from physical or chemical actions that may occur over the system life, (2) the material will not be degraded by time-integrated gamma radiation emitted by the spent fuel fission products, and (3) the small neutron flux from spontaneous fission and subcritical multiplication results in a negligible depletion of poison material over the storage period. Inclusion of evidence of satisfactory similar use of the material for a 20-year period is desirable.

Tolerances for structures, systems, and components (SSCs) material, fabrication, and assembly can be important in identifying worst case (lowest margin of safety) geometries, material compositions, and densities. The tolerances for the properties and construction of all SSCs involved in criticality analyses should be used in the analyses and must then be also identical or conservatively bounded by the tolerances shown in the definition of the ISFSI or MRS design. The analyses should be based on the most conservative combination of tolerances.

8.4.2 Stored Material Specifications

The regulatory requirements given in 10 CFR 72.124(a) identify acceptable design criteria for stored material specifications. The NRC generally considers the design criteria identified below to be acceptable to meet the criticality requirements of 10 CFR 72 for storage confinement casks.

The stored material specifications must include the ranges of properties of concern for criticality analysis (guidance on data required is provided in Standard Review Plan for Spent Fuel Dry Storage Facilities [FSRP] Section 4.4.1). Stored material characteristics of probable concern for the various criticality analyses include those listed below. These should be stated for each known type of spent fuel to be stored and for other radioactive material to be stored for which criticality analysis is appropriate. Radioactive materials which, due to their atomic properties and/or physical maximum densities of the solid material are not of criticality concern should be identified as such, as rationale for not including criticality analyses. Data identified below that are not required for the analytical approach used by the applicant should still be provided as they may be needed for confirmatory and independent analyses by the NRC (performed as part of the evaluation effort):

- fuel manufacturer identity and spent fuel design/model

- maximum U^{235} enrichment of fuel and type [e.g., 4.2% enrichment, Boiling Water Reactor (BWR)] of fuel assemblies

- the maximum fuel pin enrichment [for BWR or Pressurized Water Reactor (PWR) fuel]

- for other radioactive materials that are to be stored and may be fissionable

    - isotopes present and their densities

    - means by which densities are limited

    - geometric data on the configuration (e.g., racks, basket) holding the materials including tolerances and uncertainties, and neutron absorption material integral to the configuration

    - characteristics (materials, densities, geometries, tolerances, uncertainties) of any encapsulation used to provide confinement and structural support during handling and when within the storage confinement barrier

8.4.3 Analytical Means

The regulatory requirements given in 10 CFR 72.124(a) and (b) identify acceptable design criteria. The NRC generally considers the design criteria identified below to be acceptable to meet the criticality requirements of 10 CFR 72 for storage confinement casks:

8.4.3.1 Model Configuration

The model used for the criticality evaluation must adequately describe normal, off-normal, and accident conditions analyzed. The model must provide for the most reactive tolerance combinations and any steady state elastic deformation of the positioning (racks or basket) or plastic deformation of the structure that could result from accident events.

The dimensions and materials of the model used for the criticality analysis should be the same as those in the design definition (elsewhere in the SAR). If there are differences, the model must be shown to be conservative (result in a higher k_{eff}). The NRC accepts substitution of ordinary water for end sections and support structures of the fuel in the model. Substitution of borated water for other materials is not acceptable unless it can be shown to be conservative. Sufficient conditions must be modeled and analyzed to ensure that the highest k_{eff} have been determined and that conditions and configurations not analyzed are enveloped by those analyzed.

The model should reflect conservative assumptions in all variables. This includes (but may not be limited to) variables identified below:

- location of fissionable material within positioning basket or other framework (e.g., fuel rods would be positioned to be tight against the dividing spacing material and closest to the center of the array for vertical storage and as close to center as permitted by gravity, or potentially caused by in transit vibration for a horizontal cask position)

- fuel density

- U^{235} enrichment of fuel

- manufacturing tolerances must be assumed to be at their most conservative (i.e., maximum reactivity) value within the allowed tolerance band. No statistical combination of uncertainties is allowed for manufacturing tolerances.

- flooding in the fuel rod pellet-to-clad gap region

The NRC accepts use of a heterogeneous model of each fuel rod. If, instead, the model homogenizes the entire fuel assembly (over the volume of the assembly), the applicant must acceptably demonstrate that the homogenized model is conservative relative to a heterogeneous model. This may be done by using both homogeneous and heterogeneous models in a criticality computation or by benchmarking to an acceptable (number and relevance) set of criticality experiments.

The criticality analysis model must be described in sufficient detail, either in the SAR or supporting calculations, to show conformance to the requirements in this section.

8.4.3.2 Material Properties

The material compositions and densities must be provided for all materials used in the calculational model. The sources of the properties should be referenced. The amount and geometry of fixed poison used in the criticality analysis should be no more than the minimums to be verified by acceptance testing. Validation of the poison concentration is addressed in acceptance testing.

An appropriate set of cross-sections should be determined and identified, and the sources should be referenced. Cross-sections may be obtained with the criticality computer codes or developed independently from another source. For multigroup calculations, the spectrum of the neutron flux used to construct the group cross-sections must be similar to that of the cask. Cross-sections and the computer program must be benchmarked by comparison to experimental data.

8.4.4 Applicant Criticality Analysis

The regulatory requirements given in 10 CFR 72.40 and 10 CFR 72124(a) identify acceptable design criteria. The NRC generally considers the design criteria identified below to be acceptable to meet the criticality requirements of 10 CFR 72 for storage confinement casks:

The SAR must include criticality analyses for the most reactive cases for the items and materials that approval of the application will allow to be stored. These must be demonstrated to include or envelop the loadings and situations that have the highest values of k_{eff}.

8.4.4.1 Computer Program

The SAR must identify the computer program and cross-section used in criticality analyses. The NRC has accepted both Monte Carlo and deterministic computer codes for criticality calculations. Monte Carlo codes are generally more suited to three-dimensional geometry, and therefore, more are widely used to evaluate spent fuel cask designs. Two acceptable Monte Carlo codes are SCALE/KENO (NUREG/CR-0200 and CCC-619) and MCNP (LANL, Dec 1993). KENO is a multi-group code that is part of the SCALE sequence. MCNP permits use of continuous cross sections. The NRC has accepted use of MICROX and DTFX (deterministic computer codes).

If a multigroup treatment is used, the neutron spectrum of the cask must be appropriately considered. In addition to selecting a cross-section set collapsed with an appropriate flux spectrum, a more detailed processing of the energy-group cross-sections is also required to properly account for resonance absorption and self-shielding. The use of KENO as part of the SCALE sequence provides for such processing directly.

Some cross-section sets (e.g., HANSEN-ROACH) include data for fissile and fertile nuclides (based on a potential scattering cross-section) that can be input by the user. If a stand-alone version of KENO is used, potential scattering must be properly considered. [Note: The "working-format" library, commonly distributed with the Version 4.0 of SCALE/KENO to enable calculations of the manual's sample problems, is not acceptable for criticality calculations of actual systems (NRC Information Notice 91-26). The NRC has accepted "27 Group NDF4" cross-section library in SCALE-4.1 PC KENO Va for criticality calculations.] Multigroup cross-section sets may be used in analyses of cask models with separate regions of water and steam or variations in the boron concentration. These involve use of different flux spectra in different regions.

8.4.4.2 Multiplication Factor

Variation in results of different computations of k_{eff} for different situations and with different codes and models should be rationalized and explained. Sensitivity parametric analyses may be used to provide the required demonstration that the highest k_{eff} with confidence level of 95% (with a $k_{eff} \leq 0.95$) have been determined. Certain cases which may require a lower value than 0.95 for maximum allowable k_{eff} are discussed in Chapter 6, Section V.4.b of NUREG-1536.

For verification of Monte Carlo calculations, the number of neutron histories and convergence criteria should be appropriate. As the number of neutron histories increases, the mean value for k_{eff} should approach some fixed value, and the standard deviation associated with each mean value should decrease. Depending on the code used, a number of diagnostic calculations are generally available to demonstrate adequate convergence and adequate statistical variation. For deterministic codes, a convergence limit may be prescribed in the input. The selection of a proper convergence limit and achievement of this limit must be described and demonstrated in either the SAR or supporting criticality calculations.

8.4.4.3 Benchmark Comparisons

Computer codes for criticality calculations must be benchmarked against critical experiments. Benchmark comparisons must be documented in appropriate calculations and/or the SAR. Benchmark comparisons can validate the computer code, its use on a specific geometric configuration, the neutron cross-sections used in the analysis, and consistency in modeling. Benchmark comparisons should be made by the analyst(s) and organization that will be performing the actual criticality analysis to qualify the analyst and computer environment. The calculated k_{eff}s and confidence levels of the base criticality computations must be adjusted to include the appropriate bias (the average of the differences between results and measurement) and uncertainties determined from the benchmark comparisons.

The benchmark experiments should be relevant to the actual situation analyzed (ANSI/ANS 8.1-1983). No critical benchmark experiment will precisely match the fissile material, moderation, neutron poisoning, and configuration in the actual situation. However, the applicant can perform a proper benchmark analysis by selecting experiments that adequately represent the actual situation and fissionable material features and parameters important to reactivity. Key features and parameters that should be considered in selecting appropriate critical experiments for spent fuel include type of fuel, enrichment, hydrogen to uranium ratio or moderator to uranium ratio for graphite moderated fuel designs (rod diameter and pitch), fuel and cladding chemical composition, reflector, neutron energy spectrum, and poisoning. The applicant must justify the suitability of the critical experiments chosen to benchmark the criticality code and calculations. UCID- 21830 provides guidance for benchmarking and contains a substantial bibliography of benchmark experiments and validation testing.

Detailed guidance on determining a code bias from benchmark experiments has not been formalized. Multiple applicable benchmark experiments should be analyzed. The results of these benchmark calculations should be converted to a bias for application to the criticality computations. Simply using an average of the biases from a number of benchmark calculations is typically not considered to be sufficient, particularly if one benchmark yields results that are significantly different from the others. Benchmark comparisons must also be checked for bias trends with respect to parametric variations (such as pitch-to-rod-diameter ratio, assembly separation, reflector material, neutron absorber material, etc.). UCID- 21830 provides some guidance for this, but other methods have also been considered appropriate.

The calculated statistical uncertainties of both benchmark and cask analyses also need to be addressed for Monte Carlo codes. The uncertainties should be applied to at least the 95% confidence level and 95% probability, as a general rule, if the acceptability of the result depends on small differences between large values. A sufficient number of neutron histories can readily be used so that the treatment of these uncertainties should not significantly affect the results.

Only biases determined by benchmark comparisons that increase k_{eff} or lower the confidence level should be applied. If the benchmark calculation for a critical experiment results in a neutron multiplication that is greater than unity, it should not be used in a manner that would reduce the k_{eff} calculated for the cask. Critical experiments using a different fissionable isotope than that intended for the ISFSI or MRS should not be included in this benchmark comparison.

8.4.5 Burnup Credit in the Criticality Analysis

Unirradiated reactor fuel has a well-specified nuclide composition that provides a straightforward and bounding approach to the criticality safety analysis of transport and storage casks. As the fuel is irradiated in the reactor, the nuclide composition changes. Ignoring the presence of burnable poisons, this composition change will cause the reactivity of the fuel to decrease. Allowance in the criticality safety analysis for the decrease in fuel reactivity resulting from irradiation is typically termed burnup credit. Extensive investigations have been performed both within the United States and by other countries in an effort to understand and document the technical issues related to burnup credit. Much of this work has been considered in the development of the U.S. Department of Energy's Topical Report (TR) on Actinide-Only Burnup Credit for Pressurized Water Reactor (PWR) Spent Nuclear Fuel Packages (DOE/RW-0472).

The technical information provided in the literature and in the various TR revisions, together with the initial confirmatory analyses by the U.S. Nuclear Regulatory Commission (NRC) research program, have provided a sufficient basis for the staff to proceed with acceptance of a burnup credit approach in the criticality safety analysis of PWR spent fuel casks as discussed in the Recommendations below. Although insights gained from reviewing the TR submittals form a part of the basis for the staff's position, the NRC has not endorsed the TR or its supporting documentation. The following recommendations provide a cask-specific basis for granting burnup credit, based on actinide composition. The NRC's staff will issue additional guidance and/or recommendations as information is obtained from its research program on burnup credit and as experience is gained through future licensing activities. Except as specified in the following recommendations, the application of burnup credit does not alter the current guidance and recommendations provided by the NRC staff for criticality safety analysis of transport and storage casks.

Recommendations:

8.4.5.1 Limits for the Licensing Basis

The licensing-basis analysis performed to demonstrate criticality safety should limit the amount of burnup credit to that available from actinide compositions associated with PWR irradiation of UO_2 fuel to an assembly-average burnup value of 40 GWd/MTU or less. This licensing-basis analysis should assume an out-of-reactor cooling time of five years and should be restricted to intact assemblies that have not used burnable absorbers. The initial enrichment of the fuel assumed for the licensing-basis analysis should be no more than 4.0 wt% ^{235}U unless a loading offset is applied. The loading offset is defined as the minimum amount by which the assigned burnup loading value (see Recommendation 8.4.5.5) must exceed the burnup value used in the licensing safety basis analysis. The loading offset should be at least 1 GWd/MTU for every 0.1 wt% increase in initial enrichment above 4.0 wt%. In any case, the initial enrichment shall not exceed 5.0 wt%. For example, if the applicant performs a safety analysis that demonstrates an appropriate subcritical margin for 4.5 wt% fuel burned to the limit of 40 GWd/MTU, then the loading curve (see Recommendation 8.4.5.4) should be developed to ensure that the assigned burnup loading value is at least 45 GWd/MTU (i.e., a 5 GWd/MTU loading offset resulting from the 0.5 wt% excess enrichment over 4.0 wt%). Applicants requesting use of actinide compositions associated with fuel assemblies, burnup values, or cooling times outside these specifications, or applicants requesting a relaxation of the loading offset for initial enrichments between 4.0 and 5.0 wt%, should provide the measurement data and/or justify extrapolation techniques necessary to adequately extend the isotopic validation and quantify or bound the bias and uncertainty.

8.4.5.2 Code Validation

The applicant should ensure that the analysis methodologies used for predicting the actinide compositions and determining the neutron multiplication factor (k-effective) are properly validated. Bias and uncertainties associated with predicting the actinide compositions should be determined from benchmarks of applicable fuel assay measurements. Bias and uncertainties associated with the calculation of k-effective should be derived from benchmark experiments that represent important features of the cask design and spent fuel contents. The particular set of nuclides used to determine the k-effective value should be limited to that established in the validation process. The bias and uncertainties should be applied in a way that ensures conservatism in the licensing safety analysis. Particular consideration should be given to bias uncertainties arising from the lack of critical experiments that are highly prototypical of spent fuel in a cask.

8.4.5.3 Licensing-Basis Model Assumptions

The applicant should ensure that the actinide compositions used in analyzing the licensing safety basis (as described in Recommendation 8.4.5.1) are calculated using fuel design and in-reactor operating parameters selected to provide conservative estimates of the k-effective value under cask conditions. The calculation of the k-effective value should be performed using cask models, appropriate analysis assumptions, and code inputs that allow adequate representation of the physics. Of particular concern should be the need to account for the axial and horizontal variation of the burnup within a spent fuel assembly (e.g., the assumed axial burnup profiles), the need to consider the more reactive actinide compositions of fuels burned with fixed absorbers or with control rods fully or partly inserted, and the need for a k-effective model that accurately accounts for local reactivity effects at the less-burned axial ends of the fuel region.

8.4.5.4 Loading Curve

The applicant should prepare one or more loading curves that plot, as a function of initial enrichment, the assigned burnup loading value above which fuel assemblies may be loaded in the cask. Loading curves should be established based on a 5-year cooling time and only fuel cooled at least five years should be loaded in a cask approved for burnup credit.

8.4.5.5 Assigned Burnup Loading Value

The applicant should describe administrative procedures that should be used by licensees to ensure that the cask will be loaded with fuel that is within the specifications of the approved contents. The administrative procedures should include an assembly measurement that confirms the reactor record assembly burnup. The measurement technique may be calibrated to the reactor records for a representative set of assemblies. For an assembly reactor burnup record to be confirmed, the measurement should provide agreement within a 95 percent confidence interval based on the measurement uncertainty. The assembly burnup value to be used for loading acceptance (termed the assigned burnup loading value) should be the confirmed reactor record value as adjusted by reducing the record value by the combined uncertainties in the records and the measurement.

8.4.5.6 Estimate of Additional Reactivity Margin

The applicant should provide design-specific analyses that estimate the additional reactivity margins available from fission product and actinide nuclides not included in the licensing safety basis (as described in 8.4.5.1). The analysis methods used for determining these estimated reactivity margins should be verified using available experimental data (e.g., isotopic assay data) and computational benchmarks that demonstrate the performance of the applicant's methods in comparison with independent methods and analyses. The Organization for Economic Cooperation and Development Nuclear Energy Agency's Working Group on Burnup Credit provides a source of computational benchmarks that may be considered. The design-specific margins should be evaluated over the full range of initial enrichments and burnups on the burnup credit loading curve(s). The resulting estimated margins should then be assessed against estimates of: (a) any uncertainties not directly evaluated in the modeling or validation processes for actinide-only burnup credit (e.g., k-effective validation uncertainties caused by a lack of critical experiment benchmarks with either actinide compositions that match those in spent fuel or material geometries that represent the most reactive ends of spent fuel in casks); and (b) any potential nonconservatisms in the models for calculating the licensing-basis actinide inventories (e.g., any outlier assemblies with higher-than-modeled reactivity caused by the use of control rod insertion during burnup).

8.5 Review Procedures

The review includes evaluation of compliance with all regulatory requirements and acceptance criteria given in the FSRP and applicable other NRC documents and accepted codes.

The following provides review guidance specific to the elements of the submitted documentation within the scope of this section. This guidance is based on the required products of the review and lessons learned from prior reviews and is supplemented by that in NUREG-1536 Chapter 6.

8.5.1 Criticality Design Criteria and Features

The reviewer should ensure that the stated criteria are consistent with that used for the criticality models and computations. The reviewer should verify that the calculations determine the highest k_{eff} that might occur under all operational states under normal, off-normal, and accident conditions. The principal criticality situations for previously reviewed applications have included:

- new fuel loaded into a storage confinement cask with unborated water to top of the cask and filling adjacent annulus between the storage and transfer cask and fuel at maximum density permitted by relative sizes of grid openings and fuel rods and tolerance stack-ups

- densest and maximum fuel storage in pool

- new fuel from disintegrated rods located at bottom of vertical storage following a drop

- new fuel in densest concentration permitted by single failures of the basket structure

The reviewer should verify that technical specifications relating to criticality design have been included in the technical specification chapter of the SAR. These may have been proposed by the applicant in compliance with 10 CFR 72.26 or may result from the review and evaluation of submittals relating to those areas. The following paragraph illustrates a technical specification related to criticality evaluations which has been accepted in previous applications:

- Functional/Operating Limits and Monitoring Limits/Limiting Control Settings: The concentration of boron in the water in the pool and inserted in the confinement vessel during transfer operations or otherwise when the radioactive material within the confinement vessel is exposed, shall not be less than _____ [insert]. Water to be introduced into the cask during such operations shall be tested to confirm the acceptability of the boron concentration prior to its introduction. The pool water boron concentration shall be confirmed by sample measurement at least once every _____ hours.

8.5.2 Stored Material Specifications

The reviewer should ensure that the material specifications used and that may be presented in the criticality analyses are the same as those given for new fuel (or other stored fissionable material requiring criticality analysis). The reviewer should confirm that the sources for the specifications are acceptable. Material specifications should include all tolerances and/or uncertainties in such properties as density and isotopic enrichment.

It is recommended that the Safety Evaluation Report (SER) include consolidated table of sources used in the SAR documentation, their use, and the acceptability of that use. The sources for the stored material specifications should be included in that table.

8.5.3 Analytical Means

8.5.3.1 Model Configuration

The reviewer should verify the acceptability and appropriateness of the model used. The reviewer should examine the choice and basis of cross sections and determine if an acceptable set has been selected. The reviewer should check consistency of the models used with similar prior criticality analyses for similar situations (in 10 CFR 50, 10 CFR 71, and 10 CFR 72 licensing). The sources for models used should be included and evaluated in the SER. This can be by inclusion on the recommended consolidated listing of SAR sources in the SER.

8.5.3.2 Material Properties

The reviewer should determine the acceptability of sources for the material properties and cross sections used in the criticality analyses. The material properties and cross sections should also be consistent with accepted prior criticality analyses for systems using similar materials. Material properties should be presented in terms of the expected operating environment (e.g., temperature).

The sources for models used should be included and evaluated in the SER.

8.5.4 Applicant Criticality Analysis

8.5.4.1 Computer Program

The computer program used must be acceptable to the NRC. The reviewer should determine that the program has been used in prior applications and has been accepted. The reviewer should determine that the program is used in accordance with published instructions for its use, including selection of modeling parameters, boundary conditions, and computer program specific biases. The reviewer should determine if the source of the program has been previously accepted by the NRC.

References used for the computer program should be included and evaluated in the SER. These should be included on the recommended consolidated listing of references in the SER.

8.5.4.2 Multiplication Factor

The reviewer should determine if variations in the calculated k_{eff} and explanations for these are reasonable. The reviewer should compare these with those of prior applications for similar systems and computer programs.

8.5.4.3 Benchmark Comparisons

The reviewer should review the sources and application of the experiments used for benchmarks, determine the appropriateness of those used for the benchmark comparisons, and determine if the adjustments to the computed k_{eff}s have been appropriately calculated and applied. Research benchmark comparisons made with prior applications for data not used in the subject analysis but also applicable. The reviewer should determine if use of results of additional or alternative experiments would cause a significant change in the subject determination of adjustments to k_{eff}.

8.5.4.4 NRC Independent Criticality Analysis

The reviewer should perform independent calculations to ensure that the most reactive conditions have been addressed and that the reported k_{eff}s are conservative. This is due to the importance and complexity of the criticality evaluation. The approach to the independent calculations should be determined by consideration of the code used by the applicant, the degree of conservatism in the analysis, previously demonstrated technical expertise of the analyst, and the margins of safety in the results. A small margin of safety (i.e. difference from the maximum allowable limit of k_{eff}) or a small degree of conservatism may necessitate a more extensive analyses.

The reviewer should develop a model that is independent of the applicant's model. If the reported k_{eff} for the worst case is substantially lower than the acceptance criterion of 0.95, a simple model known to produce very conservative results may be sufficient for the independent calculations.

If the results are suspect or independent calculations disagree with the submitted analysis results, the reviewer should perform the independent calculations with a computer code different from that used by the applicant. The reviewer should use a different but acceptable cross-section set to provide a more independent confirmation.

The results of the independent analysis should be reported in the SER. The analysis itself could be an appendix to the SER. Criteria applied to review of the applicant's analysis should be used by a NRC reviewer independent of the performer of the analysis.

8.5.5 Burnup Credit

The reviewer should examine the applicant's burnup credit analysis and verify the conditions and recommendations described in section 8.4.5 were followed, and if not, the differences were described and justified in a manner acceptable to the staff. The reviewer should examine the limits for the licensing basis, code validation, licensing basis model assumptions, loading curve, assigned burnup loading value, and estimate of additional reactivity margin.

8.6 Evaluation Findings

Evaluation findings are prepared by the reviewer on satisfaction of the regulatory requirements relating to criticality. If the documentation submitted with the application fully supports positive findings for each of the regulatory requirements the statements of findings should be substantially as follows (finding numbering is for convenience in referencing within the FSRP and SER):

F8.1   The design, procedures, and materials to be stored for the proposed [ISFSI/MRS] provide reasonable assurance that the activities authorized by the license can be conducted without endangering the health and safety of the public, in compliance with 10 CFR 72.40(a)(13).

F8.2   The designs and proposed use of the [ISFSI/MRS] handling, packaging, transfer, and storage systems for the radioactive materials to be stored acceptably ensure that the materials will remain subcritical and that, before a nuclear criticality accident is possible, at least two unlikely, independent, and concurrent or sequential changes must occur in the conditions essential to nuclear criticality safety. The SAR analyses and confirmatory analysis by the NRC adequately show that acceptable margins of safety will be maintained in the nuclear criticality parameters commensurate with uncertainties in the data and methods used in calculations, and demonstrate safety for the handling, packaging, transfer and storage conditions and in the nature of the immediate environment under accident conditions; in compliance with 10 CFR 72.124(a) and 10 CFR 72.124(b).

8.7 References

NRC documents referenced are identified at Consolidated References, Section 17.

Codes, Standards, and Specifications

ANSI/ANS 8.1-1983, "Nuclear Criticality Safety in Operations with Fissionable Materials Outside Reactors," American National Standards Institute, American Nuclear Society, 1983.

ANSI/ANS 8.17-1984, "Criticality Safety Criteria for the Handling, Storage, and Transportation of LWR Fuel Outside Reactors," American National Standards Institute, American Nuclear Society, 1984.

ANSI/ANS 8.21-1995, "Use of Fixed Neutron Absorbers in Nuclear Facilities Outside Reactors," American National Standards Institute, American Nuclear Society, 1995.

Manuals and Texts

"SCALE -PC Version 4.1," CCC-619, ORNL, Radiation Shielding Information Center, December 1993.

Technical Reports

Archibald, R., Lathrop, K. D., and Mathews, D., "1DFX-A Revised Version of the IDF (DTF-IV) Sn Transport Theory Code," Gulf-GA-10820, September 1971.

Lloyd, W. R., "Determination and Application of Bias Values in the Criticality Evaluation of Storage Cask Designs," UCID-21830, Lawrence Livermore National Laboratory, January 1990.

"MCNP - A General Monte Carlo Code for Neutron and Photon Transport," LA-7396-M, LANL, July 1978.

"MCNP 4A, Monte Carlo N-Particle Transport Code System," LANL, December 1993.

Petrie, L. M., Cross, N. F., "KENO-IV, An Improved Monte Carlo Criticality Program," ORNL-4938, ORNL, November 1975.

Walti, P., and Kock, P., "MICROX- A Two Region Flux Spectrum Code for the Efficient Calculation of Group Cross Section," Gulf-GA-A10827, April 1972.

9 CONFINEMENT EVALUATION

9.1 Review Objectives

There are three review objectives for this chapter. The first is to evaluate the applicant's estimate of the amount of radionuclides that would be released to the environment under (a) normal operations and anticipated occurrences, and (b) design basis accident conditions. The estimates of releases, together with local environmental transport mechanisms (i.e., meteorology and hydrology) and distances to the controlled area boundary, are used to determine if the design meets regulatory performance standards. These specific evaluations against regulatory standards are performed in Chapters 11 (Radiation Protection Evaluation) and 15 (Accident Evaluation) of this Standard Review Plan for Spent Fuel Dry Storage Facilities (FSRP).

The second review objective is the evaluation of proposed monitoring systems. This evaluation includes monitoring systems for storage confinement systems and additional systems for measuring effluents during normal operations and accidents.

The third review objective is to evaluate systems for protection of stored materials from degradation.

Figure 9.1 presents an overview of the confinement evaluation review process. The figure shows that the confinement review draws information from both the application, as well as results of other technical reviews (e.g., structural review). The figure also shows that the results of the confinement review are both used by other technical review areas and documented in the NRC staff-prepared Safety Evaluation Report (SER).

9.2 Areas of Review

The following outline shows the areas of review addressed in Section 9.4, Acceptance Criteria, and Section 9.5, Review Procedures:

Radionuclide Confinement Analysis
    Confinement Casks or Systems
    Pool and Waste Management Facilities

Confinement Monitoring
    Storage Confinement Systems
    Effluents

Protection of Stored Materials from Degradation
    Confinement Casks or Systems
    Pool and Waste Management Systems

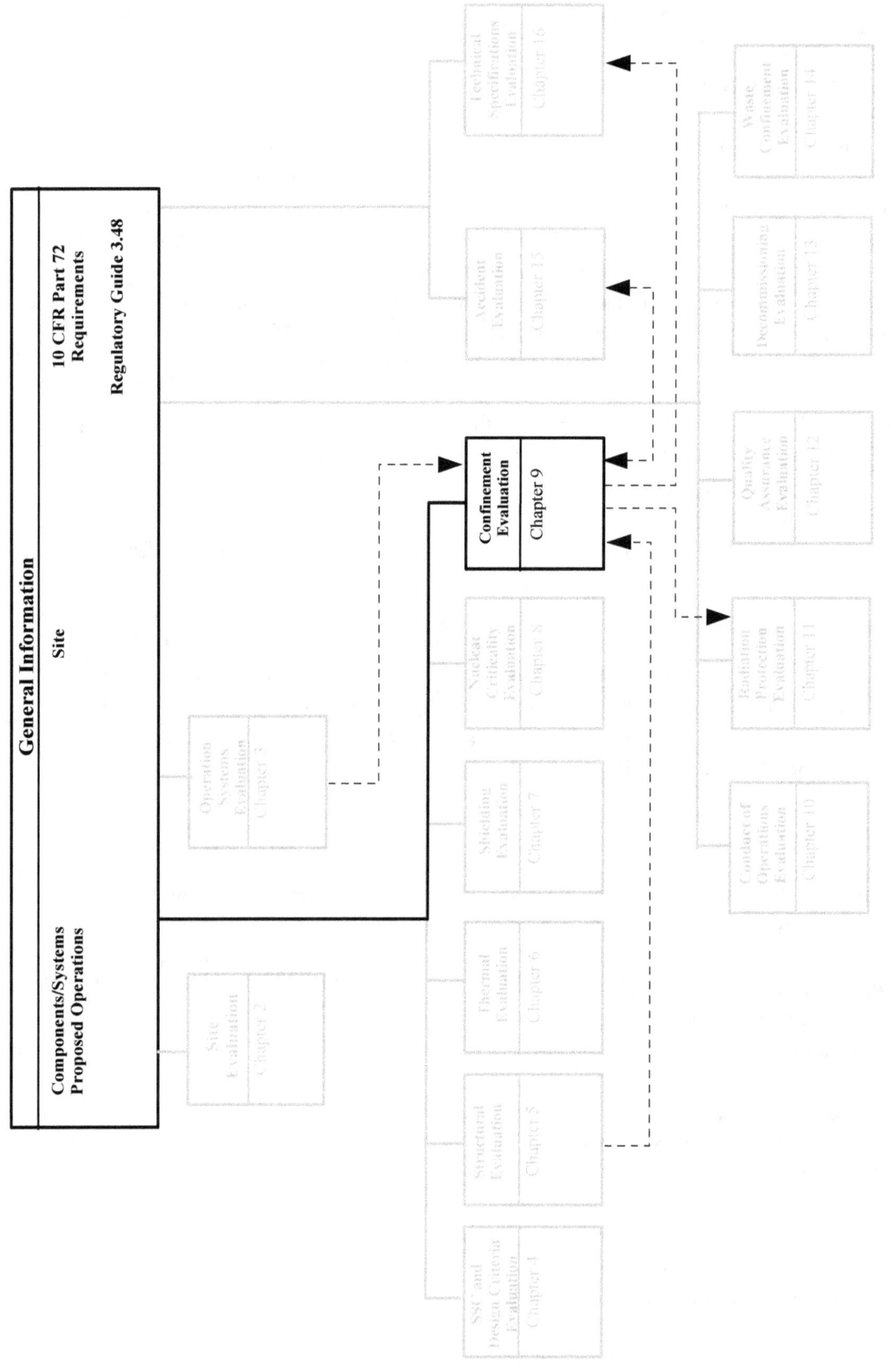

Figure 9.1 Overview of Confinement Evaluation

9.3 Regulatory Requirements

This section identifies and presents a high-level summary of Title 10 of the Code of Federal Regulations (CFR) Part 72 relevant to the review areas addressed by this chapter. The NRC staff reviewer should read the exact regulatory language. A matrix at the end of this section matches the regulatory requirements identified in this section to the areas of review identified in the previous section.

72.24 Contents of application: Technical information [Contents of SAR].
(c) "The design of the Independent Spent Fuel Storage Installations (ISFSI) or Monitored Retrievable Storage (MRS)."
(d) "Analysis and evaluation of the design and performance."
(f) "Features of ISFSI or MRS design and operating modes to reduce...radioactive waste volumes."
(g) "A description of equipment ... to maintain control over radioactive materials in gaseous and liquid effluent ."
(l)(1) "estimate the quantity ... of radionuclides expected to be released annually."

72.44 License conditions.
(c) "Technical specifications must include:"
      (1)   "Functional and operating limits and monitoring instruments and limiting control settings."
           (i) "Functional and operating limits for an ISFSI or MRS."

72.104 Criteria for radioactive materials in effluents and direct radiation from an ISFSI or MRS.
(a) "During normal operations and anticipated occurrences, the annual dose equivalent to any real individual who is located beyond the controlled area must not exceed 25 mrem to the whole body, 75 mrem to the thyroid and 25 mrem to any other organ as a result of exposure to:
      (1)   Planned discharges of radioactive materials . . . to the general environment,
      (2)   Direct radiation from ISFSI or MRS operations, and
      (3)   Any other radiation from uranium fuel cycle operations within the region. "
(b) "Operational restrictions must be established to meet as low as reasonably achievable objectives for radioactive materials in effluents and direct radiation levels associated with ISFSI or MRS operations."
(c) "Operational limits must be established for radioactive materials in effluents and direct radiation levels associated with ISFSI or MRS operations to meet the limits given in paragraph (a) of this section."

72.106 Controlled area of an ISFSI or MRS.
(b) "Any individual located on or beyond the nearest boundary of the controlled area shall not receive a dose greater than 5 rem to the whole body or any organ from any design basis accident. The minimum distance from the spent fuel or high-level radioactive waste handling and storage facilities to the nearest boundary of the controlled area shall be at least 100 meters."

72.122 Overall requirements.
(b) "Protection against environmental conditions and natural phenomena."
    (4) "If ... located over an aquifer ... measures must be taken."
(h) "Confinement barriers and systems."
    (1) "The spent fuel cladding must be protected against degradation ."
    (3) "Ventilation systems and off-gas systems must be provided."
    (4) "Storage confinement systems must have the capability for continuous monitoring."
    (5) "The high-level waste must be packaged that allows handling and retrievability."
(i) "Instrumentation and control systems"

72.126 Criteria for radiological protection.
(c) "Effluent and direct radiation monitoring"
(d) "Effluent control"

72.128 Criteria for spent fuel, high-level radioactive waste, and other radioactive waste storage and handling.
(a) "Spent fuel and high-level radioactive waste storage and handling systems"
    (1) "A capability to test and monitor components important to safety"
    (3) "Confinement structures and systems"

A matrix that shows the primary relationship of these regulations to the specific areas of review associated with this FSRP chapter is given in Table 9.1. The NRC staff reviewer should verify the matching of regulatory requirements to the areas of review presented in the matrix to ensure that no requirements are overlooked as a result of unique applicant design features.

Table 9.1 Relationship of Regulations and Areas of Review

Area of Review	10 CFR Part 72 Regulations						
	72.24	72.44	72.104	72.106	72.122	72.126	72.128
Radionuclide Confinement Analysis	●	●	●	●	●	●	●
Confinement Monitoring	●	●			●	●	●
Protection of Stored Materials from Degradation	●				●		

9.4 Acceptance Criteria

This section identifies the acceptance criteria used for the confinement review. Two types of criteria are described. The first identifies the type of descriptive and analytical information and level of detail related to confinement evaluation that should be present in the Safety Analysis Report (SAR). The second identifies particular standards that are accepted by the NRC staff when conducting confinement analysis. Compliance with the first criteria provides the information that allows the NRC staff to develop a detailed understanding of the applicant's estimate of the effectiveness of the radionuclide confinement systems under a broad spectrum of conditions.

9.4.1 Confinement Description

9.4.1.1 Confinement Casks or Systems

The application must describe the confinement system for spent fuel systems or high-level waste. The confinement design must be consistent with the regulatory requirements, as well as the applicant's "General Design Criteria" reviewed in Chapters 4 and 5 of this SRP. The NRC staff has accepted construction of the primary confinement barrier in conformance with Section III, Subsections NB or NC, of the Boiler and Pressure Vessel (B&PV) Code of the American Society of Mechanical Engineers (ASME). This code defines the standards for all aspects of construction including materials, design, fabrication, examination, testing, inspection, and certification required in the manufacture and installation of components. In such instances, the staff has relied upon Section III to define the minimum acceptable margin of safety; therefore, the applicant must fully document and completely justify any deviations from the specifications of Section III. In some cases after careful and deliberate consideration, the staff has made exceptions to this requirement.

9.4.1.2 Pool and Waste Management Facilities

A description of the confinement system for pool and waste management facilities must be presented in the SAR. Important design features associated with the control and confinement of radioactive materials include seals on closures and doors, negative pressure design, ventilation/filtration systems, charcoal beds, holdup volumes, etc. Instrumentation and control features may include detectors for monitoring releases, alarms, and control features to mitigate releases if abnormal conditions are detected.

9.4.2 Radionuclide Confinement Analysis

Confinement analysis is concerned with the release of radioactive materials to the environment for normal operations and anticipated occurrences and for accident conditions including design basis accidents. The SAR must present a clear description of the proposed confinement system as either (1) a sealed system, as is the case in most spent fuel storage systems, or (2) a vented system with off-gas treatment systems, as is often the case in pools or waste management systems. The description must state how the confinement systems would respond during anticipated occurrences or accident conditions (both design basis and less than design basis). Estimates of releases should be based on the quantity of radioactive material such as vapor pressure, particle sizes, and adsorption kinetics and equilibrium. Data sources that are used to support the physical property estimates or release quantities should be identified in the SAR.

Air and water effluents associated with normal operations must comply with average monthly concentration limits specified in Appendix B to 10 CFR 20 Sections 20.1001-20.2401. There are however, performance standards in 10 CFR Part 72 that place limits on the dose to individuals at or beyond the controlled area. Source terms developed as part of the confinement evaluation review are used for evaluating compliance with the performance standards. The assessment of performance under 10 CFR Part 72.104(a) is conducted in accordance with guidance in this Chapter and Chapter 11 (Radiation Protection Evaluation) of this FSRP. The assessment of performance relative to 72.106(b) is conducted in accordance with guidance in this Chapter and Chapter 15 (Accident Evaluation) of this FSRP.

Guidance for reviewing or evaluating source terms is presented in the following sections.

9.4.2.1 Confinement Casks or Systems

The application must identify the amount of radionuclides that would be released to the environment for normal operations, a spectrum of anticipated occurrences, and a spectrum of design basis accidents. In developing estimates of materials released to the environment, the staff uses guidance in the Cask Standard Review Plan (NUREG-1536)] and applicable ISGs, as estimates of material available for release following the failure of individual fuel pins when there are no additional forces that would move material out of the fuel pin structure. This information about material available for release must be used together with information about a specific release scenario to develop a release amount estimate. This guidance about fraction available for release is only for spent Pressurized Water Reactor (PWR) and Boiling Water Reactor (BWR) fuel with uranium in the form of UO_2.

For purposes of estimating the source term at the time of retrieval operations, NRC has accepted the assumptions that casks will have experienced an off-normal condition (e.g., 10 percent rod failure). For storage casks having closure lids that are designed and tested to be "leak tight," as defined in "American National Standard for Leakage Tests on Packages for Shipment of Radioactive Materials," ANSI N14.5-1997, confinement calculation of the doses under normal off-normal and accident conditions are unnecessary. For casks that are not tested to the leak tight standard of ANSI N14.5-1997, alternative justifications may be found acceptable to the staff.

9.4.2.2 Pool and Waste Management Facilities

Systems that do not have sealed barriers to provide confinement (i.e., transfer pools or cells, waste management facilities) may have releases to the environment under normal, off-normal, and accident conditions. The SAR must present estimates of radionuclides released to the environment for normal conditions, anticipated occurrences, and design basis accidents. The estimates must be based on evaluation of the actual design and the physical process that will move radionuclides into the environment or retain them in the storage or holding systems. If the assumptions about material available for release that are used in the SAR analysis are different from those in NUREG-1536 and applicable ISGs, such assumptions must be justified.

The SAR must include an estimate of the quantity of each of the principal radionuclides expected to be released annually to the environment in liquid and gaseous effluents produced during normal ISFSI or MRS operations. Because use of the pool facility may be intermittent, the estimated quantities of releases must be projected for the maximum usage year, typical years, and for standby (or shutdown) mode years. The estimated source term should also consider the possibility that radioactive gas from a failed sealed fuel container could be released from the facility under anticipated occurrences.

The SAR must estimate pool and waste management facility emissions resulting from anticipated occurrences (off-normal conditions), including possible emissions of radioactive gases from sealed fuel containers that may fail. The SAR must also determine any pool and waste management facility emissions which may result from design basis accidents (accident level conditions). The NRC accepts that other sources on the site can be assumed to be at normal conditions during such accident conditions unless the same initiating event affects these other sources.

Estimates of radionuclide quantities released after failure of fuel cladding can take credit for radionuclides being retained in the water in which fuel rods are immersed. The NRC accepts guidance included in Regulatory Guide 1.25, "Assumptions Used for Evaluating the Potential Radiological Consequences of a Fuel Handling Accident in the Fuel Handling and Storage Facility for Boiling and Pressurized Water Reactors," for estimation of releases from fuel rods in pools.

9.4.3 Confinement Monitoring

Confinement monitoring for ISFSI and MRS has two aspects. The first is monitoring storage confinement closure seals or overall closure effectiveness. The second is providing a system to measure radionuclides released to the environment under normal and accident conditions. This second aspect includes all areas where there is the potential for significant releases to the environment and may include storage casks, pool facilities, and waste management facilities. The SAR must present a discussion of the extent of monitoring required consistent with 10 CFR Part 72 requirements for both of these aspects of confinement monitoring.

9.4.3.1 Dry Storage Cask Confinement Systems

The applicant should describe the proposed monitoring capability and/or surveillance plans for mechanical closure seals. In instances involving welded closures, the staff has previously accepted that no closure monitoring system is required. This practice is consistent with the fact that other welded joints in the confinement system are not monitored. However, the lack of a closure monitoring system has typically been coupled with a periodic surveillance program that would enable the licensee to take timely and appropriate corrective actions to maintain safe storage conditions if closure degradation occurred. However, for storage casks having closure lids that are designed and tested to be "leak tight," as defined in "American National Standard for Leakage Tests on Packages for Shipment of Radioactive Materials," ANSI N14.5-1997, monitoring capability and/or surveillance plans are unnecessary.

To show compliance with 10 CFR Part 72.122(h)(4), cask vendors have proposed, and the staff has accepted, routine surveillance programs and active instrumentation to meet the continuous monitoring requirements. Some DCSS designs contain a component or feature whose continued performance over the licensing period has not been demonstrated to staff with a sufficient level of confidence. Therefore, the staff may determine that active monitoring instrumentation is required to provide for the detection of component degradation or failure. This particularly applies to components whose failure immediately affects or threatens public health and safety. To demonstrate compliance with 10 CFR Part 72.122(h)(4), the vendor or staff may propose a technical specification requiring such instrumentation as part of the initial use of a cask system. After initial use, and if warranted and approved by staff, such instrumentation may be discontinued or modified.

9.4.3.2 Effluents

The SAR must describe the monitoring system that provides measurement of releases under normal and accident conditions. The discussion must address all areas of the ISFSI that can release radionuclides into the environment. NRC accepts the following criteria and guidance for monitoring releases from ISFSI or MRS systems, to the extent applicable:

- Regulatory Guide 1.13, "Spent Fuel Storage Facility Design Basis"

- Regulatory Guide 1.21, "Measuring, Evaluating, and Reporting Radioactivity in Solid Wastes and Releases of Radioactive Materials in Liquid and Gaseous Effluents from Light-Water-Cooled Nuclear Power Plants"

- Regulatory Guide 4.1, "Programs for Monitoring Radioactivity in the Environs of Nuclear Power Plants"

- NUREG-0800, "Standard Review Plan," 11.5, "Process and Effluent Radiological Monitoring Instrumentation and Sampling Systems"

9.4.4 Protection of Stored Materials from Degradation

The materials that help confine the radionuclides in spent fuel and waste should be protected from degradation so that confinement effectiveness is not reduced. The SAR must identify these materials (i.e., fuel matrix and fuel cladding) and describe how these material are protected from degradation.

9.4.4.1 Confinement Casks or Systems

The primary materials in spent fuel that must be protected from degradation are the fuel matrix and fuel cladding. The applicant's SAR must describe the actions proposed to protect these materials from degradation.

The cask must provide a non-reactive environment to protect fuel assemblies against fuel cladding degradation, which might otherwise lead to gross rupture (Pacific Northwest Laboratories [PNL] 6365). Measures for providing a non-reactive environment within the confinement cask typically include drying, evacuating air and water vapor, and backfilling with a non-reactive cover gas (such as helium). For dry storage conditions, experimental data have not demonstrated an acceptably low oxidation rate for UO_2 spent fuel, over the 20-year licensing period, to permit safe storage in an air atmosphere. Therefore, to reduce the potential for fuel oxidation and subsequent cladding failure, the NRC has accepted storage designs that have cask inventories of oxidizing gases less than 1.0 gram mole per cask and an inert atmosphere (e.g., helium cover gas) for storing UO_2 spent fuel in a dry environment.

Note that other fuel types, such as graphite fuels for the high-temperature gas-cooled reactors (HTGRs), may not exhibit the same oxidation reactions as UO_2 fuels, and therefore, may not require an inert atmosphere. Applicants proposing to use atmospheres other than inert gas should discuss how the fuel and cladding will be protected from oxidation.

9.4.4.2 Pool and Waste Management Facilities

The SAR must also describe the pool and waste management facilities proposed by the applicant to prevent degradation of waste and fuel confinement materials. Pools must provide an environment that is compatible with stored materials and any elements important to safety. The SAR must give full consideration to maximum anticipated storage time for any projected corrosion. Permanent degradation of any pool confinement barrier should not occur for anticipated occurrences (off-normal events and conditions) when considering the cumulative corrosion effects over the proposed license period. The pool facility confinement barrier and liquid containment structures, systems, and components (SSCs) may experience some repairable degradation from accident-level conditions.

9.5 Review Procedures

The following review guidance relevant to the confinement evaluation is based on the required products of the review, and lessons learned from prior reviews.

9.5.1 Review of Design Features

The reviewer should review the principal design criteria and the general description of the cask presented in the SAR. All drawings, figures, and tables describing confinement features must be sufficiently detailed to stand alone. Verify that the applicant has clearly identified the confinement boundaries. This identification should include, as applicable, the confinement vessel; its penetrations, valves, seals, welds, and closure devices; and corresponding information concerning the redundant sealing.

Coordinate with the structural reviewer to ensure that the applicant has provided proper specifications for all welds and, if applicable, that the bolt torque for closure devices is adequate and properly specified.

9.5.2 Radionuclide Confinement Analysis

The procedure for review of radionuclide confinement analysis varies with the nature of the components in the ISFSI design and the certification of any of these components under 10 CFR Part 72 Subpart K. In general, the confinement analysis review involves two principle steps: (1) identification of events to be considered, and (2) evaluation of release estimates.

9.5.2.1 Identification of Release Events

The reviewer should discuss the proposed design and operations with other reviewers (e.g., structural, operations, site characteristics, etc.) to determine spectrum of events that need to be considered for the specific design and specific site. The discussions should focus on the physical condition of the confinement systems for normal operations and anticipated occurrences, and for design basis accidents. The confinement analyst should use these discussions to understand (a) the physical condition of the equipment that might serve to contain radionuclides, and (b) the forces (physical displacement, pressure differences, temperatures, etc.) that could move radionuclides into the accessible environment if the confinement system fails. The reviewer should categorize the selected events as either (a) normal operations and anticipated occurrences, or (b) design basis accidents. The reviewer-identified scenarios may be more extensive than those presented by the applicant. A specific scenario can be dismissed if the staff reviewer determines that it is less severe that another scenario being considered within the category (i.e., normal operations and anticipated occurrences or design basis accidents).

9.5.2.2 Evaluation of Release Estimates

For each of the scenarios identified and retained in the previous step, the reviewer should either (a) review the applicant's release estimate and determine that applicant's estimate is reasonable, or (b) develop independent estimates and compare them with that provided by the applicant. The method of checking or developing independent release estimates depends (as described below) on the whether the system is designed to be a sealed system and whether the design has been previously reviewed and approved by the NRC staff as part of a certification process. In preparing or reviewing the estimate, the reviewer should identify all radioactive isotopes present in the fuel or waste.

If the design is a sealed system that has not been previously certified, the confinement evaluation reviewer should discuss with the structural reviewer the response of the sealed system (primary containment system). This discussion should determine the response to the structural loads that would occur under the scenario. This discussion should result in an understanding of whether there will be any loss of containment integrity. The discussion should also result in an understanding of any forces or effects that would either promote or impede movement of radionuclides out of primary containment system. For each scenario identified and retained in step 1, the confinement evaluation reviewer should then either determine that the applicant's estimate of radionuclides released to the environment is reasonable, or develop the reviewer's own estimate of release amounts. The confinement evaluation reviewer may find the information in NUREG-1536 and applicable ISGs of value when evaluating release quantity from a low-pressure, gas-filled spent fuel canister.

If the design involves a sealed cask that has been previously certified, the confinement evaluation reviewer should review the information in the certification SER. For each scenario identified and retained in step 1, the confinement evaluation reviewer should use the information from the certification SER to either determine that the applicant's estimate of radionuclides released to the environment is reasonable, or develop the reviewer's own estimate of release amounts.

If the design involves systems and components that are not designed to be sealed, there will be releases of radioactivity under normal, as well as accident conditions. In this case, the confinement evaluation reviewer should review the design and understand the process (e.g., vapor pressure in conjunction with convective flow) that will move radioactive contamination from those areas where radionuclides are being stored or handled into the environment. The reviewer should also understand the components that are designed to reduce the flow of radionuclides into the environment (e.g., filtration systems). For each scenario identified and retained from step 1, the confinement evaluation reviewer should discuss the expected structural condition of the relevant components with the structural reviewer. This discussion should also address whether there will be additional forces or effects that would either promote or impede movement of radionuclides into the environment. The confinement evaluation reviewer should use this information to either determine that the applicant's estimate of radionuclides released to the environment is reasonable or develop the reviewer's own estimate of release amounts.

If more that one component can produce a release for the scenarios evaluated, the release estimates for each scenario must be added to produce a total for each scenario.

The NRC staff has determined that, at a minimum, the fractions of radioactive materials available for release from spent fuel, provided in Table 9.2 for pressurized-water reactor (PWR) fuel and boiling-water reactor (BWR) fuel for normal, anticipated occurrences (off-normal), and accident conditions, should be used in the confinement analysis to demonstrate compliance with 10 CFR Part 72. These fractions account for radionuclides trapped in the fuel matrix and radionuclides that exist in a chemical or physical form that is not releasable to the environment under credible normal, off-normal, and accident conditions. Other release fractions may be used in the analysis provided the applicant properly justifies the basis for their usage. For example, the staff has accepted, with adequate justification, reduction of the mass fraction of fuel fines that can be released from the cask.

The staff has accepted the following rod breakage fractions for the confinement evaluations:

    1% for normal conditions
    10% for off-normal conditions
    100% for design basis accident and extreme natural phenomena

For the source term, the NRC staff has accepted, as a minimum for the analysis, the activity from the Co^{60} in the crud, the activity from iodine, fission products that contribute greater than 0.1% of design basis fuel activity, and actinide activity that contributes greater than 0.01% of the design basis activity. In some cases, the applicant may have to consider additional radioactive nuclides depending upon the specific analysis. The total activity of the design basis fuel should be based on the cask design loading that yields the bounding radionuclide inventory (considering initial enrichment, burnup, and cool time).

The quantities of radioactive nuclides are often presented in SAR shielding descriptions, since they are generally determined during the evaluation of gamma and neutron source terms in the shielding analysis. Coordinate with the shielding reviewer to verify that the applicant has adequately developed and characterized the source term.

It is important to recognize that design basis normal or accident conditions resulting in confinement boundary failure are not acceptable. Preservation of the confinement boundary during design basis conditions is confirmed by the structural analysis. The confinement analyses demonstrate that, at the measured leakage rates, and assumed nominal meteorological conditions, the requirements of 10 CFR 72.104(a) and 10 CFR 72.106(b) can be met. Each ISFSI, whether it is a site specific or a general license, is also required to have a site specific confinement analysis and dose assessment to demonstrate compliance with these regulations.

Table 9.2* Evaluation of Release Estimates		
Variable	**Fractions Available for Release****	
	PWR AND BWR FUEL	
	Normal and Off-normal Conditions	Hypothetical Accident Conditions
Fraction of gases released due to a cladding breach, f_G†	0.3	0.3
Fraction of volatiles released due to a cladding breach, f_V†	2×10^{-4}	2×10^{-4}
Mass fraction of fuel released as fines due to cladding breach, f_F	3×10^{-5}	3×10^{-5}
Fraction of crud that spalls off cladding, f_C	$0.15^{\#}$	$1.0^{\#}$

* Values in this table are taken from U.S. Nuclear Regulatory Commission, "Containment Analysis for Type B Packages Used to Transport Various Contents," NUREG/CR-6487, November 1996.

** Except for ^{60}Co, only failed fuel rods contribute significantly to the release. Total fraction of radionuclides available for release must be multiplied by the fraction of fuel rods assumed to have failed.

† In accordance with NUREG/CR-6487, gases species include H-3, I-129, Kr-81, Kr-85, and Xe-127; volatile species include Cs-134, Cs-135, Cs-137, Ru-103, Ru-106, Sr-89, and Sr-90.

\# The source of radioactivity in crud is ^{60}Co on fuel rods. At the time of discharge from the reactor, the specific activity, S_c, is estimated to be 140 μCi/cm^2 for PWRs and 1254 μCi/cm^2 for BWRs. Total ^{60}Co activity is, this estimate times the total surface area of all rods in the cask. Decay of ^{60}Co to determine activity at the minimum time before loading is acceptable.

The reviewer should review the applicant's confinement analysis and the resulting doses for the normal, off-normal, and accident conditions at the controlled area boundary.

The analysis typically includes the following common elements:

- calculation of the specific activity (e.g., Ci/cm^3) for each radioactive isotope in the cask cavity based on rod breakage fractions, release fractions, isotopic inventory, and cavity free volume

- using the tested leak rate and conditions during testing as input parameters, calculation of the adjusted maximum seal leakage rates (Cm^3/sec) under normal, off-normal, and hypothetical accident conditions (e.g., temperatures and pressures)

- calculation of isotope specific leak rates (Q_i - Ci/sec) by multiplying the isotope specific activity by the maximum seal leakage rates for normal, off-normal, and accident conditions

- determination of doses to the whole body, thyroid, other critical organs, lens of the eye, and skin from inhalation and immersion exposures at the controlled area boundary (considering atmospheric dispersion factors - χ/Q)

Verify the applicant specified maximum allowable "as tested" seal leakage rates as a Technical Specification. Guidance on the calculations of the specific activity for each isotope in the cask and the maximum allowable helium seal leakage rates for normal, off-normal, and accident conditions can be found in NUREG/CR-6487 and ANSI N14.5-1997. Verify the minimum distance between the casks and the controlled area boundary is a design criterion and meets the 10 CFR Part 72 requirement that this distance be at least 100 meters from the ISFSI.

For the dose calculations, the staff has accepted the use of either an adult breathing rate (BR) of 2.5×10^{-4} m^3/s, as specified in Regulatory Guide 1.109, or a worker breathing rate of 3.3×10^{-4} m^3/s, as specified in EPA Guidance Report No. 11. The dose conversion factors (DCF) in EPA Guidance Report No. 11 for the whole body, critical organs, and thyroid doses from inhalation should be used in the calculation. The bounding DCFs from EPA Report No. 11 should be used for each isotope unless the applicant justifies an alternate value. No weighting or normalization of the dose conversion factors is accepted by the staff. For each isotope, the committed effective dose equivalent ($CEDE_i$ - for the internal whole-body dose) or the committed dose equivalent (CDE_i - for the internal organ dose) should be calculated as follows:

$CEDE_i$ or CDE_i (in mrem per year for normal/off-normal or mrem per accident)
$$= Q_i * DCF_i * \chi/Q * \text{B-Rate} * \text{Duration} * \text{conversion factor}[1]$$

For the contributions to the whole body, thyroid, critical organs, and skin doses from immersion (external) exposure, the DCFs in EPA Guidance Report No. 12 should be used. Again, no weighting or normalization of the dose conversion factors is accepted by the staff.

The deep dose equivalent (DDE_i - for the external whole body) and the shallow dose equivalent (SDE_i - for the skin dose) are calculated as follows:

DDE_i or SDE_i (in mrem per year for normal/off-normal or mrem per accident)
$$= Q_i * DCF_i * \chi/Q * \text{Duration} * \text{a unit conversion factor}$$

The total effective dose equivalent, $TEDE = \sum CEDE_i + \sum DDE_i$

---

[1]The conversion factor, if required, converts the input units into the desired form (e.g., mrem/year).

For a given organ, the total organ dose equivalent, $\text{TODE} = \sum \text{CDE}_i + \sum \text{DDE}_i$

The total skin dose equivalent $\text{SDE} = \sum \text{SDE}_i$

Compliance with the lens dose equivalent (LDE) limit is achieved if the sum of the SDE and the TEDE do not exceed 0.15 Sv (15 rem). This approach is consistent with guidance in ICRP-26.

In general, the staff should evaluate the applicants analyses for normal, off-normal, and accident conditions.

Normal Conditions

For normal conditions, a bounding exposure duration assumes that an individual is present at the controlled area boundary for one full year (8760 hours). An alternative exposure duration may be considered by the staff if the applicant provides justification.

Because any potential release, resulting from seal leakage, would typically occur over a substantial period of time, the staff accepts (for applications for certificates) calculation of the atmospheric dispersion factors (χ/Q) according to Regulatory Guide 1.145 assuming D-stability diffusion and a wind speed of 5 m/s.

For the likely case of an ISFSI with multiple casks, the doses need to be assessed for a hypothetical array of casks during normal conditions. Therefore, the staff anticipates that the resulting doses from a single cask will be a small fraction of the limits prescribed in 10 CFR 72.104(a) to accommodate the array and the external direct dose.

Note: If the region between redundant, confinement boundary, mechanical seals is maintained at a pressure greater than the cask cavity, the monitoring system boundaries are tested to a leakage rate equal to the confinement boundary, and the seal pressure is routinely checked and instrumentation is verified to be operable in accordance with a Technical Specification Surveillance Requirement, then the staff has accepted that no discernible leakage is credible. Therefore, calculations of dose to the whole body, thyroid, and critical organs at the controlled area boundary from atmospheric releases during normal conditions would not be required for normal conditions.

Off-normal Conditions

For off-normal conditions, the bounding exposure duration and atmospheric dispersion factors (χ/Q) are the same as those discussed above for normal conditions.

To demonstrate compliance with 10 CFR 72.104(a), the staff accepts whole body, thyroid, and critical organ dose calculations for releases from a single cask. However, the dose contribution from cask leakage should also be a fraction of the limits specified in 10 CFR 72.104(a) since the doses from other radiation sources are added to this contribution.

Accident Conditions

For hypothetical accident conditions, the duration of the release is assumed to be 30 days (720 hours). A bounding exposure duration assumes that an individual is also present at the controlled area boundary for 30 days. This time period is the same as that used to demonstrate compliance with 10 CFR 100 for reactor facilities licensed per 10 CFR 50 and provides good defense in depth since recovery actions to limit releases are not expected to exceed 30 days.

For hypothetical accidents conditions, the staff has accepted calculation of the atmospheric dispersion factors (χ/Q) of Regulatory Guide 1.145 or Regulatory Guide 1.25 on the basis of F-stability diffusion, and a wind speed of 1 m/s.

To demonstrate compliance with 10 CFR 72.106(b), the staff accepts whole body, thyroid, critical organ, and skin dose calculations for releases of radionuclides from a single cask.

The reviewer should ensure that all supportive information or documentation has been provided or is readily available. This includes, but is not limited to, justification of assumptions or analytical procedures, test results, photographs, computer program descriptions, input and output, and applicable pages from referenced documents. Reviewers should request any additional information needed to complete the review.

9.5.3 Confinement Monitoring

9.5.3.1 Confinement Casks or Systems

If applicable, assess the seals used to provide closure. Because of the performance requirements over the 20-year license period, evaluate the potential for deterioration. The NRC staff has previously accepted only metallic seals for the primary confinement. Coordinate with the thermal reviewers to ensure that the operational temperature range for the seals, specified by the manufacturer, will not be exceeded.

The NRC staff has found that casks closed entirely by welding do not require seal monitoring. However, for casks with bolted closures, the staff has found that a seal monitoring system has been needed to adequately demonstrate that seals can function and maintain a helium atmosphere in the cask for the 20-year license period. A seal monitoring system combined with periodic surveillance enables the licensee to determine when to take corrective action to maintain safe storage conditions. (Note that some fuel designs may not require an inert atmosphere in the cask. In such designs, a periodic surveillance program to check seal leak tightness may be appropriate.)

Although the details of the monitoring system may vary, the general design approach has been to pressurize the region between the redundant seals, with a non-reactive gas, to a pressure greater than that of the cask cavity and the atmosphere. The monitoring system is leakage tested to the same leak rate as the confinement boundary. Installed instrumentation is routinely checked per surveillance requirements. A decrease in pressure between these seals indicates that the non-reactive gas is leaking either into the cask cavity or out to the atmosphere. For normal operations, radioactive material should not be able to leak to the atmosphere; hence this design allows for detecting a faulty seal without radiological consequence. Note that the volume between the redundant seals should be pressurized using a *non-reactive* gas, thereby preventing contamination of the interior cover gas.

The staff has accepted monitoring systems as not important to safety and classified as Category B under the guidelines of NUREG/CR-6407. Although its function is to monitor confinement seal integrity, failure of the monitoring system alone does not result in a gross release of radioactive material. Consequently, the monitoring system for bolted closures need not be designed to the same requirements as the confinement boundary (i.e., ASME Section III, Subsections NB or NC).

Dependant on the monitoring system design, there could be a lag time before the monitoring system indicates a postulated degraded seal leakage condition. Degraded seal leakage is leakage greater than the tested rate that is not identified within a few monitoring system surveillance cycles. The occurrence of a degraded seal without detection is considered a "latent" condition and should be presumed to exist concurrently with other off-normal and design basis events (see NUREG-1536, Section 2, paragraph V.2.b.). Note that once the degraded seal condition is detected, the cask user must initiate corrective actions.

For the off-normal case, the monitoring system boundary remains intact and this condition would be bounded by the off-normal analysis. If the monitoring system would not maintain integrity under design basis accident conditions, additional safety analysis may be necessary. The staff recognizes that the possibility of a degraded seal condition is small and that the possibility of a degraded seal condition concurrent with a design basis event that breaches the monitoring system pressure boundary is very remote. However, these probabilities have not been quantified. To address this concern, the staff accepts a demonstration that the probability of occurrence of a latent, degraded seal condition concurrent with a design basis event that breaches the monitoring system boundary is acceptably low (e.g., less than 1×10^{-6} per year). Alternatively, the staff accepts a demonstration that the dose consequences of this event are within the limits of 10 CFR 72.106(b).

The reviewer should examine the specified pressure of the gas in the monitored region to verify that it is higher than both the cask cavity and the atmosphere. Coordinate with the structural and thermal reviewers to verify the pressure in the cask cavity.

The reviewer should review the applicant's analysis to verify that the total volume of gas in the seal monitoring system is such that normal seal leakage will not cause all of the gas to escape over the lifetime of the cask. In determining the proposed maximum leakage rate, the applicant should consider the volume between the redundant seals of the confinement cask, the minimum pressure to be maintained, and the length of the proposed routine recharge cycle. The reviewer should verify that the applicant specifies the calculated leakage rate as a maximum value for an acceptance test criterion in the SAR, even though the actual leakage rate of the seals is expected to be significantly lower.

For redundant seal welded closures, ensure that the applicant has provided adequate justification that the seal welds have been sufficiently tested and inspected to ensure that the weld will behave similarly to the adjacent parent material of the cask. Any inert gas should not leak or diffuse through the weld and cask material in excess of the design leak rate.

The reviewer should verify that any leakage test, monitoring, or surveillance conditions are appropriately specified in SAR, the license, and/or the Certificate of Compliance.

9.5.3.2 Pool or Waste Management Facilities

Guidance on review of monitoring systems is included in NUREG-0800, Section 11.5. The reviewer should identify those aspects of ISFSI and MRS monitoring systems for which NUREG-0800 guidance is applicable and evaluate those aspects against the NUREG criteria. Some of the primary considerations the reviewer should address are as follows:

- The location of detectors and samplers and the bases for their selection

- The equipment, piping and methods which ensure compliance with Regulatory Guide 1.21 guidelines

- The sampling frequencies, type of analyses required, analytical capabilities, and calibration methods and frequency for comparison with Regulatory Guides 1.21 and 1.97, "Instrumentation for Light-Water-Cooled Nuclear Power Plants to Assess Plant and Environs Conditions During and Following an Accident," guidelines

- Provisions for termination of releases in the event they exceed a predetermined level

- Locations of monitors, readouts, annunciators, and alarms for capability to inform operators of system performance.

9.5.4 Protection of Stored Materials from Degradation

9.5.4.1 Confinement Casks or Systems

Review procedures associated with providing a benign atmosphere within the confinement storage cask are given in the NUREG-1536 and applicable ISGs. In addition to these procedures, if the proposed confinement storage cask design is for zircalloy-clad PWR and BWR UO_2 fuels, the reviewer should determine whether (a) the design includes a specification to limit oxidizing gas concentration to less than 1.0 gram mole per cask, and (b) the proposed operation include the vacuum testing, pressure testing, and purging identified in Section 9.4.3.1.

Verify that the design and procedures provide for drying and evacuation of the cask interior as part of the loading operations, and that the design is acceptable for the pressures that may be experienced during these operations. The staff has accepted the combination of a draining procedure and a vacuum drying procedure as providing adequate assurance that the gases in the cask meet the maximum oxidizing gas criteria. The vacuum drying procedure involves a vacuum test to demonstrate that there is no water in the cask or fuel. A cask that is evacuated to less than 3 torr and, after sealing, does not have a cask pressure which increases by 1 torr over 30 minutes is considered to be free of water. The cask is then filled with an inert gas (e.g., helium of sufficient purity to preclude introduction of contaminants) and pressurized to the leak test pressure specified by ASME B&PV Code Section III, Division 1, Section NB or NC and leak tested according to ANSI 14.5. After a successful leak test, the cask is evacuated to less than 3 torr and then backfilled with helium to the design storage pressure. A cask prepared for storage by the previous procedure is acceptable as having less than 1.0 gram mole of oxidizing material per cask.

Verify that, on completion of cask loading, the gas fill of the cask interior is at a pressure level that is expected to maintain a non-reactive environment for at least the 20-year storage life of the cask interior under both normal and off-normal conditions and events. This verification can include pressure testing, seal monitoring, and maintenance for casks with seals that are not welded if these are conditions of use. The NRC has previously accepted specification of an overpressure of approximately 14 kilopascals (~2 psig) and cask leak testing as conditions of use for satisfying this requirement. In addition, if conditions of use require routine inspection of seals by the pressure testing of the cask interior, the cask fill pressure may be linked to that activity. If fuel or material other than zircalloy-clad PWR and BWR UO_2 fuel is used, the reviewer should determine if there is adequate technical basis to conclude that the fuel and cladding will not experience significant degradation over the proposed storage time.

9.5.4.2 Pool and Waste Management Facilities

For pool and waste management facilities, the reviewer should ensure that corrosion effects are considered in the design and that corrosion that is expected over the storage time will not result in confinement failures that could lead to significant releases.

9.6 Evaluation Findings

NRC staff reviewers should prepare evaluation findings regarding satisfaction of the regulatory requirements related to confinement of the radioactive material to be stored. In the case of evaluation of releases from confinement, the acceptability of releases can be determined only after reviewing the results of the dose assessment, which is addressed in Chapters 11 and 15. If the documentation submitted with the application fully supports positive findings for each of the regulatory requirements, then the findings should substantially be stated as follows (finding numbering is for convenience in referencing within the FSRP and SER):

F9.1   Section(s) _____ of the SAR describe(s) confinement structures, systems, and components (SSCs) important to safety in sufficient detail in to permit evaluation of their effectiveness.

F9.2   If the confinement system is provided by a sealed cask system, the following is applicable: the design of the [cask designation] provides redundant sealing of the confinement system closure joints by _____.

F9.3   [If applicable] The design of the [ISFSI/MRS] provides ventilation and off-gas systems that acceptably ensure the confinement of airborne radioactive particulate materials during normal or off-normal conditions, in compliance with 10 CFR 72.122(h)(3).

F9.4   The design and proposed operations of the [ISFSI/MRS] include acceptable measures that minimize the potential for transport of radioactive materials to the environment through the aquifer, in compliance with 10 CFR 72.122(b).

F9.5   The quantity of radioactive nuclides postulated to be released to the environment has been assessed as discussed above. In Section 11 of the SER, the dose from these releases will be added to the direct dose to show that the [cask designation] satisfies the regulatory requirements of 10 CFR 72.104(a) and 10 CFR 72.106(b).

F9.6   If the confinement system is provided by a sealed cask system, the following is applicable: The confinement system is monitored with a _____ monitoring system as discussed above (if applicable). No instrumentation is required to remain operational under accident conditions.

        If the confinement system is provided by a pool or an unsealed system, the following is applicable: The [ISFSI/MRS] includes the following confinement systems that are important to safety and which require monitoring over anticipated ranges for normal operation and off-normal operation: _____ [identify]. The following monitoring systems must remain operational under accident conditions: _____ [identify]. The SAR acceptably describes instrumentation and control systems that should provide these capabilities, in compliance with 10 CFR 72.122(i)(c).

F9.7   The design and proposed operations of the [ISFSI/MRS] provides adequate measures for protecting the spent fuel cladding against degradation that might otherwise lead to gross ruptures of the material to be stored, in compliance with 10 CFR 72.122(h)(1).

F9.8   The [cask/ ISFSI/MRS] confinement system has been evaluated [by appropriate tests or by other means acceptable to the Commission] to demonstrate that it will reasonably maintain confinement of radioactive material under normal, off-normal, and credible accident conditions.

F9.9   The staff concludes that the design of the confinement system of the [cask designation] is in compliance with 10 CFR Part 72 and that the applicable design and acceptance criteria have been satisfied. The evaluation of the confinement system design provides reasonable assurance that the [cask designation] will allow safe storage of spent fuel. This finding is reached on the basis of a review that considered the regulation itself, appropriate regulatory guides, applicable codes and standards, the applicant's analysis and the staff's confirmatory analysis, and accepted engineering practices.

9.7 References

NRC documents referenced are identified at Consolidated References, Section 17.

American Society of Mechanical Engineers, "ASME Boiler and Pressure Vessel Code," Section III, Sections NB and NC.

American National Standards Institute, Institute for Nuclear Materials Management, "American National Standard for Leakage Tests on Packages for Shipment of Radioactive Materials," ANSI N14.5, 1987.

International Commission on Radiation Protection, "Statement from the 1980 Meeting of the ICRP," ICRP Publication 26, Pergammon Press, New York, NewYork, 1980.

Johnson, A.B., and Gilbert, E.R., "Technical Basis for Storage of Zircalloy-Clad Spent Fuel in Inert Gasses," Pacific Northwest Laboratories, PNL-4835, September 1983.

Knoll, R.W., and Gilbert, E.R., "Evaluation of Cover Gas Impurities and Their Effects on the Dry Storage of LWR Spent Fuel," Pacific Northwest Laboratory, PNL-6365, November 1987.

Levy, I.S. *et al.,* "Recommended Temperature Limits for Dry Storage of Spent Light Water Zircalloy Clad Fuel Rods in Inert Gas," Pacific Northwest Laboratories, PNL-6189, May 1987.

U.S. Environmental Protection Agency, "Federal Guidance Report No. 11, Limiting Values of Radionuclide Intake and Air Concentration and Dose Conversion Factors for Inhalation, Submersion, and Ingestion," DE89-011065, 1988.

U.S. Environmental Protection Agency, "Federal Guidance Report No. 12, External Exposure to Radiouclides in Air, Water, and Soil," EPA 402-R-93-081, September 1993.

U.S. Nuclear Regulatory Commission, "Atmospheric Dispersion Models for Potential Accident Consequence Assessments at Nuclear Power Plants," Regulatory Guide 1.145, February 1989.

U.S. Nuclear Regulatory Commission, "Assumptions Used for Evaluating the Potential Radiological Consequences of a Fuel Handling Accident in the Fuel Handling and Storage Facility for Boiling and Pressurized Water Reactors," Regulatory Guide 1.25, March 1972.

U.S. Nuclear Regulatory Commission, "Calculations of Annual Doses to Man from Routine Releases of Reactor Effluents for the Purpose of Evaluating Compliance with 10 CFR Part 50, Appendix I," Regulatory Guide 1.109, October 1977.

10 CONDUCT OF OPERATIONS EVALUATION

10.1 Review Objective

The purpose of the review and evaluation is to ensure that the applicant has described an appropriate infrastructure to manage, test, and operate the facility, including provisions for effective training, emergency planning, and physical security programs.

An overview of the conduct of operations review process is shown in Figure 10.1. The figure shows that the confinement review draws information from the application as well as regulatory requirements.

10.2 Areas of Review

The following outline shows the areas of review addressed in Section 10.4, Acceptance Criteria, and Section 10.5, Review Procedures:

Organizational Structure
> Corporate Organization
> Onsite Organization
> Management and Administrative Controls

Preoperational Testing and Startup Operations
> Preoperational Testing Plan
> Startup Plan

Normal Operations
> Procedures
> Records

Personnel Selection, Training and Certification
> Personnel Organization
> Selection and Training of Operating Personnel
> Selection and Training of Security Guards

Emergency Planning

Physical Security and Safeguards Contingency Plans

10.3 Regulatory Requirements

This section identifies and presents a high-level summary of Title 10 of the Code of Federal Regulations (CFR) Part 72 relevant to the review areas addressed by this chapter. The NRC staff reviewer should read the exact regulatory language. A matrix at the end of this section matches the regulatory requirements identified in this section to the areas of review identified in the previous section.

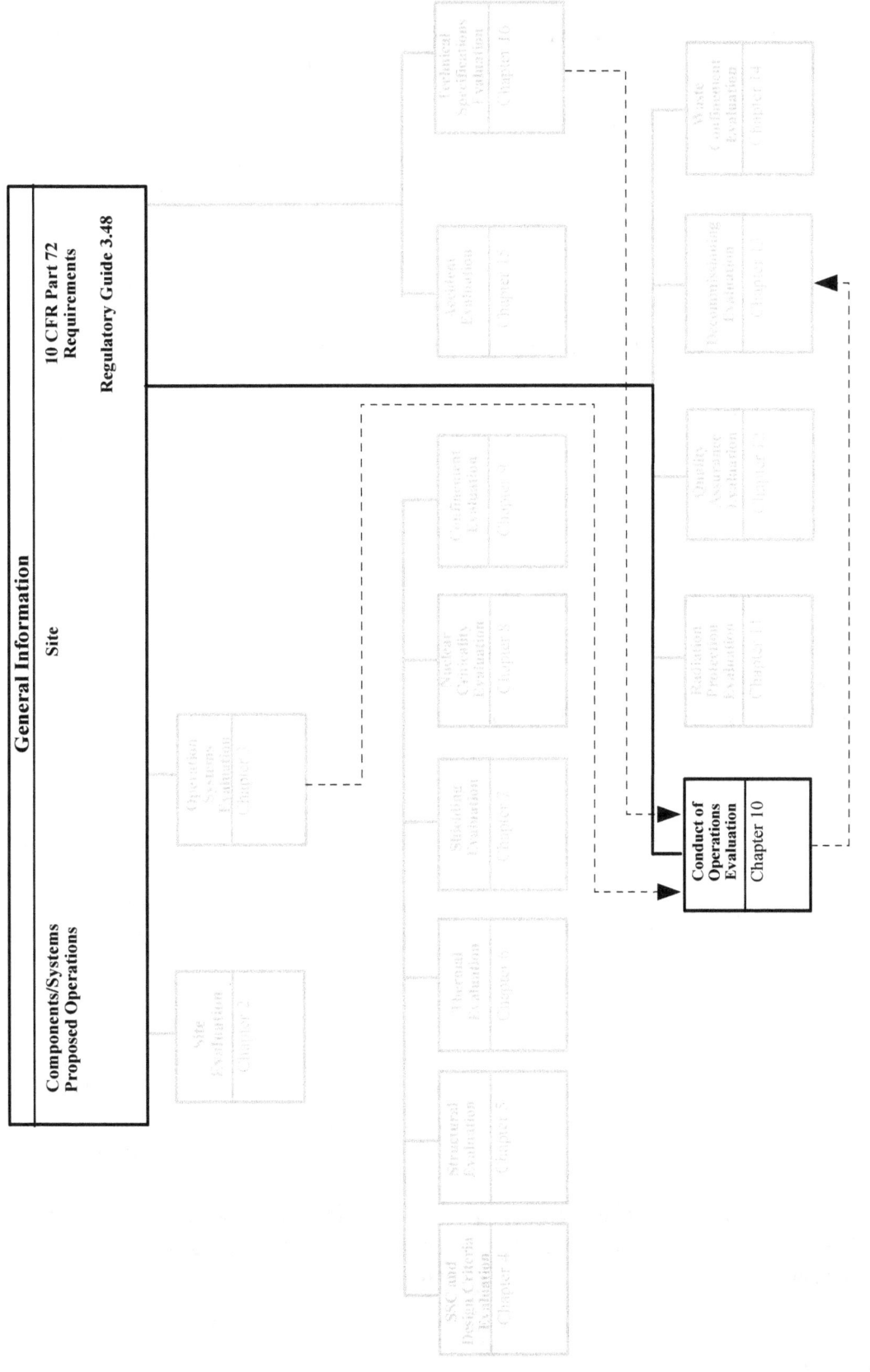

Figure 10.1 Overview of Operations Evaluation

72.24 Contents of application: Technical information.

Each application for a license under this part must include a Safety Analysis Report . . . [including] the following:

(h) "A plan for the conduct of operations, including the planned managerial and administrative controls system, and the applicant's organization, and program for training of personnel pursuant to Subpart I."

(i) "If the proposed Independent Spent Fuel Storage Installations (ISFSI) or Monitored Retrievable Storage (MRS) incorporates [structures, systems, and components (SSCs)] important to safety whose functional adequacy or reliability have not been demonstrated by prior use for that purpose or cannot be demonstrated by reference to performance data in related applications or to widely accepted engineering principles, an identification of these SSCs along with a schedule showing how safety questions will be resolved prior to the initial receipt of spent fuel or high-level radioactive waste for storage at the ISFSI or MRS."

(j) "The technical qualifications of the applicant to engage in the proposed activities, as required by Section 72.28."

(k) "A description of the applicant's plans for coping with emergencies, as required by Section 72.32."

(o) "A description of the detailed security measures for physical protection."

(p) "A description of the program covering preoperational testing and initial operations."

72.28 Contents of application: Applicant's technical qualifications.

Each application under this part must include:

(a) "The technical qualifications, including training and experience, of the applicant to engage in the proposed activities;

(b) A description of the personnel training program required under Subpart I;

(c) A description of the applicant's operating organization, delegations of responsibility and authority and the minimum skills and experience qualifications relevant to the various levels of responsibility and authority; and

(d) A commitment by the applicant to have and maintain an adequate complement of trained and certified installation personnel prior to the receipt of spent fuel or high-level radioactive waste for storage."

72.40 Issuance of license.

(a) "Except as provided in paragraph (c) of this section, the Commission will issue a license under this part upon a determination that the application for a license meets the standards and requirements of the Act and the regulations of the Commission, and upon finding that

> (4) The applicant is qualified by reason of training and experience to conduct the operations covered by the regulations in this part;
>
> (9) The applicant's personnel training program complies with Subpart I;
>
> (13) There is reasonable assurance that: (i) The activities authorized by the license can be conducted without endangering the health and safety of the public."

Subpart H - Physical Protection
Section 72.180 Physical security plan.

"The licensee shall establish a detailed plan for security measures for physical protection. . . This plan must demonstrate how the applicant plans to comply with the applicable requirements of Part 73 of this chapter and during transportation to and from the proposed ISFSI or MRS and must include the design for physical protection, the licensee's safeguards contingency plan, and the guard training plan. The plan must list tests, inspections, audits, and other means to be used to demonstrate compliance with such requirements."

Section 72.184 Safeguards contingency plan.
(a) "The requirements of the licensee's safeguards contingency plan for responding to threats and radiological sabotage must be as defined in appendix C to part 73 of this chapter. This plan must include Background, Generic Planning Base, Licensee Planning Base, and Responsibility Matrix."
(b) "The licensee shall prepare and maintain safeguards contingency plan procedures in accordance with appendix C to 10 CFR part 73 for effecting the actions and decisions contained in the Responsibility Matrix of the licensee's safeguards contingency plan."

Subpart I - Training and Certification of Personnel
72.190 Operator requirements.
"Operation of equipment and controls that have been identified as important to safety in the Safety Analysis Report and in the license must be limited to trained and certified personnel or be under the direct visual supervision of an individual with training and certification in the operation. Supervisory personnel who personally direct the operation of equipment and controls that are important to safety must also be certified in such operations."

72.192 Operator training and certification program.
"The applicant for a license under this part shall establish a program for training, proficiency testing, and certification of ISFSI or MRS personnel. This program must be submitted to the Commission for approval with the license application."

72.194 Physical requirements.
"The physical condition and the general health of personnel certified for the operation of equipment and controls that are important to safety must not be such as might cause operational errors that could endanger other in-plant personnel or the public health and safety. Any condition that might cause impaired judgment or motor coordination must be considered in the selection of personnel for activities that are important to safety."

A matrix showing the primary relationship of these regulations to the specific areas of review in this chapter is given in Table 10.1. The reviewer should independently verify the relationships in this matrix to ensure that no requirements are overlooked because of unique applicant design features.

10.4 Acceptance Criteria

Acceptance criteria for review of proposed conduct of operations include those pertaining to (a) the completeness of information submitted by the applicant describing the plan of operations, and (b) the adequacy of scope and content of the plan. The following sections provide criteria for each of these categories.

Table 10.1 Relationship of Regulations and Areas of Review

Areas of Review	10 CFR Part 72 Regulations							
	72.24	72.28	72.40	72.180	72.184	72.190	72.192	72.194
Organizational Structure	●	●	●					
Preoperational Testing and Startup Operations	●		●					
Normal Operations	●	●	●					
Personnel Selection, Training, and Certification	●	●	●			●	●	●
Emergency Planning	●		●					
Physical Security and Safeguards Contingency Plans	●		●	●	●			

10.4.1 Organizational Structure

The application must describe the organizational structure and administrative control system that will be used for the proposed ISFSI or MRS (i.e., through construction, preoperational testing and initial operations, normal operations, and decommissioning).

10.4.1.1 Corporate Organization

The application must describe the corporate organization responsible for the ISFSI or MRS installation, including organization charts and position descriptions. If the corporation is made up from two or more corporate identities, the relationship and responsibilities between each entity should be explained. The financial capabilities of the corporation to construct, operate, and decommission the installation must be demonstrated.

The application must describe the corporate functions, responsibilities, and authorities related to each aspect of the installation (design, engineering, construction, quality assurance, testing, etc.). The in-house organization and technical staff (numbers of personnel, qualifications, educational and experience backgrounds, etc.) must be described. The relationship between the applicant's in-house organization and outside contractors and suppliers, including the extent of dependence on those sources for design, construction, quality assurance and other functions, must be described.

The relationship between the corporate and onsite organizations must also be described. The nature of interaction between corporate management and the site related to health and safety, including any role in policy/procedure development, audits, inspections, and investigations, must be explained.

10.4.1.2 Onsite Organization

The application must describe site organization, including organizational charts and position descriptions with emphasis on positions that perform functions important to safety. Such positions include, but are not limited to, those with responsibilities in health physics, nuclear criticality safety, training and certification, emergency planning and response, operations, maintenance, engineering, and quality assurance.

The discussion of positions and responsibility must illustrate how these functions or aspects of these functions, including the degree of separation between the facility operations organization and other parts of the onsite organization that perform functions important to safety, are performed. The application must also identify alternates who are authorized to act in the absence of individuals assigned to key positions and identify which positions have shutdown or stop-work authority for health or safety reasons.

The application must identify minimum staffing levels for major entities within the onsite organization.

The application must identify whether the onsite organization includes a safety committee (or committees). The membership, duties, responsibilities, operating characteristics and reporting function of proposed safety committees must be described.

10.4.1.3 Management and Administrative Controls

The application must describe the proposed management and administrative control system, including provisions for:

- Administrative and general plant procedures
- A program of surveillance, testing, and inspections of items and activities important to safety
- Periodic independent audits
- Change control
- Employee training and certification programs
- Records preparation and maintenance.

Administrative procedures address the process of planning, administrative controls, and document issuance; and provide rules and instructions on personnel conduct, preparation and retention of plant documents, and interfaces among plant organizations. General plant procedures are those that prescribe the actions required to achieve safe operation and provide necessary instruction for the operation and maintenance of plant systems and equipment. The application must describe the program for preparation, review, change, and approval of procedures. The identity of the onsite organizations that use procedures and the activities or operations that are covered by such procedures, must also be specified. (See Section 10.4.3.1 below for guidance on procedures for normal plant operation.)

The applicant must describe the program of surveillance, testing, and inspection to ensure satisfactory in-service performance of items and activities important to safety. The description

must address the development and use of procedures that set forth the steps to be taken and identify the standards or criteria to be applied. The program must include provisions for:

- Pre-operational testing (see Section 10.4.2.1) to demonstrate plant operability and identify conditions adverse to safety

- Operational testing and surveillance to verify and record characteristics of plant equipment and components

- Surveillance, testing, and inspection after modification or when corrective actions have been completed.

The management control system description must also include requirements for planned and scheduled internal and external audits to evaluate the application and effectiveness of management controls, plant procedures, and other activities affecting safety. The audit program must describe audit frequency, methods for documenting and communicating audit findings, resolution of issues, and implementation of corrective actions.

The system for change control, including how change control is integrated into the management control system, must also be described. The coordination of change between potentially affected organizations (engineering, operations, maintenance, training, etc.) must be described. The application must describe how operations are shut down to effect changes and how all plant equipment and procedural changes are completed. The training of staff before resumption of operations must also be addressed.

The management system description must also include the system for maintaining records of facility operation (as addressed in Section 10.4.3.2.)

10.4.2 Preoperational Testing and Startup Operations

The Safety Analysis Report (SAR) must describe the plans for preoperational testing and initial facility (startup) operations. Guidance on the SAR informational content related to preoperational testing and startup operations is provided by Regulatory Guide 3.48, Section 9.2. The following guidance summarizes and supplements that provided by Regulatory Guide 3.48.

Classification of preoperational testing and operating startup are typically delineated by the receipt of radioactive material to be stored. In cases where an ISFSI or MRS is to be located at a site with a pre-existing pool, preoperational testing would occur before any withdrawal of the subject radioactive material from the pool for placement into storage.

The administrative procedures used for conducting the testing and startup must be described. This description must include the system to be used for preparing, approving and executing the test procedures and for evaluating, documenting, and approving test results. Provisions must be made for incorporating changes to the system or individual procedures on the basis of inadequacies in test procedures or unexpected test results. The organizational responsibilities for administering the system must be identified, and the qualifications of involved personnel must be described.

10.4.2.1 Preoperational Testing Plan

The test program description must include an identification of testing objectives and the general methods to be used to meet those objectives. The SAR must identify each item (facility, component, piece of equipment, operation) to be tested. For each physical or operational item, the following information must be provided:

- The type of test to be performed
- The expected response
- The acceptable margin of difference from the expected response
- The method of validation (if applicable)
- Appropriate corrective action for unexpected or unacceptable results.

If the proposed ISFSI or MRS contains any SSCs important to safety the functional adequacy or reliability of which has not been demonstrated by prior use or otherwise validated, the preoperational test plan must include a description and schedule showing how these safety questions will be resolved before the initial receipt of the radioactive materials to be stored.

The applicant must commit to performing a dry run (cold test) of each operation involving the radioactive material to be stored and to use the results of these tests to make necessary changes to equipment and procedures. There must also be a commitment to conduct routine full load tests of any equipment that is to carry spent fuel or high-level waste containers.

10.4.2.2 Operating Startup Plan

The operating startup plan must identify those specific operations involving the initial handling of radioactive material to be placed into storage. Although plant procedures to be used for normal operations or during steady-state conditions are not necessarily included in the operating startup plan, the evaluation of the effectiveness of those procedures are elements of the operating startup plan. For as low as is reasonably achievable (ALARA) considerations, as many of the operating startup actions as feasible must be performed during preoperational testing (i.e., before sources of exposure are present).

The operating startup plan must include the following elements:

- Tests and confirmation of procedures and exposure times involving actual radioactive sources (e.g., radiation monitoring, in-pool operations)

- Direct radiation monitoring of casks and shielding for radiation dose rates, streaming, and surface "hot-spots"

- Verification of effectiveness of heat removal features

- Documentation of results of tests and evaluations.

10.4.3 Normal Operations

Regulatory Guide 3.48, Section 9.4, provides the primary guidance related to SAR information on procedures and recordkeeping in support of normal operations. The guidance in this section summarizes and supplements the guidance provided in Regulatory Guide 3.48.

10.4.3.1 Procedures

The SAR must describe the applicant's commitment to conduct all operations that are important to safety according to written procedures and to have proposed procedures and revisions reviewed and approved by the health, safety, and quality assurance organizations that are independent of the operating management function.

The identification of proposed written procedures must include all routine and projected contingency operations. The applicant must also describe the review, change, and approval practices for all operating, maintenance, and testing procedures. This description may refer to the appropriate management controls addressed in Section 10.4.1.2.

The listing of operations requiring written procedures must include, as applicable to the ISFSI or MRS:

- All operations identified in the proposed technical specifications
- Operating, maintenance, testing, and surveillance functions important to safety.

The procedures listed must clearly indicate, by title or subject, their purpose and applicability. The applicant should identify any standards used for the preparation of these procedures.

10.4.3.2 Records

The management system for maintaining records must be described. This description may refer to the appropriate management controls addressed in Section 10.4.1.2. Although all records need not be maintained centrally, the management system must ensure that cognizance is being maintained of all records, the responsible staff, and locations.

Records stored in electronic media will generally be acceptable if the capability is maintained to produce legible, accurate, and complete records over the required retention period. The record format must include all pertinent information, such as stamps, initials, and signatures. The retention period for each type of record, because it varies depending on applicable regulatory requirements, must be specified. The management system must also provide for adequate safeguards against tampering and loss of records over the retention period.

The SAR must identify, by type, the records to be maintained. Records maintained must include:

- Construction records, as specified in applicable construction codes (e.g., American Concrete Institute [ACI] 349) and including as-built drawings and specifications, material certifications and audit trail to the incorporating SSCs, inspection records, test reports, and certifications (per 10 CFR 72.30(d)(3), 72.156, 72.174)

- As required by 10 CFR 72.30(d)(3), a list contained in a single document and updated no less than every 2 years of the following:

  - All areas designated and formerly designated as restricted areas as defined under 10 CFR 20.1003

  - All areas outside of restricted areas that require documentation under 10 CFR 72.30(d)(1) (see next entry)

- Records of spills or other abnormal occurrences involving the spread of radiation in and around the facility, equipment, or site (per 10 CFR 72.30(d)(1))

- Records of the cost estimate performed for the decommissioning funding plan or of the amount certified for decommissioning and records of the funding method used for ensuring funds, if either funding plan or certifications are used (per 10 CFR 72.30(d)(4)) (i.e., record copy of proposed decommissioning plan filed with license application, attached decommissioning funding plan, any modifications to these plans, and final decommissioning plan when prepared)

- Receipt, inventory, disposal, acquisition, and transfer of all spent fuel and high-level radioactive waste in storage, as required by 10 CFR 72.72(a) (including provisions for duplicate records storage at different locations, per 10 CFR 72.72[d])

- Records of physical inventories and current material control and accounting procedures (per 10 CFR 72.72[b] and [c])

- Operating records, including principal maintenance, alternations, or additions made

- Records of off-normal occurrences and events associated with radioactive releases

- Records of employee certification (per 10 CFR 72.44)

- Quality Assurance (QA) records (per 10 CFR 72.174)

- Environmental survey records and environmental reports

- Radiation monitor readings or records (e.g., stripcharts or electronic results)

- Radiation protection program records (per 10 CFR 20, Subpart L), including those related to:
  - Program contents, audits, and reviews
  - Radiation surveys
  - Determination of prior occupational dose
  - Planned special exposures
  - Individual (worker) monitoring results
  - Dose to individual members of the public
  - Radioactive waste disposal
  - Tests of entry control devices for very high radiation areas

- Records of changes to the physical protection plan (per 10 CFR 72.44(e) and 72.186), and other physical protection records (per 10 CFR 73.21 and 73.70)

- Records of occurrence and severity of natural phenomena (10 CFR 72.92)

- Record copies of:
  - SAR, SAR updates, FSAR (10 CFR 72.70)
  - Reports of accidental criticality or loss of special nuclear material (10 CFR 72.74 and 10 CFR 73.71)
  - Material status reports (10 CFR 72.76)
  - Nuclear material transfer reports (10 CFR 72.78)
  - Physical security plan (per 10 CFR 72.180)
  - "Other" records and reports (per 10 CFR 72.82)
  - Report of preoperational test acceptance criteria and test results
  - Written procedures

The radiation protection records required by 10 CFR 20 Subpart L must incorporate the units of curie, rad, and rem, as applicable, including multiples or subdivisions of those units (e.g., megacurie, millicurie, millirem, etc.). Where dose is part of a record, the dose quantity used on the record (e.g., total effective dose equivalent, committed effective dose equivalent, shallow dose equivalent, etc.) must be clearly indicated.

10.4.4 Personnel Selection, Training, and Certification

The application must describe the organization responsible for personnel selection, training, and certification. The program that will be established and implemented to ensure that personnel whose responsibilities include functions that are important to safety will be appropriately qualified and trained must also be described. The process of selecting and training security guards must be described.

10.4.4.1 Personnel Organization

The description must include a discussion of the organization and management of the training component, and must identify the personnel responsible for development of training programs, conducting training and retraining of employees (including new employee orientations), and maintaining up-to-date records on the status of trained personnel.

10.4.4.2 Selection and Training of Operating Personnel

The applicant must identify the functions that are important to safety and describe the qualifications for personnel performing those functions. These personnel qualifications must include:

- Minimum qualification requirements for operating, technical, and maintenance supervisory personnel
- Qualifications, in resumé form, of persons who will be assigned to managerial and technical positions.

The program description must identify the scope of operational and safety training. Operational training must include topics such as installation design and operations, instrumentation and control, methods of dealing with operating functions, decontamination procedures, and emergency procedures. Radiation safety training must include topics such as nature and sources of radiation, methods of controlling exposure and contamination, radiation monitoring, shielding, dosimetry, biological effects, and criticality hazards control.

The type and level of training to be provided for each job description (personnel classification), including specific training provided to specific job descriptions, must be listed. Alternatively, the basis used to identify the type and level of training by job description may be described.

The requirements for certification of personnel who will operate equipment and controls that are important to safety must be clearly identified. The requirements must address the physical condition and general health of personnel to be certified in accordance with 10 CFR 72.194.

The methods of testing to determine the effectiveness of the training program must be described. Effectiveness must be determined by evaluation against established objectives and criteria, and any standards used for development and implementation of the training program must be identified.

The frequency of retraining, and the nature and duration of retention of training and testing records, must be described. Retraining must be periodic and not less than every 2 years. Training records must be kept up-to-date and retained for a minimum of 3 years.

Implementation of the training program before conduct of operations involving radioactive material (i.e., preoperational training) must be described. The applicant must commit to substantial completion of staff training and certification before receipt of the radioactive material to be stored.

The applicant should identify any standards used for selection, training, and certification of personnel.

10.4.4.3 Selection and Training of Security Guards

The process by which security guards (including watchmen, armed response persons, etc.) will be selected and qualified must be described as required by 10 CFR 73.55(b)(4)(ii). This information may be submitted as part of the applicant's physical security plan, as addressed in Section 10.4.6.

The criteria used must conform to the general criteria for security personnel contained in 10 CFR 73, Appendix B. Regulatory Guide 5.20, "Training, Equipping, and Qualifying of Guards and Watchmen," provides guidance in this area.

10.4.5 Emergency Planning

If the proposed installation is not on the site of an NRC-licensed reactor, the application must include an emergency plan that complies with 10 CFR 72.32(a) (for an ISFSI) or 10 CFR 72.32(b) (for an MRS or an ISFSI that may process or repackage spent fuel). The emergency plan may be incorporated by reference into the SAR. The SAR must include descriptive information on the applicant's plans for coping with emergencies, as required by the applicable paragraph of 10 CFR 72.32.

Because it is possible that structures, systems and components important to safety, including the confinement boundary, could be damaged or fail by a means that has not been considered, the SAR must describe the applicant's ability to detect accident events or damage to SSCs caused by conditions not analyzed in the SAR.

Information provided in the SAR may be limited to the following:

*   A statement that the outline and content of the emergency plan submitted with the license application is in accordance with the requirements of 10 CFR 72.32(a) or (b), as applicable

*   Identification of types of radioactive material accidents provided for by the emergency plan (these must include or encompass all accident level events or conditions addressed in the SAR "Accident Analyses")

*   Identification of offsite response organizations provided opportunity to comment on the plan (under 10 CFR 72.32(a)(14) or (b)(14)) and summary of responses

*   Identification of organizations with whom arrangements have been made for offsite assistance (under 10 CFR 72.32(a)(15) or (b)(15)).

Regulatory Guide 3.67, "Standard Format and Content for Emergency Plans for Fuel Cycle and Materials Facilities," contains the principal guidance on preparation of emergency plans for ISFSI and MRS installations.

10.4.6 Physical Security and Safeguards Contingency Plans

The application must contain a physical security plan as required by Section 72.180 and a safeguards contingency plan as required by Section 72.184.

The security plan must describe how the applicant will comply with the applicable requirements of 10 CFR 73. The plan must provide for physical security of materials during transportation to and from the ISFSI or MRS, as well as during the storage period. The plan must establish a security organization and include:

- Physical protection design features
- Safeguard contingency plan
- Guard training plan
- Tests, inspections, audits, and other means to demonstrate compliance.

If the application is from DOE, the SAR must include: (a) a description of the physical security plan for protection against radiological sabotage (as required by 10 CFR 72 Subpart H), and (b) a certification that it will provide safeguards at the ISFSI or MRS that meet the requirements for comparable surface DOE facilities (required by 10 CFR 72.24[o]).

The safeguards contingency plan must comply with the format and content requirements of 10 CFR 73, Appendix C. An acceptable plan must contain: (a) a predetermined set of decisions and actions to satisfy stated objectives, (b) an identification of the data, criteria, procedures, and mechanisms necessary to efficiently implement the decisions, and (c) a stipulation of the individual, group, or organizational entity responsible for each decision and action.

Regulatory Guide 5.55, "Standard Format and Content of Safeguards Contingency Plans for Fuel Cycle Facilities," provides guidance on safeguards contingency plans that are specifically applicable to ISFSI- and MRS-type facilities.

10.5 Review Procedures

In performing the conduct of operations review, the reviewer should first determine that the respective elements required by Regulatory Guide 3.48 have been submitted. The review guidance provided in the following sections is predicated on the required products for the review and on lessons learned from prior reviews and should be applied in evaluating the submitted documentation.

10.5.1 Organizational Structure

10.5.1.1 Corporate Organization

When evaluating the organizational structure, the reviewer should ensure that the relationship between the corporate organizations and the site organizations is clearly defined. The submitted documentation must enable the reviewer to understand the delineation of authority and responsibilities regarding site activities. The reviewer should ensure that the frequency and scope of any audits or inspections conducted by the corporate organizations are specified.

10.5.1.2 Onsite Organization

The reviewer should have a clear understanding of the distribution of responsibility to specific parts of the site organization and ensure that the site organization and the distribution of responsibilities for functions important to safety is clearly evident. The reviewer should verify that the functions of radiation protection, nuclear criticality safety, and other safety entities are organizationally separate from the entity responsible for facility operations.

The reviewer should determine whether the onsite organization includes a safety committee (or equivalent function) with appropriate representation and responsibilities. In making this determination, the reviewer should consider whether membership includes representatives from operating and safety support organizations. The reviewer should ensure that the safety committee has appropriate review and approval authority, and procedures for systematic review of proposed operations and changes. The reviewer should ensure that the committee reports directly to the Plant Manager or other senior management.

In reviewing staffing commitments, the reviewer should consider the extent of expected operations. In cases where the full spectrum of fuel types to be stored has been identified and evaluated, and no engineering changes are expected, the level of on site technical support (e.g., in areas such as nuclear criticality safety or structural design analysis) can be lower than in cases where more analysis and review will be required.

10.5.1.3 Management and Administrative Controls

The reviewer should review the application to ensure that adequate attention has been paid to a proposed system of management and administrative controls. The reviewer should verify that each of the system elements identified in Section 10.4.1.3 have been addressed. The reviewer should emphasize the proposed system for procedures, including provisions for initial preparation, review, change, and approval.

10.5.2 Preoperational Testing and Startup Operations

The reviewer should review the preoperational testing plan to determine that it includes all of the necessary tests and provides for proper evaluation, approval, and use of the test results. The reviewer should determine that the testing descriptions, responses expected, and contingent corrective actions are appropriate for the purposes. The reviewer should seek the assistance of NRC staff with expertise in the specific topical areas covered by the tests in performing these assessments.

In determining whether the preoperational testing plan is comprehensive, the reviewer should consider the inclusion of the following types of testing and evaluation, as applicable:

*   Tests associated with construction (or reference to submitted construction specifications)

*   Preoperational testing specified in technical specifications

*   Calibration and testing of all instruments and monitors with a safety or security function

*   Tests of supplier-owned equipment to be used in functional operations (e.g., storage confinement cask haul trailer and positioning equipment) and in testing

*   Tests of physical and programmed limits on travel of lifting and transfer equipment (e.g., travel over the pool, lift heights, positioning force)

*   Load tests of rigging, spreaders, and lift points

*   Evaluations of effectiveness of procedures and consideration of potentially improved alternatives.

10.5.3 Normal Operations

10.5.3.1 Procedures

The reviewer should ensure that the applicant commits to conduct all operations that are important to safety according to written procedures. The reviewer should determine if the identified subjects for written procedures include all routine and projected contingency operations and correlate with the narrative and flowsheet descriptions of operations at the ISFSI or MRS.

10.5.3.2 Records

The reviewer should determine whether the records identified for retention include all those required by regulations (refer to listing in Section 10.4.3.2).

10.5.4 Personnel Selection, Training, and Certification

The reviewer should review proposed training for inclusion of regulatory requirements relating to personnel selection, training, certification, exercises, and training records. Determine acceptability based on satisfaction of regulatory requirements, Regulatory Guide 3.48 guidance, and evidence of experience in planning and conduct of training programs.

The reviewer should review the minimum qualifications for operating, technical, maintenance, and supervisory personnel. The reviewer will have to exercise judgment, because there are no standard minimum qualifications. Compare proposed requirements with those of other approved license applications. Generally, the minimum qualifications for these personnel include a bachelor's degree and several years experience in a related technical area that is commensurate with the level of assigned responsibility. Higher-level managers typically have the same experience requirements plus previous supervisory or management experience.

The reviewer should ensure that the application adequately addresses the implementation of the training program before initiation of operations with spent fuels or solidified high-level waste, including a commitment to complete most of staff training and certification before receipt of the radioactive material to be stored.

The reviewer should review the selection and qualification process for security personnel. The reviewer should determine that the process will ensure that security personnel will meet the requirements of 10 CFR 73.55 and will be qualified to perform each assigned security job duty in accordance with 10 CFR 73, Appendix B.

Additional guidance on training criteria and training program content can be found in:

- ANSI/ANS 8.20-1991, "Nuclear Criticality Safety Training"

- ANSI/ANS 3.1-1993, "Selection, Qualification, and Training of Personnel for Nuclear Power Plants"

- ASTM E 1168, "Guide for Radiation Protection Training for Nuclear Facility Workers"

- Regulatory Guide 1.8, "Qualification and Training of Personnel for Nuclear Power Plants"

- Regulatory Guide 1.134, "Medical Evaluation of Licensed Personnel for Nuclear Power Plants"

- Regulatory Guide 8.27, "Radiation Protection Training for Personnel at Light-Water-Cooled Nuclear Power Plants"

- Regulatory Guide 8.29, "Instruction Concerning Risks from Occupational Radiation Exposure"

- NUREG-0800, "Standard Review Plan," Section 13.2.2.

10.5.5 Emergency Planning

The reviewer should review the description of emergency planning presented in the SAR. The reviewer should ensure that the description correlates with the emergency plan submitted with the license application in accordance with 10 CFR 72.32. The reviewer should consult with the staff reviewer responsible for the emergency plan and jointly determine the acceptability of the plan. The reviewer should not consider acceptable a description of emergency planning presented in SAR if the emergency plan lacks or is determined to be deficient in any of the required elements.

The reviewer should ensure the SAR describes the licensee's ability to detect accident events caused by the assumed failure of structures, systems and components important to safety, including the confinement boundary.

10.5.6 Physical Security and Safeguards Contingency Plans

The reviewer should review the security plan against the applicable requirements of 10 CFR 73 and ensure that the plan adequately provides for each of the required elements. If the application is from DOE, the reviewer should verify that it includes a description of the physical security plan for protection against radiological sabotage (as required by 10 CFR 72 Subpart H) and a certification that it will provide safeguards at the ISFSI or MRS that meet the requirements for comparable surface DOE facilities.

The reviewer should ensure the safeguards contingency plan complies with the format and content requirements of 10 CFR 73, Appendix C, including (a) a predetermined set of decisions and actions to satisfy stated objectives, (b) an identification of the data, criteria, procedures, and mechanisms necessary to efficiently implement the decisions, and (c) a stipulation of the individual, group, or organizational entity responsible for each decision and action. The reviewer should consult Regulatory Guide 5.55 for guidance on acceptable contents and format of safeguards contingency plans applicable to ISFSI or MRS installations. However, the applicant is not required to submit the written procedures that will implement the safeguards contingency plan (although the procedures will be subject to NRC inspection on a periodic basis).

10.6 Evaluation Findings

NRC staff reviewers prepare evaluation findings regarding satisfaction of the regulatory requirements related to conduct of operations, as identified in Section 10.3. If the documentation submitted with the application fully supports positive findings for each of the regulatory requirements, the findings should substantially be stated as follows (finding numbering is for convenient referencing within the Standard Review Plan):

F10.1   The SAR includes an acceptable plan for the conduct of operations, in compliance with 10 CFR 72.24(h).

F10.2   [If appropriate] The design of the [ISFSI/MRS] includes _____ [specify the SSCs], the functional adequacy or reliability of which has not been demonstrated by prior use for the same purpose. The SAR describes acceptable planned tests and demonstration of capability in the areas of uncertainty before use, in compliance with 10 CFR 72.24(i).

F10.3   The SAR includes an acceptable description of the program covering preoperational testing and initial operations, in compliance with 10 CFR 72.24(p).

F10.4   The applicant has provided acceptable technical qualifications, including training and experience, for personnel who will be engaged in the proposed activities, in compliance with 10 CFR 72.28(a).

F10.5   The application includes an acceptable description of a personnel training program to comply with 10 CFR 72, Subpart I.

F10.6   The application includes an adequate, acceptable description of the applicant's operating organization, delegations of responsibility and authority, and the minimum skills and experience qualifications relevant to the various levels of responsibility and authority, in compliance with 10 CFR 72.28(c).

F10.7   The application includes an acceptable commitment by the applicant to have and maintain an adequate complement of trained and certified installation personnel before receipt of spent fuel or high-level radioactive waste for storage, in compliance with 10 CFR 72.28(d).

F10.8   The application provides acceptable assurance that the applicant is qualified by reason of training and experience to conduct the operations covered by the regulations in 10 CFR 72, in compliance with 10 CFR 72.40(a)(4).

F10.9   The application is considered to provide acceptable assurance with regard to the management, organization, and planning for preoperational testing and initial operations that the activities authorized by the license can be conducted without endangering the health and safety of the public, in compliance with 10 CFR 72.40(a)(13).

10.7 References

NRC documents referenced are identified at Consolidated References, Section 17.

American National Standards Institute, "Nuclear Criticality Safety Training," ANSI/ANS 8.20-1991.

American National Standards Institute, "Selection, Qualification, and Training of Personnel for Nuclear Power Plants,"ANSI/ANS 3.1-1993.

American Society for Testing and Materials, "Guide for Radiation Protection Training for Nuclear Facility Workers," ASTM E 1168 - 1995.

11 RADIATION PROTECTION EVALUATION

11.1 Review Objective

This chapter describes the requirements and considerations associated with the radiation protection evaluation of the proposed ISFSI or MRS. As used here, radiation protection refers to organizational, design, and operational elements that are primarily intended to limit radiation exposures associated with normal operations and anticipated occurrences. The evaluation of radiological consequences of accidents is addressed in Chapter 15 (Accident Evaluation).

The primary objectives of the radiation protection evaluation are to determine whether the design features and proposed operations provide sufficient assurance that:

- Radiation exposures and radionuclide releases will be maintained at levels that are as low as reasonably achievable (ALARA).

- Occupational radiation doses will not exceed the limits specified in NRC's radiation protection standards.

- Radiation doses to the public during normal conditions and anticipated occurrences will meet regulatory standards.

The interrelationship of the radiation protection review with other areas of review is shown in Figure 11.1. The figure shows that radiation protection evaluation draws upon general information in the application as well as information reviewed or developed in site evaluation, shielding evaluation, confinement evaluation, and operation systems evaluation.

11.2 Areas of Review

The following outline shows the areas of review addressed in Section 11.4, Acceptance Criteria, and Section 11.5, Review Procedures:

As Low as Reasonably Achievable Considerations
    ALARA Policies and Programs
    Design Considerations
    Operational Considerations

Radiation Protection Design Features
    Installation Design Features
    Access Control
    Radiation Shielding
    Confinement and Ventilation
    Area Radiation and Airborne Radioactivity Monitoring Instrumentation

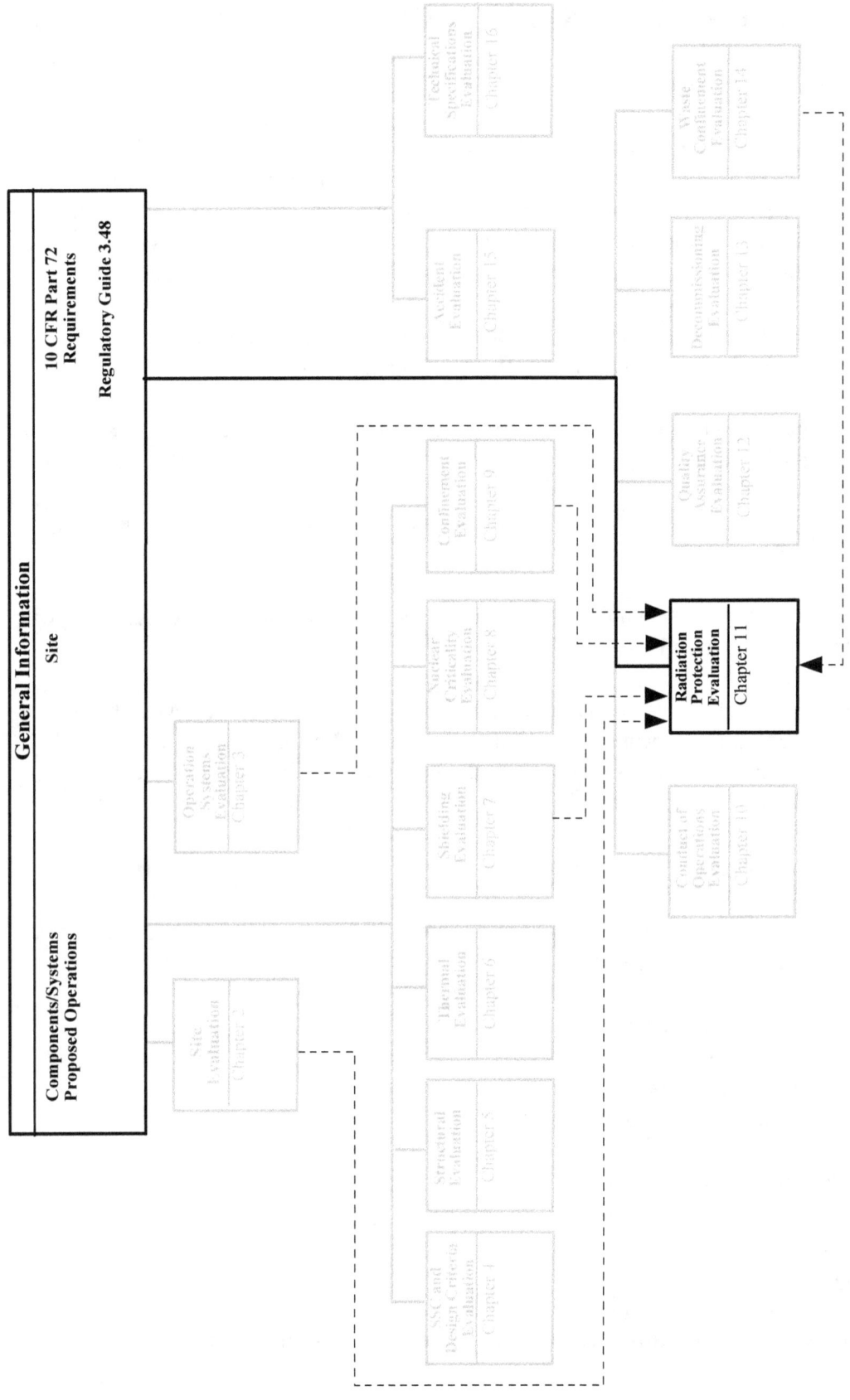

Figure 11.1 Overview of Radiation Protection Evaluation

Dose Assessment
　　Onsite Dose
　　Offsite Dose

Health Physics Program
　　Organization
　　Equipment, Instrumentation, and Facilities
　　Policies and Procedures

11.3　Regulatory Requirements

This section identifies and presents a high-level summary of Title 10 of the Code of Federal
Regulations (CFR) Part 72 relevant to the review areas addressed by this chapter. The NRC staff
reviewer should read the exact regulatory language. Virtually the entire contents of 10 CFR 20,
Standards for Protection Against Radiation, are also applicable to this review. A matrix at the
end of this section matches the regulatory requirements identified in this section to the areas of
review identified in the previous section.

20.1101　Radiation protection programs.
(a) "Each licensee shall develop, document, and implement a radiation protection program."
(b) "The licensee shall use . . . controls based upon sound radiation protection principles to
achieve . . . as low as reasonably achievable (ALARA)."
(c) "The licensee shall periodically (at least annually) review the radiation protection program."
(d) "A constraint on air emissions of radioactive material to the environment . . . shall be
established by licensees. "

20.1201　Occupational dose limits for adults.
(a) "The licensee shall control the occupational dose to individual adults, except for planned
special exposures under Section 20.1206, to the following dose limits:
　　(1) An annual limit, which is the more limiting of
　　　　(i) The total effective dose equivalent being equal to 5 rem (0.05 Sv), or
　　　　(ii) The sum of the deep-dose equivalent and the committed dose equivalent to
　　　　any individual organ or tissue other than the lens of the eye being equal to
　　　　50 rem (0.5 Sv).
　　(2) The annual limits to the lens of the eye, to the skin, and to the extremities, which are:
　　　　(i) An eye dose equivalent of 15 rem (0.15 Sv), and
　　　　(ii) A shallow-dose equivalent of 50 rem (0.50 Sv) to the skin or to any extremity."

20.1301　Dose limits for individual members of the public.
(a) "Each licensee shall conduct operations so that
　　(1) The total effective dose equivalent to individual members of the public from the
　　licensed operation does not exceed 0.1 rem (1 mSv) in a year, exclusive of the dose
　　contributions from background radiation, any medical administration the individual has
　　received, voluntary participation in medical research programs, and the licensee's disposal
　　of radioactive material into sanitary sewerage in accordance with Section 20.2003.

(2) The dose in any unrestricted area from external sources does not exceed 0.002 rem (0.02 mSv) in any one hour."

(b) "If the licensee permits members of the public to have access to controlled areas, the limits for members of the public continue to apply to those individuals."

(d) "In addition to the requirements of this part, a licensee subject to the provisions of EPA's generally applicable environmental radiation standards in 40 CFR Part 190 shall comply with those standards."

20.1302 Compliance with dose limits for individual members of the public.
(a) "The licensee shall make...surveys of radiation levels in unrestricted and controlled areas and radioactive materials in effluents released to unrestricted and controlled areas to demonstrate compliance with the dose limits for individual members of the public in Section 20.1301."

20.1406 Minimization of contamination.
"Applicants . . . shall describe...how facility design and procedures for operation will minimize . . . contamination . . . facilitate decommissioning, and minimize . . . generation of radioactive waste."

20.1501 General.
(a) "Each licensee shall make or cause to be made, surveys that
      (1) May be necessary for the licensee to comply with the regulations in this part."

20.1701 Use of process or other engineering controls.
"The licensee shall use controls (e.g., containment or ventilation) to control the concentrations of radioactive material in air."

20.1702 Use of other controls.
"When it is not practicable to apply process or other engineering controls . . . the licensee shall . . . increase monitoring and limit intakes by one or more of the following means:
(a) Control of access;
(b) Limitation of exposure times;
(c) Use of respiratory protection equipment; or
(d) Other controls."

72.24 Contents of application: Technical information [Contents of the SAR].
(e) "The means for controlling and limiting occupational radiation exposures.
      (l) A description of the equipment . . . to maintain control over radioactive materials in gaseous and liquid effluent. The description must include:
            (i) An estimate of the quantity of each of the principal radionuclides expected to be released annually."

72.104 Criteria for radioactive materials in effluents and direct radiation from an ISFSI or MRS.
(a) "During normal operations and anticipated occurrences, the annual dose equivalent to any real individual who is located beyond the controlled area must not exceed 25 mrem to the whole body, 75 mrem to the thyroid and 25 mrem to any other organ as a result of exposure to:
      (1) Planned discharges of radioactive materials . . . to the general environment,

(2) Direct radiation from ISFSI or MRS operations, and
(3) Any other radiation from uranium fuel cycle operations within the region."

(b) "Operational restrictions must be established to meet as low as reasonably achievable objectives for radioactive materials in effluents and direct radiation levels associated with ISFSI or MRS operations."
(c) "Operational limits must be established for radioactive materials in effluents and direct radiation levels associated with ISFSI or MRS operations to meet the limits given in paragraph (a) of this section."

72.106 Controlled area of an ISFSI or MRS.
(b) "Any individual located on or beyond the nearest boundary of the controlled area shall not receive a dose greater than 5 rem to the whole body or any organ from any design basis accident. The minimum distance from the spent fuel or high-level radioactive waste handling and storage facilities to the nearest boundary of the controlled area shall be at least 100 meters."

72.126 Criteria for radiological protection.
(a) "Exposure control. Radiation protection systems must be provided for all areas and operations where onsite personnel may be exposed to radiation or airborne radioactive materials. Structures, systems, and components for which operation, maintenance, and required inspections may involve occupational exposure must be designed, fabricated, located, shielded, controlled, and tested so as to control external and internal radiation exposures to personnel. The design must include means to:
>   (3) Control access to areas of potential contamination or high radiation within the ISFSI or MRS;
>   (4) Measure and control contamination of areas requiring access;
>   (5) Minimize the time required to perform work in the vicinity of radioactive components
>   (6) Shield personnel from radiation exposure."
(c) "Effluent and direct radiation monitoring.
>   (2) Areas containing radioactive materials must be provided with systems for measuring the direct radiation levels in and around these areas."
(d) "Effluent control. The ISFSI or MRS must be designed to provide means to limit to levels as low as reasonably achievable the releases of radioactive materials in effluents during normal operations. Analyses must be made to show that releases to the general environment during normal operations and anticipated occurrences will be within the exposure limit given in Section 72.104."

40 CFR 190.10 Standards for normal operations.
"Operations covered by this subpart shall be conducted in such a manner as to provide reasonable assurance that:
(a) The annual dose equivalent does not exceed 25 millirem to the whole body, 75 millirem to the thyroid, and 25 millirem to any other organ of any member of the public as the result of exposures to planned discharges of radioactive materials, radon and its daughters excepted, to the general environment from uranium fuel cycle operations and to radiation from these operations."

40 CFR 191 (Note: applicable requirements are essentially identical to those of 40 CFR 190.10)

A matrix that shows the primary relationship of these regulations to the specific areas of review associated with this FSRP chapter is given in Table 11.1. The NRC staff reviewer should verify the matching of regulatory requirements to the areas of review presented in the matrix to ensure that no requirements are overlooked as a result of unique applicant design features.

Table 11.1 Relationship of Regulations and Areas of Review

Areas of Review	10 CFR Part 20 Regulations							
	20.110 1	20.120 1	20.130 1	20.130 2	20.140 6	20.150 1	20.170 1	20.170 2
As Low As Reasonably Achievable	●				●	●		●
Radiation Protection Design Features	●				●		●	●
Dose Assessment	●	●	●	●				
Health Physics Program	●			●	●	●		●

Table 11.1 Relationship of Regulations and Areas of Review (continued)

Areas of Review	10 CFR Part 72 Regulations			
	72.24	72.104	72.106	72.126
As Low As Reasonably Achievable	●	●		●
Radiation Protection Design Features	●	●		●
Dose Assessment		●	●	
Health Physics Program	●			●

11.4 Acceptance Criteria

This section describes the acceptance criteria used for review of radiation protection features and programs. These criteria are organized according to the areas specified in Section 11.2. The reviewer should note that some overlap exists between acceptance criteria for radiation protection and those related to shielding (Chapter 7), confinement (Chapter 9), and site-generated waste (Chapter 14).

11.4.1 As Low As Reasonably Achievable (ALARA) Considerations

This element should include a description of the proposed program for maintaining exposures to workers and the public ALARA. An ALARA policy statement and the manner in which equipment design and layout and operational features contribute to the ALARA objective should be provided in the license application.

11.4.1.1 ALARA Policy and Program

As a minimum, the policy, program, and activities for ensuring that radiation exposures will be ALARA should include the elements described below. Acceptable guidance on the development of an ALARA program is provided in Regulatory Guide 8.10, Revision 1R, "Operating Philosophy for Maintaining Occupational Radiation Exposures As Low As Reasonably Achievable."

Policy statement

The application should include a written policy that states management's commitment to maintain exposures to workers and the public ALARA. The policy should include provisions that:

- No practice involving radiation exposure will be undertaken unless its use produces a net benefit.

- All exposures will be kept ALARA with technological, economic, and social factors considered.

- Exposures to individuals will not exceed the limits recommended for the appropriate circumstances.

- Supervisors will integrate appropriate radiation protection controls into all work activities.

- Individuals will be appropriately instructed in the ALARA program.

- There will be strict compliance with all regulatory requirements regarding procedures, radiation exposures, and releases of radioactive materials.

- A comprehensive program will be maintained to ensure that both individual and collective doses are ALARA.

ALARA program organization

This element should include the organizational structure of the ALARA program and the responsibilities and activities of ALARA personnel.

ALARA program elements

Information should be provided to document how implementation of the program will ensure that ALARA objectives are achieved. ALARA program elements should include the use of:

- Procedures and engineering controls to minimize dose (detailed in Section 11.4.1.2).

- Tracking of individual doses to identify trends and causes and use of these data in the development of alternative procedures that can result in lower doses (detailed in Section 11.4.1.3).

- Periodic training and exercises for management, radiation workers, and other site workers in radiation protection, operating procedures, and emergency response (see Section 11.4.1.3).

11.4.1.2 Design Considerations

The applicant's discussion of facility design and layout should demonstrate consideration of ALARA principles. The design criteria (presented in Section 3 of the SAR) should include ALARA criteria, and the documentation should identify choices between otherwise comparable alternatives affected by ALARA considerations. Regulatory Guide 8.8, "Information Relevant to Ensuring that Occupational Radiation Exposure at Nuclear Power Stations Will Be As Low As Is Reasonably Achievable," should be used for ALARA design guidance, although specific alternative approaches may be used if clearly indicated. Examples of ALARA design considerations include the following:

- Engineered design features that minimize the amount of time that maintenance, health physics, or inspection personnel must stay in restricted areas

- Provisions for use of remotely operated or robotic equipment such as welders, wrenches, radiation monitors, etc.

- Use of closed-circuit television to monitor for possible blockage of air cooling passages to perform inspections, etc.

- Provisions for remote placement and use of temporary shielding.

- Incorporation of materials and design features that minimize the potential for accumulation of radioactive materials or surface contamination and that facilitate decontamination and decommissioning.

- Incorporation of proven ALARA design alternatives used at other ISFSIs, pool facilities, or waste management facilities.

- Placement of occupiable areas (e.g., office, security, or laboratory facilities) away from radiation sources.

- Provisions for ALARA and health physics training facilities and equipment.

11.4.1.3 Operational Considerations

The description of proposed operations should reflect incorporation of ALARA principles in operational procedures. Detailed plans and procedures should be developed in accordance with Regulatory Guides 1.33, "Quality Assurance Program Requirements," 8.8, and 8.10, and should consider, to the extent practical:

*   Tradeoffs between requirements for increased monitoring or maintenance activities (and the increased exposures that would result) and the potential hazards associated with reduced frequency of these activities.

*   Performance of cask preparation efforts (for loading) away from the pool or dry transfer facility.

*   Sequencing the placement of spent fuel in a manner that maximizes shielding by storage casks or structures.

*   Dry runs to develop proficiency in procedures involving radiation exposures, to determine exposures likely to be associated with specific procedures, and to consider alternative procedures to minimize exposures.

*   Inclusion of tested contingency procedures for potential off-normal occurrences.

*   Consideration of ALARA operational alternatives based on experience with other ISFSIs, pool facilities, or waste management facilities.

*   Operations research on procedures, types of tools and instruments, and personal protective equipment to minimize exposure times or the effects of exposure.

11.4.2 Radiation Protection Design Features

This section identifies elements that indicate whether adequate attention has been paid to radiation protection in the design of the ISFSI or MRS. There is considerable overlap between radiation protection and ALARA criteria in this regard. Applicable guidance on criteria and SAR contents for ISFSI or MRS radiation protection design features is provided in Regulatory Guide 3.48, "Standard Format and Content for the Safety Analysis Report for an Independent Spent Fuel Storage Installation (Dry Storage)" (Section 7.1.2), and NUREG-0800, Sections 12.3 - 12.4, "Radiation Protection Design Features."

11.4.2.1 Installation Design Features

Installation design features for radiation protection are listed separately below as they apply to minimizing offsite and onsite exposures.

Features that specifically minimize offsite exposures include:

- Siting considerations -- Site location away from population centers to the extent feasible, consistent with other factors.

- Controlled area/perimeter distance -- Site the ISFSI or MRS-controlled area to maintain distance to the perimeter of the site and locations of public occupancy.

- Transfer route -- Locate transfer routes for ISFSI or MRS containers to maintain distance from the site perimeter.

- Effluent discharges and impacts -- Incorporate consideration of natural and manmade contours, existing or planned rerouting of natural surface water, and points at which surface water exits the site relative to residences and public use areas; use cutoffs, drains, well points, or other means to control water flow.

Features that minimize onsite exposures include:

- Transfer route -- Locate transfer routes for ISFSI or MRS containers to or from the storage area and the handling areas (intermodal transfer points, or wet or dry transfer facility) to minimize the route between the handling and storage facilities, provide for minimal other traffic on the route, remain within the single controlled area, and maintain distance from the site perimeter.

- Multiple restricted areas -- Incorporate use of multiple restricted areas within the controlled area to provide control of access to areas with radiation levels that would pose unacceptable risks or exposures to workers within those areas.

- Controlled area/perimeter distance -- Provide separation of radioactive material handling and storage functions from other functions on the site; provide distance between radioactive material and both the boundary of the controlled area and onsite work stations outside the restricted area.

11.4.2.2 Access Control

Control of access to controlled and restricted areas is performed for reasons related to radiation protection and for safeguards and security purposes. This section addresses control of access for purposes of limiting exposure to external radiation and radiological contamination hazards.

The description of the ISFSI or MRS installation design should include (with consideration of the "information to be protected" provisions of 10 CFR 73.21(b)) the following access control elements:

- Site layout to scale showing ISFSI or MRS controlled area (per 10 CFR 72.106) and any traversing right(s)-of-way.

- Description of barrier used to preclude ready access to the controlled area.

- Location and summary description of gate and/or overlook stations.

The criteria used to designate restricted areas (or zones within restricted areas) should be identified. Descriptions should be provided for all protective features designed to limit access to restricted areas, including physical barriers, locked entryways, and audible or visible alarm signals. Continuous direct or electronic surveillance used to prevent unauthorized entry should also be described.

Restricted areas may require further designation as high or very high radiation areas according to 10 CFR 20.1601 and 1602, respectively. Guidance on access control features applicable to these areas is provided in Regulatory Guide 8.39, "Control of Access to High and Very High Radiation Areas of Nuclear Plants."

Restricted areas may be further divided to identify areas where the potential for contamination exists. Criteria used to designate contamination control areas (including airborne radioactivity areas) should also be identified. Access control features applicable to contamination control areas may include:

- Incorporation of access control facilities into the building design or provisions for temporary or mobile-type facility(ies) immediately adjacent to the confinement barrier of the potentially contaminated area.

- Male and female change rooms, including lavatories and showers; provisions for protective equipment and garments; stations for monitoring hands, feet, and whole body; and threshold stations for removal of booties on leaving the area

- Shower and lavatory water collection, storage, and provisions for routing of potentially contaminated water.

Drawings should document that appropriate measures are provided for collection of possibly contaminated wash water and that leakage of possibly contaminated liquid onto or into the ground is precluded. Wash water may include liquids temporarily stored pending sampling and sample analysis before release to the sanitary sewer (in accordance with 10 CFR 20.2003) or collection for handling and treatment as radioactive waste.

11.4.2.3 Radiation Shielding

Provisions for effective shielding must be incorporated into the ISFSI or MRS design as an integral part of the ALARA and radiation protection programs to protect the public and workers against direct radiation. Guidance for conducting detailed engineering evaluations aimed at determining the performance and effectiveness of the proposed shield design is beyond the scope of this section but is provided in Chapter 7, Shielding Evaluation. However, criteria that can be specifically evaluated to determine whether the proposed shielding and installation designs satisfy dose rate and ALARA requirements are addressed in Section 11.4.3. The radiation protection review also uses the dose rates from the shielding review in combination with radionuclide emission rate estimates (from confinement and site-generated waste reviews) to ensure that the combined dose rates (i.e., from all sources and pathways) meet the acceptance criteria, as described in Section 11.4.3.

11.4.2.4 Confinement and Ventilation

Confinement refers to the ability of the ISFSI or MRS to prevent the release of radioactive materials from areas where these materials are normally contained to areas where they are not normally contained, and ultimately, to the surrounding environment. Confinement barrier systems may be sealed, as in the case of most ISFSIs, or vented with off-gas treatment systems, as in the case with storage pools or waste management systems. For the latter, intake and exhaust filters and dampers, as well as portions of ducts and stacks of the ventilation systems function as elements of the confinement system. Confinement and ventilation function to provide protection of personnel against radiation exposures associated with releases of radioactive materials under normal conditions, anticipated occurrences, and accidents.

Detailed evaluation of confinement is beyond the scope of this section, but is provided in Chapter 9, Confinement Evaluation, of this FSRP. The radiation protection review uses the emission rate estimates of the confinement analysis with shielding and site-generated waste review findings to determine whether the combined dose rates meet regulatory criteria, as described in FSRP Section 11.4.3.

11.4.2.5 Area Radiation and Airborne Radioactivity Monitoring Instrumentation

The locations, types, capabilities, and parameters of fixed area radiation monitors and continuous airborne monitoring instrumentation (as required by Regulatory Guide 3.48, Section 7.3.4) must be detailed in the drawings and specifications defining the ISFSI or MRS design. The NRC accepts for an ISFSI or MRS the criteria for fixed area radiation monitors and continuous airborne monitoring instrumentation as provided in:

* ANSI N13.1-1993, "Guide to Sampling Airborne Radioactive Materials in Nuclear Facilities," as it relates to principles for obtaining valid samples of airborne radioactive materials, and acceptable methods and materials for gas and particle sampling.

* ANSI/ANS-HPSSC-6.8.1-1981, "Location and Design Criteria for Area Radiation Monitoring Systems for Light Water Reactors," as it relates to the criteria for locating fixed continuous area gamma radiation monitors and for design features and ranges of measurement.

* NUREG-0800, "Standard Review Plan," Section 11.5, "Process and Effluent Radiological Monitoring Instrumentation and Sampling Systems."

* Regulatory Guides 8.5, "Criticality and Other Interior Evacuation Signals," as it relates to use of interior evacuation alarm signals, and 8.25, "Air Sampling in the Workplace," as it relates to use of fixed and portable air samplers in the workplace.

Classification of auxiliary power for monitoring instrumentation as "emergency" (important to safety) or "standby" (not important to safety) should correspond to the classification of the instrumentation itself. Some discriminators for classifying instrumentation and auxiliary power as important to safety are shown below:

• If data provided by the monitoring system can have an immediate and determining effect on personal actions and operations to maintain compliance with the basic safety criteria, including prevention of unacceptable worker doses, then monitoring instrumentation should be classified as emergency.

• If any of the following exist, then the instrumentation and its auxiliary power are probably not important to safety:

   - Instrumentation data are not provided real-time to a central control room or, if provided, do not trigger an alarm that results in actions that should preclude or mitigate unacceptable consequences.

   - Instrumentation does not trigger an alarm necessary to avoid unacceptable worker exposures at its location when a threshold is reached.

   - Data are collected only periodically.

   - No normal, off-normal, or accident-level events or conditions can result in changes in the monitored phenomena that can jeopardize satisfaction of the basic safety criteria.

11.4.3 Dose Assessment

From the evaluations described in the shielding evaluation (Chapter 7), estimated dose rates should be provided for representative points within the restricted areas as well as on and beyond the perimeter of the controlled area. Additionally, the confinement evaluation (Chapter 9) and site-generated waste management evaluation (Chapter 14) should have produced estimates of radionuclide concentrations in effluents. The radiation protection review includes a dose assessment that incorporates findings of each of those reviews, as applicable. The major elements of the dose assessment and the applicable acceptance criteria are described below.

11.4.3.1 Onsite Doses

Individual and collective dose rates should be calculated for all onsite areas at which workers will be exposed to elevated dose rates (e.g., greater than 2 mrem/hr) or airborne radioactivity concentrations. Radiation doses should be based on direct exposure and radionuclide inhalation and should be computed for workers performing specific ISFSI or MRS functions, including routine, contingency, maintenance, or repair procedures, or other activities that can occur in elevated dose-rate areas. Individual and collective doses should also be determined for onsite functions outside the ISFSI or MRS restricted area associated with transportation and intermodal transfer of the radioactive materials to be stored.

The SAR should include estimated occupancy time for personnel involved in these functions, including the maximum expected total hours per year for any individual and total person-hours per year for all personnel. The annual collective doses associated with each major function and each radiation area should be estimated, and the bases, models, and assumptions used in arriving at these values should be identified.

If the material to be put into dry storage is in holding or storage in a pool, the doses associated with preparation for the loading operations should be based on assumptions of maximum radioactive material in the pool and in the storage location (unless the SAR analysis acceptably demonstrates that alternative conditions are justified).

All individual doses to workers should be well below the dose limits specified in 10 CFR 20.1201. Collective doses should be consistent with the objectives contained in the applicant's ALARA program. The information provided by the applicant must allow for the determination of compliance with these criteria. In general, the following information will allow for such a determination:

- Collective and individual doses associated with all operations involved with placing one full storage confinement cask in storage position are identified and listed according to associated function.

- Annual collective and individual doses are estimated by multiplying the single-cask dose by the maximum annual rate for placing casks into storage. This estimation assumes that the same personnel will be involved in the same operations for each cask. If the doses exceed those allowed by 10 CFR 20.1201(a), the planned conduct of operations (Chapter 10) should commit to conditions (staffing plan, monitoring, etc.) that would ensure that 10 CFR 20.1201(a) dose limits are not exceeded.

- Estimates of annual doses for operation of the ISFSI for material in storage and material in wet holding or wet storage should be provided for comparison with maximum allowable doses given in 10 CFR 20.1201.

- The SAR should include discussion of sensitivity of the doses to assumptions and uncertainties, including the use of conservative assumptions.

11.4.3.2 Offsite Doses

The dose rate beyond the controlled area boundary should not exceed 2 mrem in any one hour from all licensed activities at the site (10 CFR 20.1301). The maximum projected annual dose at or beyond the controlled area boundary should not exceed 25 mrem to the whole body, 75 mrem to the thyroid, or 25 mrem to any other organ (10 CFR 72.104). Demonstration of compliance with 10 CFR 72.104 dose limits is an indication that the dose limits of U.S. Environmental Protection Agency regulations 40 CFR 190 and 191 also will be met.

The collective dose should be determined as the sum of the products of individual doses in each of 16 compass sectors around the installation and the number of population members in each sector. Sectors should be centered between the arcs having radii of 1.5, 3, 5, 6.5, and 8 km (about 1, 2, 3, 4, and 5 miles). The dose should be based on all important exposure pathways (direct radiation, airborne releases, etc.) and modes of exposure (external exposure, inhalation, etc.) and should be specified as whole-body or effective. In addition, the organ receiving the highest dose should be identified. Dose calculations must consider direct radiation and discharges of radioactive material under both normal conditions and anticipated occurrences as well as contributions from other uranium fuel-cycle facilities within the region. The methodology applied must be acceptable to the NRC.

The following considerations also apply:

- The direct radiation dose rate should be calculated on the basis of the maximum quantity of radioactive material permitted by the ISFSI or MRS license.

- The dose assessment should assume that the radioactive material is distributed in such a manner as to produce the highest perimeter dose rate, unless such arrangements are specifically precluded by proposed operational considerations.

- The effective dose to any member of the public resulting from airborne emissions of radioactive material should conform to the ALARA constraint level of 10 mrem/yr specified by 10 CFR 20.1101(d).

- The applicant's environmental monitoring program should be designed to provide exposure and concentration data for those pathways that lead to the highest potential external and internal doses.

11.4.4 Health Physics Program

11.4.4.1 Organization

Guidance on the organization and planning for health physics (radiation protection) activities at an ISFSI or MRS is contained in Regulatory Guides 8.8, 8.10, and others. The ISFSI or MRS management organization should identify an individual with clearly designated responsibility for health physics. To avoid the potential for conflict of interest, this individual's reporting line should not go through the manager responsible for operations. The position should be maintained for the life of the facility, including all decontamination and decommissioning (D&D) operations.

11.4.4.2 Equipment, Instrumentation, and Facilities

Health physics program equipment, instrumentation, and facilities should be described in the SAR. The need for specific health physics components depends on the nature of the installation, for example, whether it includes a pool facility or whether some laboratory functions are performed at offsite facilities. In any case, portable and laboratory equipment and instrumentation should include:

- Personal monitoring devices for external dosimetry, including provisions for dosimeter processing by a National Voluntary Laboratory Accreditation Program-accredited dosimetry service.

- Handheld/portable radiation meters and detectors for performing radiation and contamination surveys; an appropriate number of instruments should be available for each type of survey to be performed (e.g., Geiger-Mueller (G-M) survey instruments for contamination surveys and personnel "frisking," ionization chambers for exposure rate surveys, neutron detectors for neutron flux or dose rate surveys, etc.).

- Portable air sampling equipment.

- Facilities for internal radiation monitoring, including whole-body counters, thyroid counters, bioassay sample analysis equipment, etc.

- Personnel protective equipment (including respirators certified by National Institute for Occupational Safety and Health/Mine Safety and Health Administration).

- Decontamination equipment and facilities, including spill control materials, shower, eyewash, changing facilities, etc.

Health physics facilities can be in permanent structures, temporary buildings, or trailers. Facilities should be located outside restricted areas and, if practicable, away from areas with elevated dose rates. Exceptions can include facilities for storing items that need to be readily available within restricted or elevated dose rate areas, as well as personnel decontamination, shower, and changing facilities. Health physics facilities should be identified on the plot drawing of the installation and should be described to an extent that acceptably demonstrates the applicant's understanding of the associated requirements and functions.

The following regulatory guides and industry standards provide information, recommendations, and guidance on various aspects of health physics equipment, instrumentation, and facilities. The NRC considers these sources as acceptable in describing a basis for implementing activities to comply with applicable regulatory requirements:

- ANSI/ANS N13.1, "Guide to Sampling Airborne Radioactive Materials in Nuclear Facilities."

- Regulatory Guides 8.4, "Direct-Reading and Indirect-Reading Pocket Dosimeters"; 8.6, "Standard Test Procedures for Geiger-Mueller Counters"; 8.14, "Personnel Neutron Dosimeters"; 8.25, "Air Sampling in the Workplace"; and 8.28, "Audible Alarm Dosimeters."

- NUREG-0800, Section 12.5, "Operations Radiation Protection Program."

11.4.4.3 Policies and Procedures

10 CFR Part 20.1101 requires that licensees "develop, document, and implement a radiation protection program commensurate with the scope and extent of licensed activities." The application should describe the radiation protection program, including details on all health physics-related policies and procedures, to be implemented at the ISFSI or MRS. The applicant should commit to reviewing the program for content and implementation at least annually. A listing of major program elements, along with the parameters that should be described under each element, is given in Table 11.2. Applicable regulatory criteria and guidance documents are also listed.

In addition to the regulatory guides identified in Table 11.2, applicable guidance and criteria for health physics procedures relevant to ISFSI or MRS operations are contained in the following:

- ANSI/ANS N13.2, "Guide to Administrative Practices in Radiation Monitoring."

- ANSI/ANS N13.6, "Practice for Occupational Radiation Exposure Record Systems."

- ASTM E 1167, "Guide for Radiation Protection Program for Decommissioning Operations."

- ASTM E 1168, "Guide for Radiation Protection Training for Nuclear Facility Workers."

- HPS N 13.30-1996, "Performance Criteria for Radiobioassay" (An American National Standard).

- HPS N 13.32-1995, "Performance Testing of Extremity Dosimeters" (An American National Standard).

- HPS N 13.41-1997, "Criteria for Performing Multiple Dosimetry" (An American National Standard).

- HPS N 13.42-1997, "Internal Dosimetry Program for Mixed Fission and Activation Products" (An American National Standard).

- National Council on Radiation Protection (NCRP) 59, "Operational Radiation Safety Program."

- NCRP 71, "Operational Radiation Safety Training."

- National Safety Council (NSC), "Accident Prevention Manual for Industrial Operations."

- NUREG-0800, Section 12.5.

11.5 Review Procedures

This section provides review guidance specific to submitted documentation that falls within the scope of radiation protection-related procedures. The radiation protection review includes evaluation of compliance with all regulatory requirements and acceptance criteria given in the FSRP and other applicable NRC documents and accepted codes. The reviewer should always assume that such a comprehensive nature of the review applies, even though it is not further detailed or repeated in this section.

11.5.1 As Low As Reasonably Achievable

Determine that an ALARA philosophy will be applied to most functions associated with construction, operation, and eventual D&D of a nuclear facility. Determine if ALARA philosophies and program goals are evident throughout the SAR, in the description of equipment, facility designs, and operational procedures. In conjunction with the review, and through discussions with reviewers of other topics, determine whether the ALARA policies are additionally reflected in the other topics of the SAR.

Ensure that the applicant's ALARA policy and program includes a written policy statement that expresses management's commitment to maintain exposures to workers and the public ALARA. Determine that the policy includes the elements identified in Section 11.4.1. Review the proposed ALARA program organization and ensure that the organizational structure is identified, including responsibilities and activities of ALARA personnel. Review the ALARA program content, and ensure that it includes provisions for (a) procedures and engineering controls to minimize dose, (b) tracking and trend analysis of individual doses, and (c) periodic training and exercises for management, radiation workers, and other site workers in radiation protection, operating procedures, and emergency response.

NUREG-0800, Section 12.1, "Assuring that Occupational Radiation Exposures Are As Low As Is Reasonably Achievable," and Regulatory Guide 8.8 provide other guidance that may be applicable to the review of an applicant's proposed ALARA program for an ISFSI or MRS.

11.5.1.1 Design Considerations

Ensure that the facility design and layout demonstrate consideration of ALARA principles. Ensure that the design criteria (presented in Section 3 of the SAR) include ALARA criteria. Determine whether the documentation provided identifies choices between otherwise comparable alternatives affected by ALARA considerations. Evaluate the design and layout for consideration of the factors identified in FSRP Section 11.4.1.2.

Table 11.2 Policy and Procedural Elements of the Health Physics Program

Item	Description	Criteria
Radiation surveys	Method, frequency, and plans for conducting radiation surveys, records of surveys	10 CFR 20.1501(a), Section 20.2103
ALARA plans	Plans developed to ensure occupational exposures will be ALARA	10 CFR 20.1101(b) Regulatory Guide 8.8, "Information Relevant to Ensuring that Occupational Radiation Exposures at Nuclear Power Stations are as Low as Reasonably Achievable" Regulatory Guide 8.10, "Operating Philosophy for Maintaining Occupational Radiation Exposures as Low as Reasonably Achievable"
Access control	Physical and administrative functions (e.g., personnel monitoring) for controlling access to and limiting stay times in restricted and controlled areas	10 CFR 20.1601, Section 1602, Section 1702 10 CFR 72.126(a)(3) Regulatory Guide 8.38, "Control of Access to High and Very High Radiation Areas of Nuclear Power Plants"
External exposure monitoring	Monitoring criteria, types of dosimeters, collection frequency, processing, review of results (including how results are used for operational planning), record-keeping, and reporting of results	10 CFR 20.1501(c), Section 1502 Regulatory Guide 8.4, "Direct-Reading and Indirect-Reading Pocket Dosimeters" Regulatory Guide 8.14, "Personnel Neutron Dosimeters" Regulatory Guide 8.28, "Audible Alarm Dosimeters" Regulatory Guide 8.34, "Monitoring Criteria and Methods to Calculate Occupational Radiation Doses"
Internal exposure monitoring	Types of monitoring (whole-body counts, lung counts, urinalysis, etc.), monitoring criteria, procedures for estimating dose from bioassay results, record-keeping, and review and reporting of results	10 CFR 20.1204, Section 1502 Regulatory Guide 8.9, "Acceptable Concepts, Models, Equations, and Assumptions for a Bioassay Program" Regulatory Guide 8.26, "Applications of Bioassay for Fission and Activation Products" Regulatory Guide 8.34, "Monitoring Criteria and Methods to Calculate Occupational Radiation Doses"
Air sampling and analysis	Methods and procedures for air sampling and analysis, evaluation and control of airborne radioactivity, requirements/procedures for special air sampling	10 CFR 20.1204(a)(1), Sections 1501(a)(2)(ii), 1502, 1701, 1702, 1703(a)(3)(i) Regulatory Guide 8.25, "Air Sampling in the Workplace"
Respiratory protection program	Policy statement on respirator usage; respirator certification, fit-testing, and usage; medical surveillance of respirator users	10 CFR 20.1703 Regulatory Guide 8.15, "Acceptable Programs for Respiratory Protection"
Radiation protection training	Requirements for initial and refresher training, contents (topics), health physics-related qualification of workers, etc.	Regulatory Guide 8.27, "Radiation Protection Training for Personnel at Light-Water-Cooled Nuclear Power Plants" Regulatory Guide 8.29, "Instruction Concerning Risks from Occupational Radiation Exposure"

Item	Description	Criteria
Pregnant worker protection	Provisions to inform female workers of fetal protection requirements, to monitor fetal dose, and to provide alternatives to minimize fetal dose	10 CFR 20.1208 Regulatory Guide 8.13, "Instruction Concerning Prenatal Radiation Exposure"
Instrument quality assurance	Requirements and procedures for calibration, maintenance, and care of radiation detection, monitoring, and dosimetry instruments	10 CFR 20.1501(b) Draft Regulatory Guide OP 032-5, "Test and Calibration of Radiation Protection Instrumentation"
Record-keeping	Preparing of records for health physics program contents and audits, surveys, calibrations, personnel monitoring results, etc.	

11.5.1.2 Operational Considerations

Determine that the descriptions of proposed operations adequately demonstrate that ALARA principles have been incorporated into operational procedures. Ensure that plans and procedures have been developed in accordance with applicable guidance (refer to Table 11.1) and that the considerations detailed in FSRP Section 11.4.1.3 have been addressed adequately.

11.5.2 Radiation Protection Design Features

NUREG-0800, Section 12.3 - 12.4, provides acceptable guidance for review of ISFSI and MRS radiation protection design features. This section addresses the review procedures that apply to installation design, access control, shielding, confinement, and area radiation and airborne radioactivity monitoring instrumentation.

11.5.2.1 Installation Design Features

Review the installation design features and ensure that (a) installation features for which credit is taken in radiation protection analyses are clearly identified on the site drawing; and (b) the license application includes commitments to construct those site improvements that have been credited in analyses to show compliance with regulatory requirements or ALARA goals.

11.5.2.2 Access Control

Review the description and provisions for access control and verify that: (a) necessary and desirable personal protective measures have been incorporated in facility and operational planning; (b) provisions reflect facility designer appreciation of potential dose rates and contamination levels in the pool (or dry transfer) facilities and waste management facilities; and (c) provisions for access control are incorporated into SAR descriptions of ALARA or other radiological protection features and in the physical protection planning.

11.5.2.3 Shielding Design, Use, and Effectiveness

Examine the applicant's input to the computer program used for the shielding analysis, input to the program, and reasonableness of results (Chapter 7). Discuss the use of the shield evaluation results in developing projected dose rates with shield design reviewers. The radiation protection reviewer should then determine whether the findings of the shielding analysis can be used in conjunction with other applicable information to verify the SAR estimates of external radiation dose associated with radioactive materials handling and storage during routine ISFSI or MRS operations. In general, such a determination can be made if (a) the shield design review has confirmed the dose rates estimates in the SAR (or has resulted in independent estimates) and (b) conservative assumptions regarding the amount and locations of radioactive material at the ISFSI or MRS have been used to determine direct dose rates.

If independent computations of external dose rates are determined to be required to verify the probable accuracy of estimates in the SAR, coordinate the performance of these computations with the shield design reviewer. In determining the need or identifying an appropriate level of effort for confirmatory calculations, consider how the estimated dose rates compare with those of similar casks that have been reviewed previously. Consider the amount of conservatism applied in the shielding analysis as well as conditions in which the actual personnel doses will be monitored and controlled to meet Part 20 limits and ALARA objectives.

Evaluate whether the proposed shielding use is consistent with the applicant's design objectives relative to keeping radiation doses ALARA. Include the applicant's plans, if any, for use of temporary or portable shielding, remote handling, or other protective features in this evaluation.

11.5.2.4 Confinement and Ventilation

The confinement analysis (Chapter 9) includes an assessment of the applicant's estimates of radionuclide releases to the environment, or alternatively, development of new estimates. The analysis of site-generated waste confinement and management (Chapter 14) addresses radionuclide releases from site-generated wastes. Those analyses include confinement and ventilation aspects applicable to sealed casks (for which releases are usually minimal) and to systems and components that are not designed to be sealed.

The radiation protection review of confinement and ventilation has two components. The first is to evaluate information from the confinement and site-generated waste analyses and to determine if the radionuclide release estimates and other site-specific information are adequate for estimating onsite and offsite doses, as described in Section 11.5.3. The second is to evaluate the personnel protection features of pool and waste management facility ventilation systems. This part of the review should identify how the confinement and ventilation system components and controls function to:

- Maintain radiation exposures ALARA.
- Prevent spread of radioactive materials and contamination between areas.
- Limit the spread of radioactive materials within the ventilation system(s).
- Interface with process off-gases (e.g., waste treatment, cask venting).

In reviewing the protective features, ensure that confinement and ventilation systems conform to the applicable guidance of NUREG-0800 (Section 11.3), and Regulatory Guides 1.140, "Design, Testing, and Maintenance Criteria for Normal Ventilation Exhaust System Air Filtration and Adsorption Units of Light-Water-Cooled Containment Isolation Provisions for Fluid Systems," and 1.143, "Design Guidance for Radioactive Waste Management Systems, Structures, and Components Installed in Light-Water-Cooled Nuclear Power Plants."

11.5.2.5 Area Radiation and Airborne Radioactivity Monitoring Instrumentation

NUREG-0800, Section 11.5, includes acceptable guidance for conducting reviews of ISFSI or MRS area radiation and airborne radioactivity monitoring instrumentation. Evaluate the applicant's description of the fixed area radiation monitors and continuous airborne monitoring instrumentation as well as the placement of these monitors. Review the criteria and methods used for determining alarm setpoints. Review the information provided on the auxiliary and emergency power supply. Evaluate information and specifications on instrument range, sensitivity, accuracy, energy dependence, calibration methods and frequency, recording devices, readouts, and alarms.

11.5.3 Dose Assessment

This section provides review guidance on the assessment of dose to onsite (worker) personnel and to persons located beyond the controlled area boundary for compliance with applicable regulatory criteria.

11.5.3.1 Onsite Dose

Use all relevant information to estimate the total individual and collective doses received by ISFSI or MRS workers and to determine whether applicable dose and ALARA criteria are met.

This onsite dose evaluation includes the following steps:

- Review the estimated annual occupancy times, including the maximum expected total hours per year for any individual and total person-hours per year for all personnel for each radiation area, including storage areas, during normal operation and anticipated operational occurrences.

- Ensure that the estimated annual doses are based on the maximum number of casks placed into storage in one year and include both direct radiation and inhalation of airborne radionuclides.

- Ensure that descriptions of procedures that involve worker exposure are compatible with the occupancy times and proximities assumed for the dose estimates.

- Ensure that the individual and collective dose estimates are based on reasonable assumptions regarding presumption of skill level, extent of care taken in nuclear-safety-related operations, presence of supervisor and quality control, and other factors that tend to increase doses.

• Ensure that the dose calculational methodology is consistent with methods and models approved by the NRC.

Perform independent estimates of onsite collective dose if the assessment indicates that SAR estimates are not bounding. Clearly identify assumptions or models that differ from those in the SAR, and discuss whether the collective dose estimates support the applicant's considerations related to keeping occupational exposures ALARA.

Determine whether annual individual doses meet the dose limits of 10 CFR 20.12019(a). If the estimated doses approach or exceed these limits, the planned conduct of operations (Chapter 10) must commit to conditions (staffing plan, monitoring, etc.) that ensure that individual doses will be controlled and that dose limits will not be exceeded.

11.5.3.2 Offsite Dose

Estimate offsite doses that will result from ISFSI or MRS operations. Make three principal determinations for comparison to acceptance criteria: (1) annual collective (person-rem) dose to surrounding population; (2) annual dose to the maximally exposed individual; and (3) maximum hourly dose rate in unrestricted area. For each of these determinations, ensure that the applicant's environmental monitoring program is designed to provide the exposure and concentration data needed to assess those pathways that lead to the highest potential external and internal doses. Ensure that the dose contributions from any other nuclear fuel cycle facilities within an 8-km (5-mile) radius are also addressed.

Collective Dose to Surrounding Population

In reviewing annual collective dose attributable to direct radiation and facility effluents, ensure that the assumptions and models that were used to estimate doses have duly considered the following factors:

• Site layout.
• Land use, topography and population data.
• Direct radiation dose rates as a function of distance and direction.
• Site meteorological data.
• Radioactive material release rates, dispersion, and deposition.

Ensure that collective dose has been determined for the population within an 8-km (5-mile) radius and that the dose considers all important exposure pathways (direct radiation, airborne releases, etc.) and modes of exposure (external exposure, inhalation, etc.). Assess the increment by which the collective dose would be increased by the presence of any other nuclear fuel-cycle facilities (existing or projected) within an 8-km (5-mile) radius of the proposed ISFSI or MRS. Ensure that the computational models or equations and assumptions used are acceptable to the NRC.

Perform independent dose estimates if the offsite collective dose review indicates that the SAR dose estimates are not bounding, have high uncertainty levels, or are not acceptable for other reasons. In cases where the SAR methods are deemed acceptable, perform a limited number of confirmatory or spot-check calculations. For all independent or confirmatory calculations, clearly identify any assumptions or models that differ from those in the SAR.

Review the effluent and environmental monitoring program and ensure (1) that it identifies the pathways that lead to the highest potential external and internal exposures of the offsite population; and (2) that the program can yield information that can be used to estimate collective dose with reasonable accuracy. Finally, discuss whether the collective dose estimates support the applicant's considerations related to keeping radionuclide releases and offsite doses ALARA.

Dose to Maximally Exposed Real Individual

Determine whether the highest offsite dose received by a real individual is less than the limits specified in 10 CFR 72.104(a). Many of the same factors considered in the collective dose assessment are applicable to this review.

Evaluate the applicant's assessment of direct dose rates and radionuclide concentrations beyond the controlled area boundary for normal operations and anticipated occurrences. Identify the location of the offsite individual likely to receive the highest dose from direct radiation and facility effluents.

Assess the annual whole-body dose equivalent, as well as the dose equivalent to the thyroid and to the organ (other than the thyroid) receiving the highest dose equivalent. Total effective dose equivalent may be calculated as a surrogate for whole-body dose equivalent.

As in the collective dose assessment, perform independent calculations as necessary or confirmatory calculations to verify the applicant's results. Clearly identify any assumptions or models that differ from those in the SAR.

Assess the increment by which the whole-body dose would be increased by the presence of any other nuclear fuel-cycle facilities (existing or projected) within an 8-km (5-mile) radius of the proposed ISFSI or MRS. Ensure that the combined annual dose equivalent from the ISFSI or MRS and other nuclear facilities does not exceed 25 mrem to the whole body, 75 mrem to the thyroid, or 25 mrem to any other organ.

Assess the total effective dose equivalent that can result from airborne emissions only, and ensure that this dose does not exceed the 10 mrem/yr ALARA constraint level of 10 CFR 20.1101(d). Review applicant's determination that the maximum dose rate in any unrestricted area resulting from external sources does not exceed 0.002 rem in any one hour, and determine whether the distance to the nearest boundary of the controlled area is sufficient to ensure compliance with this dose rate standard.

Review the effluent and environmental monitoring program. Ensure that it identifies the pathways that lead to the highest potential external and internal exposures of offsite individuals from ISFSI or MRS operations and that the program can yield information that can be used to estimate this dose with reasonable accuracy.

11.5.4 Health Physics Program

Evaluate the administrative organization of the applicant's health physics program. Determine that the program describes the authority, responsibility, experience, and qualifications of the personnel responsible for the health physics program.

Review the applicant's description of the portable and laboratory equipment and instrumentation for performing radiation and contamination surveys, sampling airborne radioactive material, monitoring area radiation, and monitoring personnel during normal operations, anticipated operational occurrences, and accident conditions. Ensure that an appropriate number of instruments will be available for each type of radiation survey to be performed (e.g., G-M survey instruments for contamination surveys and personnel "frisking," ionization chambers for exposure rate surveys, neutron detectors for neutron flux or dose rate surveys, etc.). Review the calibrating survey equipment, area radiation monitors, continuous airborne monitors, effluent monitors, and laboratory equipment.

Review the applicant's plans and procedures to ensure that provisions have been made for:

- Controlling, storing, and moving radioactive material.

- Physical and administrative measures aimed at ensuring that occupational doses are ALARA (also addressed in FSRP Section 11.5.1).

- Retaining records of personnel dosimetry results, surveys, training, instrument calibration, and decommissioning.

Other review procedures that may be applicable to health physics programs at an ISFSI or MRS are provided in NUREG-0800, Sections 11.5, 12.1, 12.5, and 13.2.2.

11.6 Evaluation Findings

Evaluation findings are prepared by the reviewer upon determination that the regulatory requirements related to radiation protection, as identified in Section 11.3, have been satisfied. Some of these determinations (e.g., with respect to radiation levels or airborne radionuclide releases) can be made only after evaluating the results of reviews performed under other FSRP sections. If the documentation submitted with the application fully supports positive findings for each of the regulatory requirements, then the statements of findings should be substantially as follows (finding numbering is for convenience in referencing within the FSRP and SER).

> F11.1 The design and operating procedures of the [ISFSI/MRS] provide acceptable means for controlling and limiting occupational radiation exposures within the limits given in 10 CFR 20 and for meeting the objective of maintaining exposures ALARA, in compliance with 10 CFR 72.24(e).

F11.2   The SAR and other documentation submitted in support of the application provide acceptable and reasonable assurance that the activities authorized by the license can be conducted without endangering the health and safety of the public, in compliance with 10 CFR 72.40(a)(13).

F11.3   [If appropriate] The proposed [ISFSI/MRS] is to [be on the same site as/near other] nuclear facilities, _____ [identify]. The cumulative effects of the combined operations of these facilities will not constitute an unreasonable risk to the health and safety of the public, in compliance with 10 CFR 72.122(e).

F11.4   The SAR provides analyses showing that releases to the general environment during normal operations and anticipated occurrences will be within the exposure limits given in 10 CFR 72.104.

F11.5   The design of the [ISFSI/MRS] provides suitable shielding for radiation protection under normal and accident conditions, in compliance with 10 CFR 72.128(a)(2).

11.7 References

NRC documents referenced are identified at Consolidated References, Section 17.

American National Standards Institute, "Location and Design Criteria for Area Radiation Monitoring Systems for Light Water Reactors," ANSI/ANS-HPSSC-6.8.1-1981.

American National Standards Institute, Institute for Nuclear Materials Management, "Guide to Sampling Airborne Radioactive Materials in Nuclear Facilities," ANSI N13.1, 1993.

American National Standards Institute, "Guide to Administrative Practices in Radiation Monitoring," ANSI/ANS N13.2, 1969 (Reaffirmed in 1989).

American National Standards Institute,"Practice for Occupational Radiation Exposure Record Systems," ANSI/ANS N13.6, 1972 (Reaffirmed in 1989).

American Society for Testing and Materials, "Guide for Radiation Protection Program for Decommissioning Operations," ASTM E 1167, 1987.

American Society for Testing and Materials, "Guide for Radiation Protection Training for Nuclear Facility Workers," ASTM E 1168.

Health Physics Society, "Performance Criteria for Radiobioassay," HPS N 13.30, 1996 (An American National Standard).

Health Physics Society, "Performance Testing of Extremity Dosimeters," HPS N 13.32-1995 (An American National Standard).

Health Physics Society, "Criteria for Performing Multiple Dosimetry," HPS N 13.41-1997 (An American National Standard).

Health Physics Society, "Internal Dosimetry Program for Mixed Fission and Activation Products," HPS N 13.42-1997 (An American National Standard).

National Council on Radiation Protection, "Operational Radiation Safety Program," NCRP Report No. 59, 1978.

National Council on Radiation Protection, "Operational Radiation Safety - Training," NCRP Report No. 71, 1983.

National Safety Council (NSC), "Accident Prevention Manual for Industrial Operations."

12 QUALITY ASSURANCE EVALUATION

12.1 Review Objective

The purpose of the review is to determine whether the applicant for a license to store spent fuel or high-level waste has a quality assurance (QA) program that complies with the requirements of 10 CFR Part 72, Subpart G. The basis for that determination is a review and evaluation of the applicant's QA program submitted as a part of the application in accordance with 10 CFR 72.24(n). The results of the review and evaluation are documented in the Safety Evaluation Report (SER).

This scope of review does not include specific procedures and instructions that implement the QA program. The staff may approve a QA program that is a high-level description listing the procedures and controls to be developed prior to initiation of any design or other quality activities to meet the requirements of 10 CFR Part 72, Subpart G. After the license is issued, the staff performs inspections to determine whether the QA program has been implemented effectively. The licensee is required to take corrective actions to resolve all deficiencies identified by the staff during the inspections of critical activities affecting quality.

An overview of the QA review process is shown in Figure 12.1. The figure shows that the review draws information from the application and other relevant review sections.

12.2 Areas of Review

The following outline shows the areas of review addressed in Section 12.4, Acceptance Criteria, and Section 12.5, Review Procedures:

QA Organization
Assignment of Functional Responsibilities
Responsibility for QA
Organization for Performing QA Functions

QA Program
Scope of Program
Implementation of Program
Review of Program

Design Control
Design Process
Design Interfaces
Design Verification
Design Changes

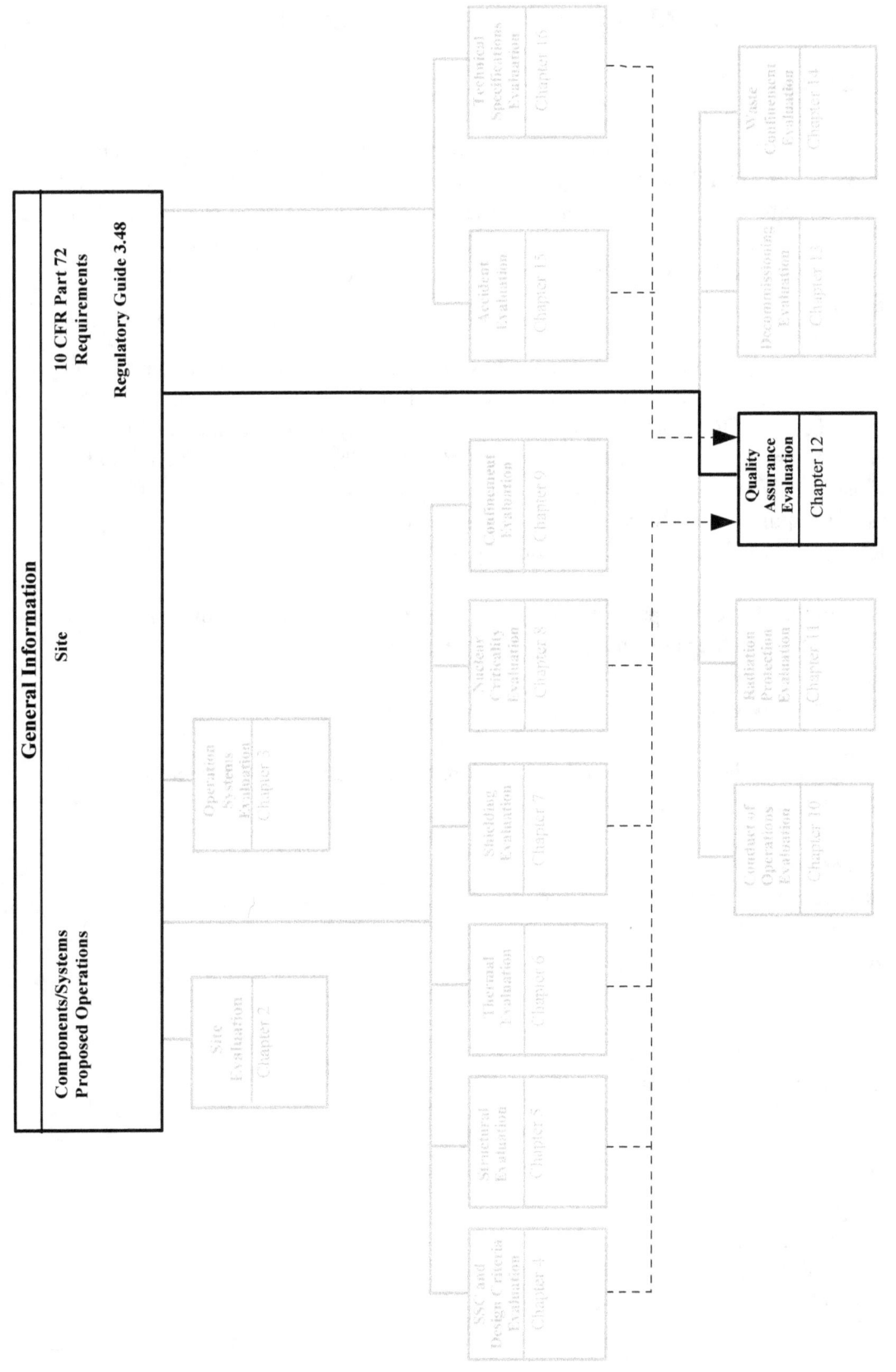

Figure 12.1 Overview of Quality Assurance Evaluation

Procurement Document Control
    Control Process
    Contractor QA Programs

Instructions, Procedures, and Drawings
    Development Process
    Utilization

Document Control
    Scope of Control Process
    Control Process

Control of Purchased Material, Equipment, and Services
    Control Process
    Acceptance of Services Only
    Commercial-Grade Items
    Assessments of Effectiveness

Identification and Control of Materials, Parts, and Components
    Identification and Control Process
    Stored Items
    Items with Limited Lifetimes

Control of Special Processes
    Control Process
    Qualification

Licensee Inspection
    Scope of Inspection Program
    Implementation of Program

Test Control
    Scope of Test Program
    Test Control Process

Control of Measuring and Test Equipment
    Control Process
    Commercial Devices

Handling, Storage, and Shipping Control
    Control Process
    Tools and Equipment
    Markings

Inspection, Test, and Operating Status

Inspection and Test Status Process
Operating Status Process

Nonconforming Materials, Parts, or Components
Control Process
Notification

Corrective Action
Initiation Process
Correction Process

QA Records
Scope of Records Program
Implementation

Audits
Scope of Audit Program
Implementation

12.3 Regulatory Requirements

This section identifies and presents a high-level summary of Title 10 of the Code of Federal Regulations (CFR) Part 72 relevant to the review areas addressed by this chapter. The NRC staff reviewer should read the exact regulatory language. A matrix at the end of this section matches the regulatory requirements identified in this section to the areas of review identified in the previous section.

72.24 Contents of application: Technical information.
(n) "A description of the quality assurance program that satisfies the requirements of Subpart G."

72.122 Overall requirements.
(a) "Quality Standards. [Structures, Systems, and Components] SSCs important to safety must be designed, fabricated, erected, and tested to quality standards commensurate with the importance to safety of the function to be performed."

72.140 Quality assurance requirements.
(a) "This subpart describes quality assurance requirements...that are important to safety."
(b) "Each licensee shall establish, maintain, and execute a quality assurance program satisfying each of the applicable criteria."
(c) "Prior to receipt of spent fuel at the [Independent Spent Fuel Storage Installation] ISFSI or spent fuel and high-level radioactive waste at the [Monitored Retrievable Storage] MRS, each licensee shall obtain Commission approval of its quality assurance program."
(d) "A Commission-approved quality assurance program which satisfies the applicable criteria of Appendix B to Part 50 of this chapter...will be accepted as satisfying the requirements of paragraph (b) of this section."

72.142 Quality assurance organization.

"The licensee shall be responsible for the establishment and execution of the quality assurance program...clearly establish and delineate in writing the authority and duties of persons and organizations....The quality assurance functions are:

(a) Assuring that an appropriate quality assurance program is established and effectively executed and

(b) Verifying...activities affecting the functions that are important to safety have been correctly performed...organization performing quality assurance functions shall report to a management level that ensures...the required authority and organizational freedom."

72.144 Quality assurance program.

(a) "The licensee shall establish, at the earliest practicable time...a quality assurance program...document the quality assurance program...and shall carry out the program."

(b) "The licensee...shall provide control over activities affecting the quality of the identified SSCs to an extent commensurate with the importance to safety... The licensee shall ensure that activities affecting quality are accomplished under suitably controlled conditions.

(c) "Base the requirements and procedures of its quality assurance program on:

    (1) The impact of malfunction or failure of the item on safety;

    (2) The design and fabrication complexity or uniqueness of the item;

    (3) The need for special controls and surveillance over processes and equipment;

    (4) The degree to which functional compliance can be demonstrated by inspection or test; and

    (5) The quality history and degree of standardization of the item."

(d) "Provide for indoctrination and training of personnel performing activities affecting quality."

72.146 Design control.

(a) "Establish measures to ensure that applicable regulatory requirements and the design basis...are correctly translated into specifications, drawings, procedures, and instructions."

(b) "Establish measures for the identification and control of design interfaces and for coordination among participating design organizations...The licensee shall apply design control measures to items such as: criticality physics, radiation, shielding, stress, thermal, hydraulic, and accident analyses; compatibility of materials; accessibility for in-service inspection, maintenance, and repair; features to facilitate decontamination; and delineation of acceptance criteria for inspections and tests."

(c) "Subject design changes...to design control measures commensurate with those applied to the original design."

72.148 Procurement document control.

"Establish measures to assure that applicable regulatory requirements, design bases, and other requirements are included or referenced in the documents for procurement...the licensee shall require contractors or subcontractors to provide a quality assurance program."

72.150 Instructions, procedures, and drawings.

"Prescribe activities affecting quality by documented instructions, procedures, or drawings....
The instructions, procedures, and drawings must include appropriate quantitative or qualitative
acceptance criteria."

72.152 Document control.
"Establish measures to control the issuance of documents."

72.154 Control of purchased material, equipment, and services.
(a) "Establish measures to ensure that purchased material, equipment and services... conform to
the procurement documents."
(b) "Have available documentary evidence that material and equipment conform to the
procurement specifications prior to installation or use."
(c) "Assess the effectiveness of the control of quality by contractors and subcontractors."

72.156 Identification and control of materials, parts, and components.
"Establish measures for the identification and control of materials, parts, and components...
traceable to the item as required, throughout fabrication, installation, and use of the item."

72.158 Control of special processes.
"Establish measures to ensure that special processes... are controlled and accomplished by
qualified personnel using qualified procedures."

72.160 Licensee inspection.
"Establish and execute a program for inspection of activities affecting quality... must be
performed for each work operation where necessary to assure quality."

72.162 Test control.
"Establish a test program to ensure that all testing required... is identified and performed in
accordance with written test procedures."

72.164 Control of measuring and test equipment.
"Establish measures to ensure that tools, gauges, instruments, and other measuring and testing
devices used in activities affecting quality are properly controlled, calibrated, and adjusted."

72.166 Handling, storage, and shipping control.
"Establish measures to control... the handling, storage, shipping, cleaning, and preservation of
materials and equipment to prevent damage or deterioration."

72.168 Inspection, test, and operating status.
(a) "Establish measures to indicate, by the use of markings such as stamps, tags, labels, routing
cards, or other suitable means, the status of inspections and tests performed upon individual
items of the ISFSI or MRS. These measures must provide for the identification of items which
have satisfactorily passed required inspections and tests where necessary to preclude inadvertent
bypassing of the inspections and tests."
(b) "Establish measures to identify the operating status of SSCs of the ISFSI or MRS, such as
tagging valves and switches to prevent inadvertent operation."

72.170 Nonconforming materials, parts, or components.
"Establish measures to control materials, parts, or components that do not conform to the licensee's requirements to prevent their inadvertent use or installation."

72.172 Corrective action.
"Establish measures to ensure that conditions adverse to quality... are promptly identified and corrected. In the case of a significant condition adverse to quality, the measures must ensure that the cause of the condition is determined and corrective action is taken to preclude repetition."

72.174 Quality assurance records.
"Maintain sufficient records to furnish evidence of activities affecting quality.... Records must be identifiable and retrievable."

72.176 Audits.
"Carry out a comprehensive system of planned and periodic audits to verify compliance with all aspects of the quality assurance program and to determine the effectiveness of the program."

A matrix that shows the primary relationship of these regulations to the specific areas of review associated with this chapter is given in Table 12.1. The NRC staff reviewer should verify the matching of regulatory requirements to the areas of review presented in the matrix to ensure that no requirements are overlooked as a result of unique applicant design features.

12.4 Acceptance Criteria

The applicant's QA program must describe the program that is established, will be maintained, and will be executed for the design, fabrication, construction, testing, operation, modification, and decommissioning of the SSCs of the ISFSI or MRS that are important to safety. The QA program description must identify the items important to safety and include information about managerial and administrative controls to ensure safe operation of the ISFSI or MRS.

The applicant's QA program must be structured to apply QA measures and controls to all activities and items in proportion to their importance to safety (graded approach). A graded application of QA requires applicant justification and reviewer acceptance. The graded approach for the application of QA must be described adequately. The QA program must identify the activities and items that are important to safety and the degree of their importance. The highly important-to-safety activities and items must have a high-level of control, while those less important may have a lower level of control. An applicant may choose to apply the highest level of QA and control to all activities and items. From that point on, the assignment of QA levels of control to be used within the QA program must be based on a graded application to the 18 criteria listed.

QUALITY ASSURANCE

SECTION 12

For each of the activities and items identified as important to safety, the applicant must identify and define the level of application for each of the following QA programmatic elements. The attributes listed for each topic must be applied collectively only in the most stringent application of the QA program. Less stringent application of requirements may be effected by modifying or eliminating some attributes from selected topics.

The applicant's QA program and associated QA program controls and implementing procedures regarding activities performed must be in place before activities begin.

Defining a process involves establishing authorities and assigning responsibilities. Implementing a process involves issuing instructions and procedures.

The acceptance criteria are organized to reflect the 18 sections of 10 CFR 72, Subpart G, that correspond to the 18 basic requirements of NQA-1, 1983 Edition. Each acceptance criteria section shows the title of the relevant area of review in Section 12.2. Notes following each acceptance criterion identify the source(s) of substantive content of the criterion.

12.4.1 QA Organization

12.4.1.1 Assignment of Functional Responsibilities

The applicant must present organization charts describing functional responsibilities that:

- Establish clear lines of authority and responsibility for activities important to safety to be performed by applicant organizational elements and by contractors, whether onsite or offsite (10 CFR 72.142; NQA-1/Part I/Sec II Basic 1, 1S-1 Par 2, 3.1, 3.2; NUREG-0800; NUREG-1536).

- Cover activities to achieve quality objectives, i.e., performance in accordance with specified requirements of activities such as site design, purchasing, fabricating, constructing, handling, shipping, receiving, storing, cleaning, erecting, assembling, installing, inspecting, testing, operating, maintaining, repairing, modifying, and decommissioning (10 CFR 72.24(n), 72.140(a); NQA-1, Part I, Sec II Basic 1, 1S-1 Par 2.1(a)).

- Cover QA functions, that is, ensuring that an appropriate QA program is established and effectively executed, and verifying, by procedures such as checking, auditing, and inspection, that work activities have been correctly performed and physical characteristics and quality of material and components adhere to predetermined requirements (10 CFR 72.140(a),(b), 72.142(a),(b); NQA-1/Part I/Sec II, 1S-1 Par 2.1(b)).

Table 12.1 Relationship of Regulations and Areas of Review

Areas of Review	10 CFR Part 72 Requirements						
	72.24	72.122	72.140	72.142	72.144	72.146	72.148
QA Organization	●	●	●	●			
QA Program	●	●	●		●		
Design Control	●	●	●			●	
Procurement Document Control	●	●	●				●
Instructions, Procedures, and Drawings	●	●	●				
Document Control	●	●	●				
Control of Purchased Material, Equipment, and Services	●	●	●				
Identification and Control of Materials, Parts, and Components	●	●	●				
Control of Special Processes	●	●	●				
Licensee Inspection	●	●	●				
Test Control	●	●	●				
Control of Measuring and Test Equipment	●	●	●				
Handling, Storage, and Shipping Control	●	●	●				
Inspection, Test, and Operating Status	●	●	●				
Nonconforming Materials, Parts, or Components	●	●	●				
Corrective Action	●	●	●				
QA Records	●	●	●				
Audits	●	●	●				

Table 12.1 Relationship of Regulations and Areas of Review (continued)

Areas of Review	10 CFR Part 72 Requirements						
	72.150	72.152	72.154	72.156	72.158	72.160	72.162
QA Organization							
QA Program							
Design Control							
Procurement Document Control							
Instructions, Procedures, and Drawings	●						
Document Control		●					
Control of Purchased Material, Equipment, and Services			●				
Identification and Control of Materials, Parts, and Components				●			
Control of Special Processes					●		
Licensee Inspection						●	
Test Control							●
Control of Measuring and Test Equipment							
Handling, Storage, and Shipping Control							
Inspection, Test, and Operating Status							
Nonconforming Materials, Parts, or Components							
Corrective Action							
QA Records							
Audits							

Table 12.1 Relationship of Regulations and Areas of Review (continued)

Areas of Review	10 CFR Part 72 Requirements						
	72.164	72.166	72.168	72.170	72.172	72.174	72.176
QA Organization							
QA Program							
Design Control							
Procurement Document Control							
Instructions, Procedures, and Drawings							
Document Control							
Control of Purchased Material, Equipment, and Services							
Identification and Control of Materials, Parts, and Components							
Control of Special Processes							
Licensee Inspection							
Test Control							
Control of Measuring and Test Equipment	●						
Handling, Storage, and Shipping Control		●					
Inspection, Test, and Operating Status			●				
Nonconforming Materials, Parts, or Components				●			
Corrective Action					●		
QA Records						●	
Audits							●

12.4.1.2 Responsibility for QA

The applicant must:

• Assign to elements of its own organization the responsibility for establishment and execution of its QA program (10 CFR 72.142, first paragraph; NQA-1/Part I/Sec II, 1S-1 Par 2.2; NUREG-0800; NUREG-1536) and oversight and evaluation of work to establish and execute its QA program that is delegated to others (10 CFR 72.142, first paragraph; NQA-1/Part I/Sec II, 1S-1 Par 2.2; NUREG-0800; NUREG-1536)

• Assign to a high-level of applicant line management the responsibility for promulgating corporate or company QA policies, goals, and objectives; maintaining a continuing involvement in QA matters; and approving procedures for resolution of disputes between applicant organizational elements about QA activities (10 CFR 72.142, first paragraph; NQA-1/Part I/Sec II, 1S-1 Par 2.2; NUREG-0800; NUREG-1536)

• Establish lines of communication among that line position, intermediate levels of line management, contractors involved in activities impacted by the QA program, and QA management positions (10 CFR 72.142, first paragraph; NQA-1/Part I/Sec II, 1S-1 Par 2.2; NUREG-0800; NUREG-1536).

12.4.1.3 Organization for Performing QA Function

The applicant's organization for performing QA functions must:

• Include a management position having authority and responsibility for developing the QA program (e.g., QA Manager, Director, or Vice President). This management position must (a) have appropriate qualification requirements (e.g., education, experience, technical competence) (10 CFR 72.142, last paragraph; NUREG-0800; NUREG-1536), (b) be at an organizational level at least as high as that of the highest level line manager having direct responsibility for cost and scheduling of work activities to attain quality objectives (10 CFR 72.142, last paragraph; NQA-1/Part I/Sec II Basic 1; NUREG-0800; NUREG-1536), (c) be independent of responsibility for cost and scheduling of work activities to attain quality objectives and have no other duties that would prevent full attention to QA matters (10 CFR 72.142, last paragraph; NQA-1/Part I/Sec II Basic 1; NUREG-0800; NUREG-1536), and (d) have an assigned staff (without responsibilities for work activities to impart quality) to develop and implement procedures for verifying conformance of work activities and results of work activities to specified requirements, or authority and a budget to contract for such services (10 CFR 72.142, last paragraph; NUREG-0800).

• Include individuals assigned responsibility for ensuring effective execution of any portion of the QA program who have sufficient authority, access to work areas, and organizational freedom to (a) identify quality problems; (b) initiate, recommend, or provide solutions to quality problems through designated channels; (c) verify implementation of solutions; and (d) stop unsatisfactory work and ensure that further processing, delivery, installation, or use is controlled until proper disposition of a nonconformance, deficiency, or unsatisfactory condition (10 CFR 72.142, last paragraph; NQA-1/Part I/Sec II Basic 1; NUREG-0800; NUREG-1536).

12.4.2 QA Program

12.4.2.1 Scope of Program

The applicant must establish a QA program that:

- Identifies SSCs and activities that are important to safety, major organizations participating in the program, and functions of those organizations (10 CFR 72.24(n), 72.140(a),(b), 72.144(a); NQA-1/Part I/Sec I Par 2, 3).

- Applies to (a) siting, designing, purchasing, fabricating, constructing, handling, shipping, receiving, storing, cleaning, erecting, assembling, installing, inspecting, testing, operating, maintaining, repairing, modifying, and decommissioning of the identified SSCs; and (b) managerial and administrative controls to ensure safe operation of the facility, both before issuance of a license and throughout the life of licensed activity (10 CFR 72.24(n), 72.140(a),(b), 72.144(a); NQA-1/Part I/Sec I/Par 2; NUREG-0800; NUREG-1536).

- Covers activities that provide confidence that an SSC will perform satisfactorily in service (QA), including activities that determine that physical characteristics and quality of materials or components adhere to predetermined requirements (quality control) (10 CFR 72.140(a); NQA-1/Part I/Sec II Basic 2).

12.4.2.2 Implementation of Program

The applicant must have written statements of QA policies, goals, and objectives, and procedures and instructions that:

- Are under controlled distribution (both original versions and subsequent changes) (10 CFR 72.144(a); NUREG-1536).

- Inform responsible organizations and individuals of the applicant's QA policies, goals, and objectives, and make compliance with QA policies and procedures mandatory (10 CFR 72.142(a), 72.144(a); NUREG-0800; NUREG-1536).

- Identify criteria of 10 CFR 72, Subpart G, and any other specific provisions that apply to applicant activities in progress (10 CFR 72.24(n), 72.140(a); 72.144(a); NQA-1/Part I/ Basic 1).

- Relate QA procedures to the criteria of 10 CFR 72, Subpart G, and other specific provisions that apply to applicant activities (10 CFR 72.142(a), 72.144(a); NUREG-1536).

- Prescribe actions by persons and organizations engaged in activities affecting safety that will result in compliance with requirements of the applicable criteria of 10 CFR 72 to an extent that is commensurate with their importance to safety (NQA-1, Sec I Par 3; NQA-1/Part I/Sec II Basic 2) and ensure conformance to the approved design of each ISFSI or MRS (10 CFR 72.144(b)).

- Prescribe suitably controlled conditions for performing activities affecting quality, including (a) satisfying prerequisites for the given activity, establishing suitable environmental conditions (e.g., adequate cleanliness), and using appropriate equipment; (b) providing special controls, processes, test equipment, tools, and skills needed to attain the required quality; and (c) verifying quality by inspection and test (NQA-1/Part I/Sec II Basic 2).

- Provide a level of detail in requirements and procedures for performing work that is appropriate to the complexity and proposed use of the structures, systems, or components affected, including (a) the impact of malfunction or failure of the item on safety; (b) the design and fabrication complexity or uniqueness of the item; (c) the need for special controls and surveillance over processes and equipment; (d) the degree to which functional compliance can be demonstrated by inspection or test; and (e) the quality history and degree of standardization of the item (10 CFR 72.144(c); NQA-1/Part I/Sec II Basic 2).

- Provide for indoctrination and training of personnel performing activities affecting quality including (a) instructing personnel responsible for performing activities affecting quality in the purpose, scope, and implementation of quality-related manuals, instructions, and procedures; (b) training and qualifying personnel performing activities affecting quality in the principles and techniques of the activity being performed; (c) certifying qualified personnel in accordance with relevant codes and standards; (d) maintaining proficiency of personnel performing activities affecting quality by retraining, reexamining, and recertifying; and (e) maintaining records of completed training and qualification (NQA-1/Part I/Sec II Basic 2; NUREG-1536)

12.4.2.3 Review of Program

The applicant must have assigned responsibilities and have provided, prior to implementation, written instructions and procedures for:

- Review, by the applicant, of the status and adequacy of the QA program through frequent contact with the program through reports, meetings, and audits; and periodic formal assessments in which performance is documented and necessary corrective and follow up actions are identified (10 CFR 72.144(d); NQA-1/Part I/Sec II Basic 2; NUREG-0800; NUREG-1536).

- Review by management of other organizations participating in the QA program of that part of the program they are executing (10 CFR 72.144(d); NQA-1/Part I/Sec II Basic 2; NUREG-0800; NUREG-1536).

12.4.3 Design Control

The applicant must have responsibilities assigned and, prior to implementation, instructions and procedures issued for the design process, design interfaces, design verification, and design changes.

12.4.3.1 Design Process

Instructions and procedures must provide for:

•   Translation of applicable regulatory requirements and the design basis into specifications, drawings, procedures, and instructions (10 CFR 72.146(a); NQA-1/Part I/Sec II Basic 3, 3S-1 Par 2, 3; NUREG-1536).

•   Actions to ensure that appropriate quality standards are specified and included in design documents (10 CFR 72.146(a); NQA-1/Part I/Sec II 3S-1 Par 3; NUREG-1536) and deviations from standards are controlled (10 CFR 72.146(a); NQA-1/Part I/Sec II 3S-1 Par 2; NUREG-1536).

•   Actions to ensure selection and review for suitability of application of materials, parts, equipment (including commercial-grade items and computer systems), and processes that are essential to the functions of the SSCs important to safety (NQA-1/Part I/Sec II 3S-1 Par 3; NUREG-1536).

•   Application of the above actions to criticality physics, radiation, shielding, stress, thermal, hydraulic, and accident analyses; compatibility of materials; accessibility for in-service inspection, maintenance, and repair; features to facilitate decontamination; and delineation of acceptance criteria for inspections and tests (10 CFR 72.146(a)).

12.4.3.2 Design Interfaces

Instructions and procedures must provide for:

•   Actions to identify and control design interfaces and to provide for coordination among participating design organizations (10 CFR 72.146(b); NQA-1/Part I/Sec II 3S-1 Par 6; NUREG-0800; NUREG-1536).

•   Actions to establish written procedures among participating design organizations for the review, approval, release, distribution, and revision of documents involving design interfaces (10 CFR 72.146(b); NQA-1/Part I/Sec II 3S-1 Par 6).

12.4.3.3 Design Verification

Instructions and procedures must provide for:

- Actions to verify and check the adequacy of design, by methods such as design reviews or alternate or simplified calculational methods, or by a suitable testing program (10 CFR 72.146(b); NQA-1/Part I/Sec II 3S-1 Par 4).

- Verifying or checking of designs by individuals or groups other than those who were responsible for the original design (and normally other than the designer's immediate supervisor) and who have a level of skill at least equal to that of the original designer to (a) confirm that the design of a structure, system, or component is suitable for its intended purpose by critical reviews of design inputs, assumptions, design methods, incorporation of inputs into the design, design outputs, and design interfaces; and (b) verify correctness of design calculations or analyses by calculations or analyses using alternate methods (10 CFR 72.146(b); NQA-1/Part I/Sec II 3S-1 Par 4; NUREG-0800; NUREG-1536).

- Ensuring that a test program used to verify the adequacy of a specific design feature, in lieu of other verifying or checking processes, includes suitable qualification testing of a prototype or example unit under the most adverse design conditions (10 CFR 72.146(b); NQA-1/Part I/Sec II 3S-1 Par 4.2.3; NUREG-0800; NUREG-1536).

12.4.3.4 Design Change

Instructions and procedures must provide for:

- Subjecting design changes, including field changes, to design control measures commensurate with those applied to the original design (10 CFR 72.146(c); NQA-1/Part I/Sec II 3S-1 Par 5; NUREG-0800; NUREG-1536).

- Obtaining NRC approval of changes in the conditions specified in the license (10 CFR 72.146(c)).

12.4.4 Procurement Document Control

12.4.4.1 Control Process

The applicant must have responsibilities assigned and, prior to implementation, instructions and procedures issued for ensuring that documents for procurement of material, equipment, or services issued by the applicant or its contractors or subcontractors, and changes thereto:

- Include a statement of work to be performed by the supplier (10 CFR 72.148; NQA-1/Part I/Sec II 4S-1 Par 2.1).

- Include or reference applicable regulatory requirements, design bases, and other requirements that are necessary to ensure adequate quality of purchased items or services. (10 CFR 72.148; NQA-1/Part I/Sec II 4S-1 Par 2.2; NUREG-0800; NUREG-1536).

- Specify technical requirements (may be done by reference to specific drawings, specifications, codes, standards, regulations, procedures, or instructions) (10 CFR 72.148; NQA-1/Part I/Sec II 4S-1, Par 2.2; NUREG-0800; NUREG-1536).

- Identify test, inspection, and acceptance requirements of the purchaser (10 CFR 72.148; NQA-1/Part I/Sec II 4S-1 Par 2.2; NUREG-0800; NUREG-1536).

- Provide for access to the supplier's facilities and records for inspection or audit by the purchaser or purchaser's representative (10 CFR 72.148; NQA-1/Part I/Sec II 4S-1 Par 2.4; NUREG-1536).

- Identify documentation to be submitted for information, review, or approval by the purchaser (10 CFR 72.148; NQA-1/Part I/Sec II 4S-1 Par 2.5; NUREG-1536).

- Prescribe retention times and disposition requirements for QA records that are to be maintained by the supplier (10 CFR 72.148; NQA-1/Part I/Sec II 4S-1 Par 2.5; NUREG-1536).

- Include purchaser's requirements for reporting and approving disposition of nonconformances (10 CFR 72.148; NQA-1/Part I/Sec II 4S-1 Par 2.6).

- Require identification of appropriate spare and replacement parts or assemblies and delineation of the technical and QA-related data required for ordering these parts or assemblies (10 CFR 72.148; NQA-1/Part I/Sec II 4S-1 Par 2.7).

- Are reviewed prior to transmission to the supplier to ensure that they include appropriate provisions for assuring that items or services will meet specified requirements (10 CFR 72.148; NQA-1/Part I/Sec II 4S-1 Par 3; NUREG-1536).

12.4.4.2 Contractor QA Programs

The applicant must have responsibilities assigned and, prior to implementation, instructions and procedures issued for requiring, to the extent necessary, that contractors or subcontractors adhere to a QA program consistent with the applicable provisions of 10 CFR 72, Subpart G (10 CFR 72.148; NQA-1/Part I/Sec II 4S-1 Par 2.3; NUREG-0800).

12.4.5 Instructions, Procedures, and Drawings

12.4.5.1 Development Process

The applicant must have responsibilities assigned and, prior to implementation, instructions and procedures issued for obtaining or producing instructions, procedures, and drawings that:

- Prescribe how activities affecting quality are to be performed.

- Are appropriate to the circumstances under which the activities are performed.

- Include acceptance criteria for determining that important activities have been accomplished satisfactorily, (e.g., quantitative criteria such as dimensions, tolerances, and operating limits, and/or qualitative criteria such as legible, "well-defined borders," or "smooth to the touch").

- Are reviewed and concurred to by the applicant's QA organization if the instructions or procedures prescribe performance of inspections, tests, calibration, or special processes (10 CFR 72.150; NQA-1/Part I/Sec II Basic 5; NUREG-0800; NUREG-1536).

12.4.5.2 Utilization

The applicant must have responsibilities assigned and, prior to implementation, instructions and procedures issued for:

- Distributing controlled copies of instructions, procedures, or drawings important to safety to those performing the activity (see Section 12.4.6 control process)

- Enforcing adherence to the provisions of the instructions, procedures, and drawings (10 CFR 72.150; NQA-1/Part I/Sec II Basic 5).

12.4.6 Document Control

12.4.6.1 Scope of Control Process

The applicant must have responsibilities assigned and, prior to implementation, instructions and procedures issued for preparing, issuing, and changing documents that:

- Specify quality requirements or

- Prescribe activities affecting quality, including, at a minimum, design specifications; design and fabrication drawings; procurement documents; QA procedures; design criteria documents; instructions and procedures for fabrication, inspection, and tests; as-built documentation; QA procedures; and nonconformance reports (10 CFR 72.152; NQA-1/Part I/Sec II Basic 6; NUREG-1536).

12.4.6.2 Control Process

The applicant must have responsibilities assigned and, prior to implementation, instructions and procedures issued for:

- Identifying documents that prescribe activities affecting quality or specify quality requirements.

- Preparing, reviewing, and approving such documents.

- Preparing, reviewing, and approving changes to such documents, with review and approval by the same organizations that performed the original review and approval unless other organizations are specifically designated to do so.

- Distributing such documents and changes to locations where they are to be used (10 CFR 72.152; NQA-1/Part I/Sec II 6S-1 Par 2, 3; NUREG-0800; NUREG-1536).

12.4.7 Control of Purchased Material, Equipment, and Services

12.4.7.1 Control Process

The applicant must have responsibilities assigned and, prior to implementation, instructions and procedures issued for ensuring that purchased material, equipment, and services conform to relevant procurement documents by:

- Evaluating potential suppliers' capability to provide acceptable products and services, prior to award of a procurement order or contract, by assigning qualified technical and QA personnel to make evaluations. The extent of investigation should be commensurate with importance to the safety of the item or service being procured and should document results of supplier evaluations for use in current and future procurement actions (10 CFR 72.154(a); NQA-1/Part I/Sec II Basic 7, 7S-1 Par 3; NUREG-1536). The investigation must be performed by either: (a) reviewing records of past performance of suppliers in providing items or services of the type being procured or items with similar technical and quality requirements or (b) surveying supplier facilities and QA programs to determine current capability to supply items or services meeting applicant's technical and quality requirements.

- Conducting surveillance of supplier activities during fabrication, inspection, testing, and shipment of materials, components, assemblies, parts, and other products that are important to safety by: (a) identifying technical and quality requirements conformance of which to purchase orders or contracts cannot be confirmed satisfactorily by inspections or tests upon receipt; (b) identifying supplier processes or activities in which conformance to those requirements can be confirmed by inspecting, witnessing, or verifying results; (c) preparing plans specifying methods to be used in inspecting, witnessing, or verifying results of each supplier process or activity selected for surveillance; acceptance criteria for the process or activity; schedules for surveillance activities; and extent of documentation of the surveillance; (d) informing suppliers of the purpose and scheduling of surveillance activities and requirements for notification to the applicant when processes or activities are nearing hold or witness points (by reference to provisions of procurement documents or changes to them if original documents are not adequate); (e) assigning qualified technical and QA personnel to perform surveillance activities in accordance with the prepared plans and schedules; and (f) performing follow up surveillance activities when needed to verify correction of deficiencies found in scheduled surveillance activities

(10 CFR 72.154(a); NQA-1/Part I/Sec II Basic 7, 7S-1, Par 8; NUREG-0800; NUREG-1536).

- Preparing or obtaining documentary evidence of the quality of products and services that are important to safety, including: (a) reports of results of surveillance activities; (b) supplier certificates of conformance and other documents identifying specific procurement requirements that have been met for products or services delivered; (c) supplier certificates of conformance and other documents identifying specific procurement requirements (e.g., codes, standards, specifications) that have been met for products or services delivered; and (d) applicant or supplier documents identifying specific procurement requirements that have not been met and describing the disposition of nonconforming items (e.g., accept as is, repair, rework, scrap, return to vendor) (10 CFR 72.154(b); NQA-1/Part I/Sec II Basic 7, 7S-1 Par 6, 8; NUREG-1536).

- Retaining or having available the documentary evidence of quality of products and services that are important to safety for the life of ISFSI or MRS (10 CFR 72.154(b)).

- Performing receiving inspections upon delivery of purchased items and materials to ensure that: (a) items or materials received are correctly identified and are the items or materials specified by relevant purchasing documents; (b) items or materials received conform to predetermined acceptance requirements (e.g., dimensions, weight, color, condition, accompanying documents) before they are released for use; and (c) items or materials that do not conform to predetermined acceptance requirements or require further inspections or tests before acceptance are identified and placed under appropriate controls (see Sections 12.4.14 and 12.4.15) (10 CFR 72.154(a); NQA-1/Part I/Sec II Basic 7, 7S-1 Par 8; NUREG-0800; NUREG-1536).

12.4.7.2 Acceptance of Services Only

The applicant must have responsibilities assigned and, prior to implementation, instructions and procedures issued for accepting services such as third-party inspection; engineering and consulting services; and installation, repair, overhaul, or maintenance work by:

- Technical verification of data produced.

- Surveillance or audit of the activity.

- Review of certifications and reports submitted for evidence of conformance to procurement document requirements (10 CFR 72.154(a); NQA-1/Part I/Sec II 7S-1 Par 8.3).

12.4.7.3 Commercial-Grade Items

The applicant must have responsibilities assigned and, prior to implementation, instructions and procedures issued for:

- Identifying commercial-grade items in approved design output documents.

- Conducting supplier evaluation when warrented by complexity and importance to safety.

- Identifying commercial-grade items in purchase documents by the manufacturer's published product description (e.g., catalog number).

- Inspecting and testing items received to ensure conformance to the manufacturer's published requirements.

- Documenting receipt and acceptability of the item (10 CFR 72.154(a),(b); NQA-1/Part I/Sec II 7S-1, Par 10).

12.4.7.4 Assessments of Effectiveness

The applicant must have responsibilities assigned and, prior to implementation, instructions and procedures issued for assessing the effectiveness of quality control by contractors and subcontractors at intervals consistent with the importance, complexity, and quantity of the product or services (10 CFR 72.154(c); NQA-1/Part I/Sec II 7S-1, Par 5; NUREG-0800; NUREG-1536).

12.4.8 Identification and Control of Materials, Parts, and Components

12.4.8.1 Identification and Control Process

The applicant must have responsibilities assigned and, prior to implementation, instructions and procedures issued for ensuring that only correct and accepted materials, parts, and components are used or installed, by:

- Identifying purchased materials, parts, or components upon receipt by tagging, marking, or labeling; physical separation; documents traceable to the items; or other means that provide adequate identification (10 CFR 72.156; NQA-1/Part I/Sec II Basic 8, 8S-1 Par 2; NUREG-1536).

- Maintaining identification of purchased items during storage, subdivision for issue, and use or installation by means mentioned in the previous bullet or by procedural controls (10 CFR 72.156; NQA-1/Part I/Sec II Basic 8, 8S-1 Par 2; NUREG-1536).

- Identifying materials produced, items fabricated onsite at the time that they are produced or fabricated, and items assembled onsite at the time assembly is begun, by means similar to that described above for purchased items in the first bullet above; and maintaining identification of such materials and items through use or installation by means similar to that described for purchased items (10 CFR 72.156; NQA-1/Part I/Sec II Basic 8, 8S-1 Par 2; NUREG-1536).

- Providing for traceability of items (when required by codes, standards, or specifications) to: (a) applicable specification and grade of material; (b) heat, batch, lot, part, or serial number; and (c) specified inspection, test, or other records such as drawings, purchase orders, deviation reports, or reports of nonconformances and their disposition (10 CFR 72.156; NQA-1/Part I/Sec II Basic 8, 8S-1 Par 3.1; NUREG-1536).

12.4.8.2 Stored Items

The applicant must have responsibilities assigned and, prior to implementation, instructions and procedures issued for maintaining identification of items in prolonged storage or storage under adverse conditions by:

- Protecting markings and identification records of items in storage from deterioration from environmental exposure or adverse storage conditions.

- Restoring or replacing markings or identification records that are damaged by aging or storage conditions (10 CFR 72.156; NQA-1/Part I/Sec II Basic 8, 8S-1 Par 3.3).

12.4.8.3 Items With Limited Lifetimes

The applicant must have responsibilities assigned and, prior to implementation, instructions and procedures issued for:

- Identifying items with limited calendar or operating life or cycles (e.g., certain batteries, chemical products, mechanical relays, and control switches).

- Establishing and maintaining records of shelf or operating life or cycles remaining.

- Preventing issue of items whose shelf life has expired.

- Preventing further use of items that have reached the end of their operating life or cycles (10 CFR 72.156; NQA-1/Part I/Sec II Basic 8, 8S-1 Par 3.2).

12.4.9 Control of Special Processes

12.4.9.1 Control Process

The applicant must have responsibilities assigned and, prior to implementation, instructions and procedures issued for:

- Performing all processes affecting quality of items or services in accordance with instructions, procedures, drawings, checklists, travelers, or other appropriate means that specify or reference applicable codes and standards and acceptance criteria for the process (10 CFR 72.150; NQA-1/Part I/Sec II Basic 9, 1S-1 Par 2, 3.2; also see NQA-1/Part II/Subpart 2.18, Subpart 2.21).

- Identifying special processes (i.e., those meeting one or both of the following conditions): (a) process results that are highly dependent on control of the process or skill of the operators, or both; and (b) quality of results that cannot be readily determined by inspection or test of the product (e.g., welding, heat treating, and nondestructive testing) (10 CFR 72.158; NQA-1/Part I/Sec II Basic 9; NUREG-0800; NUREG-1536).

- Preparing appropriate instructions for each special process that include or reference requirements for qualifying procedures, personnel, and equipment (10 CFR 72.158; NQA-1/Part I/Sec II Basic 9, 9S-1 Par 3; NUREG-0800; NUREG-1536).

- Performing special processes in accordance with those instructions and applicable codes, standards, specifications, criteria, and other special requirements (10 CFR 72.158; NQA-1/Part I/Sec II Basic 9, 9S-1 Par 3; NUREG-1536).

12.4.9.2 Qualification

The applicant must have responsibilities assigned and, prior to implementation, instructions and procedures issued for:

- Qualifying procedures, personnel, and equipment to be used in special processes in accordance with applicable codes, standard, and specifications.

- Maintaining records of such qualifications (10 CFR 72.158; NQA-1/Part I/Sec II Basic, 9S-1 Par 3.1.1, 3.3; NUREG-0800; NUREG-1536).

12.4.10 Licensee Inspection

12.4.10.1 Scope of Inspection Program

The applicant must have responsibilities assigned and, prior to implementation, instructions and procedures issued for verifying conformance of items and activities to specified requirements by planning and conducting:

- Receiving and pre-service inspections.

- In-process inspections.

- Final inspections.

- In-service inspections, including inspections of modifications, repairs, and replacements (10 CFR 72.160; NQA-1/Part I/Sec II Basic 10, 10S-1 Par 6,7,8; also see NQA-1/Part II/Subpart 2.4).

12.4.10.2 Implementation of Program

The applicant must have responsibilities assigned and instructions and procedures issued for:

•   Preparing inspection instructions, procedures, and checklists that: (a) prescribe frequency of inspections or identify occasions on which testing is required, including inspections prescribed by documents establishing mandatory hold points, and modification, repair, or replacement of items; (b) specify sampling procedures if acceptability of batch materials or groups of items is to be based on inspection of a sample from the batch or group; (c) identify characteristics of items and activities to be inspected to verify conformance to original design and inspection requirements (or acceptable alternatives for modifications, repairs, and replacements); (d) describe means of inspection (e.g., examinations, measurements, tests) of each item and activity to be inspected; (e) describe methods of monitoring processing methods, equipment, and personnel if inspection of processed items is impossible or disadvantageous, or if a combination of inspection and monitoring is required, for an item or activity; (f) establish acceptance criteria for inspections and monitoring; and (g) provide for recording names of inspectors and data recorders used, and objective evidence of results of inspections and monitoring (10 CFR 72.160; NQA-1/Part I/Sec II Basic 10, 10S-1 Par 5; NUREG-1536).

•   Performing inspections in accordance with the instructions, procedures, and checklists, and using inspection personnel who: (a) are qualified to perform the inspection task; (b) did not perform or supervise the work being inspected; and (c) do not report directly to supervisors who are responsible for the work (10 CFR 72.160; NQA-1/Part I/Sec II Basic 10, 10S-1 Par 3; NUREG-1536).

•   Qualifying and certifying personnel to be used in inspections or monitoring in accordance with requirements of applicable codes, standards, and specifications (10 CFR 72.160; NQA-1/Part I/Sec II Basic 10, 10S-1 Par 3.2; NUREG-1536).

12.4.11 Test Control

12.4.11.1 Scope of Test Program

The applicant must have responsibilities assigned and, prior to implementation, instructions and procedures issued for verifying conformance of items (including safety-related computer programs) and activities to specified requirements by planning and conducting: prototype qualification tests, production tests, proof tests before installation, construction tests, pre-operational tests, and operational tests (10 CFR 72.162; NQA-1/Part I/Sec II Basic 11, 11S-1 Par 2; also see NQA-1/Part II/Subpart 2.4, Subpart 2.7).

12.4.11.2 Test Control Process

The applicant must have responsibilities assigned and, prior to implementation, instructions and procedures issued for:

- Identifying testing required to demonstrate that SSCs important to safety can perform satisfactorily in service,.

- Preparing test procedures that include or reference documents containing: (a) test objectives; (b) prerequisites for the test (e.g., condition, state, or configuration of the item or facility; required environmental conditions; instrumentation and personnel required); (c) applicable design, procurement document, and facility license requirements; (d) instructions and procedures for performing the test and recording results; and (e) instructions and procedures for evaluating test results (10 CFR 72.162; NQA-1/Part I/Sec II Basic 11, 11S-1 Par 3; also see NQA-1/Part II/Subpart 2.4, Subpart 2.7; NUREG-0800; NUREG-1536).

- Performing testing and evaluating results in accordance with the instructions and procedures (10 CFR 72.162; NQA-1/Part I/Sec II Basic 11, 11S-1 Par 3, 4; 11S-2; also see NQA-1/Part II/Subpart 2.4, Subpart 2.7; NUREG-0800; NUREG-1536).

- Preparing test records that identify, at a minimum, the item tested, date of test, names of test personnel or data recorder, type of observation, results and acceptability, action taken in connection with any deviations noted, and persons evaluating test results (10 CFR 72.162; NQA-1/Part I/Sec II Basic 11, 11S-1 Par 5 (see 11S-2 for computer programs); NUREG-0800; NUREG-1536).

12.4.12 Control of Measuring and Test Equipment

12.4.12.1 Control Process

The applicant must have responsibilities assigned and, prior to implementation, instructions and procedures issued for:

- Selecting measuring and test equipment (MTE) for use in processes, inspections, and tests that: (a) is of the type appropriate for measuring specified physical characteristics of items being processed, inspected, or tested (e.g., appropriate instruments, tools, gauges, fixtures, reference and transfer standards, nondestructive inspection and test equipment, or other devices) (10 CFR 72.164; NUREG-0800; NUREG-1536); and (b) has range, accuracy, and tolerance sufficient to determine conformance of specified physical characteristics to specified requirements (10 CFR 72.164; NQA-1/Part I/Sec II 12S-1 Par 2).

- Identifying each item of MTE in such a way that it can be: (a) distinguished from all others of similar type (e.g., by permanent markings such as serial numbers or by applied labels); and (b) traced to calibration data for it (e.g., by tagging, labeling, and/or documentation) (10 CFR 72.164; NQA-1/Part I/Sec II Basic 12; NUREG-0800; NUREG-1536).

- Prescribing methods and frequency (or occasions, such as immediately before use) for calibration and adjustment of MTE, on the basis of type of equipment, stability characteristics, required accuracy, intended use, and other conditions affecting measurements (10 CFR 72.164; NQA-1/Part I/Sec II Basic 12, 12S-1 Par 3.2; NUREG-0800; NUREG-1536).

- Performing calibration, adjustment, and maintenance of MTE on prescribed frequencies or occasions, or when its accuracy is suspect, against: (a) certified equipment having known, valid relationships to nationally recognized standards; or (b) documented bases for calibration if no nationally recognized standards exist (10 CFR 72.164; NQA-1/Part I/Sec II Basic 12, 12S-1 Par 3.1; also see Part II/Subpart 2.16; NUREG-1536).

- Indicating status (acceptable or not) of MTE "as found" and after each calibration, adjustment, or maintenance, by tagging or documentation traceable to the MTE (10 CFR 72.164; NQA-1/Part I/Sec II Basic 12, 12S-1 Par 3.2; NUREG-1536).

- Tagging or segregating out-of-calibration MTE to prevent its use until it has been re-calibrated (10 CFR 72.164; NQA-1/Part I/Sec II Basic 12, 12S-1 Par 3.2; NUREG-1536).

- Evaluating and documenting validity of previous inspection or test results, and acceptability of items previously inspected or tested when MTE is found to be out of calibration (10 CFR 72.164; NQA-1/Part I/Sec II Basic 12, 12S-1 Par 3.2; NUREG-0800; NUREG-1536).

- Repairing or replacing equipment found to be consistently out of adjustment (10 CFR 72.164; NQA-1/Part I/Sec II Basic 12, 12S-1 Par 3.2).

- Handling and storing such MTE so that its accuracy is maintained (10 CFR 72.164; NQA-1/Part I/Sec II Basic 12, 12S-1 Par 4).

- Maintaining records of calibration status that are traceable to the MTE (10 CFR 72.164; NQA-1/Part I/Sec II Basic 12, 12S-1 Par 5; NUREG-0800; NUREG-1536).

12.4.12.2 Commercial Devices

Commercial equipment such as rules, tape measures, and levels need not be subjected to the above control process if normal commercial equipment provides adequate accuracy. (10 CFR 72.164; NQA-1/Part I/Sec II 12S-1, Par 3.3).

12.4.13 Handling, Storage, and Shipping Control

12.4.13.1 Control Process

The applicant must have responsibilities assigned and, prior to implementation, instructions and procedures issued for:

* Preparing and issuing work and inspection instructions, drawings, specifications, shipment instructions, or other documents or procedures prescribing activities to prevent damage or deterioration of items during handling, storage, and shipping (10 CFR 72.166; NQA-1/Part I/Sec II Basic 13, 13S-1 Par 2; NUREG-0800; NUREG-1536).

* Identifying requirements for special equipment (e.g., containers, shock absorbers, accelerometers), special protective environments (e.g., inert gas, moisture content, temperatures), or special procedures for particular items (10 CFR 72.166; NQA-1/Part I/Sec II Basic 13, 13S-1 Par 3.1, 3.2; NUREG-0800; NUREG-1536).

* Conducting cleaning, preserving, handling, storing, packing, and shipping activities in accordance with the prepared instructions, procedures, and special requirements, by using appropriately- trained and qualified personnel and appropriate tools and equipment (10 CFR 72.166; NQA-1/Part I/Sec II Basic 13, 13S-1 Par 3.4; NUREG-0800; NUREG-1536).

12.4.13.2 Tools and Equipment

The applicant must have responsibilities assigned and, prior to implementation, instructions and procedures issued for:

* Identifying special handling tools and equipment that are required for safe handling of items.

* Inspecting and testing such tools and equipment in accordance with specified procedures and at specified intervals to verify that they are adequately maintained (10 CFR 72.166; NQA-1/Part I/Sec II Basic 13, 13S-1 Par 3.3; NUREG-1536).

12.4.13.3 Markings

The applicant must have responsibilities assigned and, prior to implementation, instructions and procedures issued for marking or labeling items being handled, shipped, or stored to identify the items and any special environments or controls they require (10 CFR 72.166; NQA-1/Part I/Sec II Basic 13, 13S-1 Par 4; NUREG-1536).

12.4.14 Inspection, Test, and Operating Status

12.4.14.1 Inspection and Test Status Process

The applicant must have responsibilities assigned and, prior to implementation, instructions and procedures issued for:

- Indicating the status of inspections and tests of individual items of the facility by markings such as stamps, tags, labels, and routing cards, or in documents accompanying or traceable to the item.

- Preventing installation, use, or further processing of an item that has not passed prerequisite inspections and tests, by tagging, labeling, or segregating, and by procedural controls.

- Specifying authority and procedures for applying or removing inspection and test status stamps, tags, markings, and labels (10 CFR 72.168(a); NQA-1/Part I/Sec II Basic 14; NUREG-0800; NUREG-1536).

12.4.14.2 Operating Status Process

The applicant must have responsibilities assigned and, prior to implementation, instructions and procedures issued for:

- Preventing inadvertent use or operation of a structure, system, or component of the facility by indicating its operating status on tags or markings on control panels, switches, and other locations where its use or operation can be initiated.

- Specifying authority and procedures for applying or removing operating status tags and markings (10 CFR 72.168(b); NQA-1/Part I/Sec II Basic 14; NUREG-0800; NUREG-1536).

12.4.15 Nonconforming Materials, Parts, or Components

12.4.15.1 Control Process

The applicant must have responsibilities assigned and, prior to implementation, instructions and procedures issued for:

- Identifying materials, parts, or components that do not conform to the applicant's requirements (10 CFR 72.170; NQA-1/Part I/Sec II Basic 15, 15S-1 Par 2; NUREG-0800; NUREG-1536).

- Segregating such items and materials in designated hold areas, or by other precautions if segregation is not practicable to prevent their inadvertent use or installation (10 CFR 72.170; NQA-1/Part I/Sec II Basic 15, 15S-1 Par 3, 4.1; NUREG-0800; NUREG-1536).

- Identifying individuals or groups with authority to perform evaluations and approve disposition of nonconforming items and materials (e.g., use-as-is, reject, repair, rework) (10 CFR 72.170; NQA-1/Part I/Sec II Basic 15, 15S-1 Par 4.2).

- Conducting evaluations in accordance with approved instructions and procedures using personnel with: (a) demonstrated competence in the area they are evaluating; (b) an adequate understanding of the requirements; and (c) access to pertinent background information (10 CFR 72.170; NQA-1/Part I/Sec II Basic 15, 15S-1 Par 4.3).

- Applying design control measures to items dispositioned "use-as-is" or "repair" commensurate with those applied in the original design (10 CFR 72.170; NQA-1/Part I/Sec II Basic 15, 15S-1 Par 4.4; NUREG-1536).

- Documenting the nonconformance and disposition, with information in the documentation, including: (a) identification of the nonconforming item or material by heat, batch, lot, part, or serial number; (b) identification of the specification or requirement that was not met; (c) description of the nonconformance; (d) disposition of the nonconforming item or material; (e) technical justification for "use-as-is" or "repair" dispositions; and (f) individual or group authorizing the disposition (10 CFR 72.170; NQA-1/Part I/Sec II Basic 15, 15S-1 Par 4.4; NUREG-0800; NUREG-1536).

- Conducting reinspections and retests of items that are repaired or reworked against the original acceptance criteria unless the disposition of the nonconforming item established alternate acceptance criteria (10 CFR 72.170; NQA-1/Part I/Sec II Basic 15, 15S-1 Par 4.5; NUREG-0800; NUREG-1536).

12.4.15.2 Notification

The applicant must have responsibilities assigned and, prior to implementation, instructions and procedures issued for notifying affected organizations of nonconforming items (10 CFR 72.170; NQA-1/Part I/Sec II Basic 15).

12.4.16 Corrective Action

12.4.16.1 Initiation Process

The applicant must have responsibilities assigned and, prior to implementation, instructions and procedures issued for:

- Identifying conditions adverse to quality, such as failures, malfunctions, deficiencies, deviations, defective material and equipment, and nonconformances.

- Identifying a subset of such conditions that are significant conditions adverse to quality.

- Reporting significant conditions adverse to quality to appropriate levels of management (10 CFR 72.172; NQA-1/Part I/Sec II Basic 16; NUREG-0800; NUREG-1536).

12.4.16.2 Correction Process

The applicant must have responsibilities assigned and, prior to implementation, instructions and procedures issued for:

- Determining the cause of significant conditions adverse to quality.

- Taking action to prevent their recurrence.

- Documenting action taken and reporting it to appropriate levels of management for review and approval.

- Following up to verify implementation of corrective actions (10 CFR 72.172; NQA-1/Part I/Sec II Basic 16; NUREG-0800; NUREG-1536).

12.4.17 QA Records

12.4.17.1 Scope of Records Program

The applicant must have responsibilities assigned and, prior to implementation, instructions and procedures issued for:

- Generating, supplying, or maintaining, by or for the applicant, records of the quality of SSCs and activities important to safety (10 CFR 72.174; NQA-1/Part I/Sec II Basic 17, 17S-1 Par 2.2; NUREG-1536).

- Generating, supplying, or maintaining records including: (a) design records; (b) procurement records; (c) records of configuration of SSCs important to safety, "as built" and as changed; (d) records of conformance to operations requirements and constraints; (e) records of qualification of personnel, procedures, and equipment; (f) records of inspections, tests, surveillances, audits, and assessments; and (g) records of nonconformances (for material, items, processes and activities), disposition of nonconforming items and materials, and actions to correct nonconformances in process and activities (10 CFR 72.174; NQA-1/Part I/Sec II Basic 17, 17S-1 Par 2.2; NUREG-1536).

12.4.17.2 Implementation

The applicant must have responsibilities assigned and, prior to implementation, instructions and procedures issued for:

- Specifying in design specifications, procurement documents, operational procedures, or other documents the types of records to be generated, supplied, or maintained by or for the applicant (10 CFR 72.174; NQA-1/Part I/Sec II Basic 17, 17S-1 Par 2.2; NUREG-1536).

- Preparing records (individual records or groups of records assembled into record packages), including: (a) identifying the item, process, or activity to which the record applies; (b) assigning unique identification to each record; and (c) validating (e.g., stamping, initialing, signing and dating) or otherwise authenticating records (e.g., by a statement from the responsible individual or organization) (10 CFR 72.174; NQA-1/Part I/Sec II Basic 17, 17S-1 Par 2.3, 2.6; NUREG-1536).

- Distributing and handling records, including: (a) identifying the individual having custody of each record by a system of receipt control; and (b) specifying measures to be taken by the individual having custody to protect them from damage or loss (10 CFR 72.174; NQA-1/Part I/Sec II Basic 17, 17S-1 Par 3; NUREG-1536).

- Classifying records (lifetime or nonpermanent) and indexing records (at a minimum, by retention times and location within the records system) (10 CFR 72.174; NQA-1/Part I/Sec II Basic 17, 17S-1 Par 2.4, 2.7; NUREG-1536).

- Storing, preserving, and protecting records in storage, including: (a) providing storage facilities meeting requirements of applicable codes, standards, and regulations; and (b) specifying procedures for receiving, storing, preserving, and protecting records in such storage facilities (10 CFR 72.174; NQA-1/Part I/Sec II Basic 17, 17S-1 Par 4; NUREG-0800; NUREG-1536).

- Retrieving records, including identification of personnel who may have access to the applicant's record files, and procedures for retrieving records maintained by suppliers or contractors (10 CFR 72.174; NQA-1/Part I/Sec II Basic 17, 17S-1 Par 5).

- Disposition of records, including ensuring that disposition of records is governed by the most stringent regulatory requirements that apply to the records (this may be an agency other than the NRC) and ensuring that supplier's nonpermanent records are not disposed of until the following conditions are met: (a) items are released for shipment and a Code Data Report is signed, or a Code Symbol Stamp is affixed; (b) regulatory requirements are satisfied; (c) operational status permits; (d) warranty consideration is satisfied, and (e) purchaser's requirements are satisfied (10 CFR 72.174; NQA-1/Part I/Sec II Basic 17, 17S-1 Par 6).

12.4.18 Audits

12.4.18.1 Scope of Audit Program

The applicant's audit program must address planning and performance of audits to:

- Verify compliance with specifications and other requirements for activities affecting quality.

- Determine the effectiveness of the QA program.

- Verify that supplier's and contractor's QA programs comply with applicable requirements of 10 CFR 72 and meet acceptance criteria equivalent to those for such requirements in this Standard Review Plan (SRP) (10 CFR 72.176; NQA-1/Part I/Sec II Basic 18, 18S-1; NUREG-0800; NUREG-1536).

12.4.18.2 Implementation

The applicant must have responsibilities assigned and instructions and procedures issued for:

- Initiating audits early enough to ensure effective QA of activities during the early stages of the installation life cycle (e.g., design, procurement) (10 CFR 72.176, 72.154; NQA-1/Part I/Sec II 18S-1 Par 4; NUREG-0800; NUREG-1536).

- Scheduling audits at a frequency commensurate with the status and importance of the activity audited (10 CFR 72.176; NQA-1/Part I/Sec II 18S-1 Par 2; NUREG-0800; NUREG-1536).

- Preparing written audit plans for each audit that identify the audit scope, requirements, audit personnel, activities to be audited, organizations to be notified, applicable documents, schedule, and written procedures or checklists (10 CFR 72.176; NQA-1/Part I/Sec II 18S-1 Par 3.1).

- Selecting and assigning auditors who are members or under direction of the applicant QA organization and who are independent of any direct responsibilities for performance of the activities they audit (10 CFR 72.176; NQA-1/Part I/Sec II 18S-1 Par 3.2, 3.3; NUREG-1536).

- Performing audits with an audit team led by a certified lead auditor in accordance with written procedures and checklists for the audit (10 CFR 72.176; NQA-1/Part I/Sec II 18S-1 Par 3.2, 3.3; NUREG-1536).

- Preparing an audit report for each audit and requesting a response from the audited organization, additionally requiring that the aduit report be signed by the audit team leader and include: (a) a description of the audit scope; (b) identification of the auditors; (c) identification of persons contacted during audit activities; (d) a summary of audit results, including a statement about the effectiveness of the QA program elements that were audited; and (e) a description of each adverse audit finding in sufficient detail to enable the audited organization to take corrective action (10 CFR 72.176; NQA-1/Part I/Sec II Basic 18, 18S-1 Par 5; NUREG-1536).

- Evaluating and concurring in corrective action planned by the audited organization and taking follow up action to verify that agreed-upon corrective action is performed as scheduled (10 CFR 72.176; NQA-1/Part I/Sec II Basic 18, 18S-1 Par 6, 7; NUREG-1536).

- Maintaining audit plans, audit reports, written replies, and records of completion of corrective action as QA records (10 CFR 72.176; NQA-1/Part I/Sec II Basic 18, 18S-1 Par 8).

12.5 Review Procedures

The objective of this review is to determine whether that the applicant has developed and described a QA program for design, fabrication, construction, testing, operations, modification, and decommissioning that complies with the requirements of 10 CFR Part 72, Subpart G, and has committed to comply with this program. If the QA program is graded, the staff should be able to conclude that the structure of the graded program is acceptable and that the highest levels of QA are applied to those SSCs that are most important to safety.

The applicant's QA program must have sufficient detail for NRC staff to evaluate it against the acceptance criteria. It is possible that aspects of the QA program are described in various portions of the application (e.g., structural analysis sections of the Safety Analysis Report (SAR) and a separate QA plan submittal). The staff should review all of this information and verify its internal consistency. This staff should use this information when evaluating the program against the acceptance criteria of Section 12.4. The staff reviewer should examine the scope of the program and verify that it applies to design, fabrication, construction, testing, operations, modifications, and decommissioning. The reviewer should evaluate the program against the acceptance criteria identified in Section 12.4 or a functional equivalent.

Having a graded QA program is considered acceptable. If the applicant presents a graded QA program, the staff reviewer should review the applicant's justification for the nature of the gradation. The staff should verify that any gradation applies the highest levels of QA to those SSCs that are most important to safety. In making determinations about the application of QA to those SSCs, the staff reviewer should coordinate with the NRC project manager and use those SSCs that the NRC staff has determined are important to safety.

During the review of the application, the staff should evaluate the performance of the QA program related to design control. The applicant's design control program and associated QA program controls, as well as the implementing procedures for design control, must be in place before design activities begin.

After the license is issued, the staff performs inspections to determine whether the procedures for design control have been effectively implemented and executed. The licensee is required to take corrective actions to resolve all deficiencies identified by the staff during inspections of design activities.

The lead QA reviewer should consider obtaining assistance from NRC staff who are involved in specific technical review areas to help evaluate the applicant's program description.

Upon completion of their review, staff reviewers should document their findings in an SER that summarizes conduct of the review, what material in the application forms the basis for acceptance, and any recommendations for modifications in the application that are required before the application can be accepted.

The QA program can be determined to be acceptable if the following evaluation criteria are met:

- The description of the QA program meets the acceptance criteria in Section 12.4 for each QA criterion applicable to activities to be performed.

- The QA program covers activities affecting SSCs important to safety.

- The organizations and persons performing QA functions have the independence and authority to perform their functions without undue influence from those directly responsible for costs and schedules.

12.6 Evaluation Findings

The reviewer prepares evaluation findings on satisfaction of the regulatory requirements relating to QA shown in Section 12.3. If the reviewer concludes that information provided with the application, along with additional information provided in response to NRC requests, shows that the QA program meets the acceptance requirements shown in Section 12.3, findings of the following type should be included in the staff's SER (finding numbering is for convenience in referencing within the FSRP and SER):

F12.1   Based upon a detailed review and evaluation of the QA program description contained in the [Safety Analysis Report] for [ISFSI or MRS installation], the staff concludes that:

- The QA program describes requirements, procedures, and controls that, when properly implemented, comply with the requirements of 10 CFR 72, Subpart G.

- The QA program covers activities affecting SSCs important to safety as identified in the Safety Analysis Report.

- The organizations and persons performing QA functions have the independence and authority to perform their functions without undue influence from those directly responsible for costs and schedules.

- The licensee's description of the QA program is in compliance with applicable NRC regulations and industry standards, and the QA program can be implemented for the (specify: design, fabrication and construction, operation, decommissioning) phases of the installation's life cycle.

12.7 References

NRC documents referenced are identified at Consolidated References, Section 17.

American Society of Engineers, "Quality Assurance Program Requirements for Nuclear Facilities," ANSI/ASME NQA-1-1983.

13 DECOMMISSIONING EVALUATION

13.1 Review Objective

The primary objective of the review is to ensure that the applicant's provisions for eventual decontamination and decommissioning of the independent spent fuel storage installation (ISFSI) or monitored retrievable storage (MRS) give reasonable assurance of adequate protection of public health and safety. The review examines the design and operational features intended to facilitate eventual decommissioning, and the proposed decommissioning plan and associated financial assurance and recordkeeping requirements.

The overview of the decommissioning evaluation process given in Figure 13.1 shows that the decommissioning evaluation draws information from the application and from the results of design criteria evaluation and the conduct of operations evaluation.

13.2 Areas of Review

The following outline shows the areas of review addressed in Section 13.4, Acceptance Criteria, and Section 13.5, Review Procedures:

Design Features
Operational Features
Decommissioning Plan

13.3 Regulatory Requirements

This section identifies and presents a high-level summary of Title 10 of the Code of Federal Regulations (CFR) Part 72 relevant to the review areas addressed by this chapter. The NRC staff reviewer should read the exact regulatory language. The decommissioning of an ISFSI or MRS at the end of its useful life must also comply with the decommissioning criteria of 10 CFR Part 20, Subpart E, "Radiological Criteria for License Termination." A matrix at the end of this section matches the regulatory requirements identified in this section to the areas of review identified in the previous section.

72.24 Contents of application: Technical information
"Each application for a license under this part ... must consist of the following:
(g) An identification and justification for the selection of those subjects that will be probable license conditions and technical specifications.
(n) A description of the quality assurance program.
(q) A description of the decommissioning plan."

72.30 Financial assurance and recordkeeping for decommissioning
(a) "Each application under this part must include a proposed decommissioning plan."
(b) "The proposed decommissioning plan must also include a decommissioning funding plan."

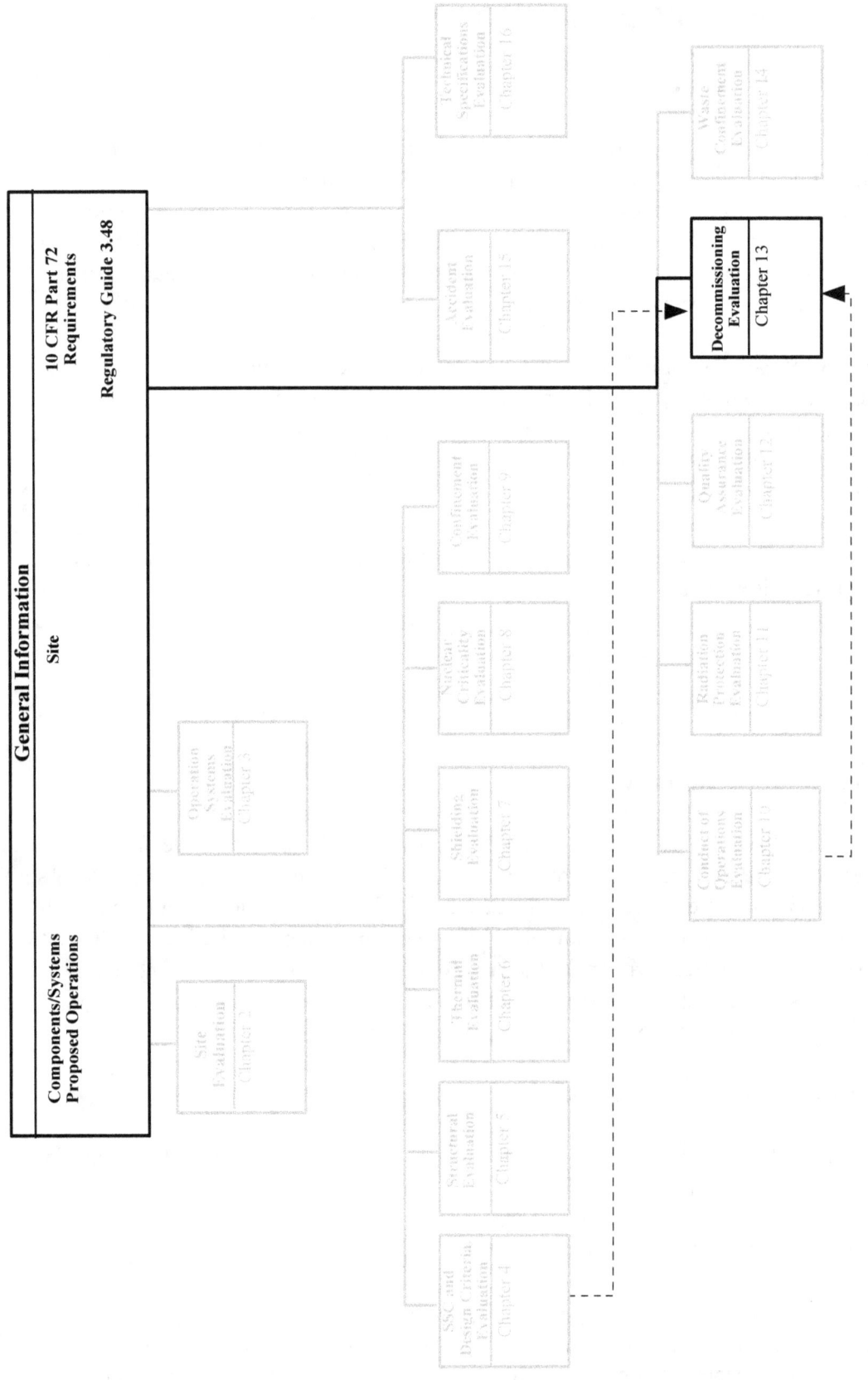

Figure 13.1 Overview of Decommissioning Evaluation

(d) "Each person licensed under this part shall keep records of information important to the decommissioning of a facility in an identified location until the site is released for unrestricted use.... Information the Commission considers important to decommissioning consists of --

> (1) Records of spills or other unusual occurrences involving the spread of contamination in and around the facility, equipment, or site
>
> (2) As-built drawings and modifications of structures and equipment in restricted areas
>
> (3) A list contained in a single document and updated no less than every 2 years of the following:
>
> > (i) All areas designated and formerly designated as restricted areas as defined under 10 CFR 20.1003; and
> >
> > (ii) All areas outside of restricted areas that require documentation under 72.30(d)(1).
>
> (4) Records of the cost estimate performed for the decommissioning funding plan."

Subpart F - General Design Criteria

72.130. Criteria for decommissioning

"The ISFSI or MRS must be designed for decommissioning. Provisions must be made to facilitate decontamination of structures and equipment, minimize the quantity of radioactive wastes and contaminated equipment, and facilitate the removal of radioactive wastes and contaminated materials at the time the ISFSI or MRS is permanently decommissioned."

A matrix showing the primary relationship of these regulations to the specific areas of review in this chapter is given in Table 13.1. The reviewer should independently verify the relationships in this matrix to ensure that no requirements are overlooked because of unique applicant design features.

Table 13.1 Relationship of Regulations and Areas of Review

Areas of Review	10 CFR Part 72 Regulations	
	72.24	72.30
Design Features	●	●
Operational Features	●	●
Decommissioning Plan	●	●

13.4 Acceptance Criteria

The ISFSI must be decommissioned at the end of service life, and every effort must be made to terminate the license and release the ISFSI site for unrestricted use according to the requirements of 10 CFR Part 20, Subpart E, "Radiological Criteria for License Termination." The requirements related to eventual decommissioning the ISFSI or MRS applicable at the time of initial licensing are satisfied if the applicant adequately addresses the acceptance criteria for design features, operational features, and decommissioning plan.

13.4.1 Design Features

The application must identify the design features included in the design of the ISFSI or MRS that will facilitate decontamination and decommissioning. This information may be in the Safety Analysis Report (SAR) or in the decommissioning plan.

Design features include surfaces that are less susceptible to contamination (or activation) and are readily decontaminated, as well as shielding to minimize any occupational exposure associated with decontamination. Design features also include equipment to facilitate the decontamination and removal of air circulation and filtration systems, and components of waste treatment and packaging systems.

13.4.2 Operational Features

The application must identify the operational features that will facilitate eventual decontamination and decommissioning of the ISFSI or MRS. Such features include minimizing contamination buildup on components, maintaining accurate records of spills or other unusual occurrences involving the spread of contamination, and maintaining accurate as-built drawings or suitable substitutions. This information is in either the SAR or the decommissioning plan, and includes technical specifications or aspects of the proposed quality assurance (QA) program.

13.4.3 Decommissioning Plan

The application must include a proposed decommissioning plan as required by 10 CFR 72.30. The plan must describe the proposed practices and procedures for (a) the decontamination of the site and facilities, and (b) the disposal of residual radioactive materials after the stored spent fuel or high-level waste has been removed. Design features of the ISFSI or MRS that facilitate its decommissioning at the end of its useful life must be identified and discussed.

The plan should provide reasonable assurance that the proposed decontamination and decommissioning of the ISFSI or MRS will adequately protect public health and safety and will leave the site suitable for unrestricted use. A site is considered acceptable for unrestricted use if (a) residual radioactivity has been reduced to levels that are as low as is reasonably achievable (ALARA), and (b) compliance with other radiological criteria of 10 CFR 20.1402 can be demonstrated.

The decommissioning plan submitted with the license application need not comply with the form and content requirements of Regulatory Guide 3.65, "Standard Format and Content of Decommissioning Plans for Licensees Under 10 CFR Parts 30, 40, and 70." Regulatory Guide 3.65 provides guidance on the content and format of final decommissioning plans submitted at the time of license termination.

As part of the decommissioning plan, the application must contain a funding plan, which, in turn, includes a cost estimate for the decommissioning and a financial assurance mechanism that will ensure availability of funds in the amount of the cost estimate.

Guidance on the format and content of the financial assurance mechanism and the means for cost estimating are provided in Regulatory Guide 3.66, "Standard Format and Content of Financial Assurance Mechanisms Required for Decommissioning Under 10 CFR 30, 40, 70 and 72." A legal, executed copy of the financial assurance mechanism must be provided. Acceptance criteria are provided in NUREG-1337, Rev. 1, "Standard Review Plan (SRP) for the Review of Financial Assurance Mechanisms for Decommissioning Under 10 CFR Parts 30, 40, 70, and 72."

The funding plan must be signed (i.e., certified) by an individual authorized to make financial commitments for the applicant.

13.5 Review Procedures

13.5.1 Design Features

The reviewer should first ensure that the application identifies, discusses, and justifies the design features and choices as they relate to decommissioning the ISFSI or MRS, as required by 10 CFR 72.24(g), 72.30, and 72.130.

The reviewer should determine whether the design satisfactorily facilitates decommissioning. The design can be considered to meet this requirement if (a) provisions are incorporated where feasible and economic, and (b) design choices that support decommissioning were selected over competing alternatives, or an acceptable rationale for not adopting the most favorable alternatives is provided.

In determining that the design facilitates decommissioning, the reviewer should consider the extent to which the applicant has selected design features which have characteristics favorable to decommissioning. Examples of favorable design features are:

- Selection of materials and processes to minimize waste production

- Minimize mass of shielding materials subject to activation

- Facilitate future demolition and removal by use of modular design and inclusion of lifting points (with anticipation of the size containers that may be used for transportation and permanent disposal)
- Selection of materials compatible with projected decommissioning and waste processing

- Use of minimum surface roughness finishes on structures, systems, and components (SSCs)

- Use of selected coatings to preclude penetration into porous materials of radioactive gas, condensate, or deposited aerosols (if probably present), to permit future decontamination by surface treatment

- Incorporation of features to contain leaks and spills

•   Consideration of current industry technology for waste production minimization.

In performing these design reviews, the reviewer should also ensure that the design features have adequately considered health and safety, including provisions to maintain occupational and public radiation exposures ALARA during decommissioning.

13.5.2 Operational Features

The reviewer should review the SAR and decommissioning plan for operational features that facilitate eventual decommissioning and minimize the associated impacts. The reviewer should verify that the applicant has committed to a plan to keep records of spills or other unusual occurrences until the license is terminated. Records should include information on contamination that may have spread to inaccessible areas, as in the case of seepage into porous materials like concrete. Records must include any known information on identification of nuclides, quantities, forms, and concentrations. The reviewer should verify that the applicant has a plan to maintain records of as-built drawings and modifications (or suitable substitute records if drawings are not available) of structures and equipment in restricted areas.

The reviewer should consult with the reviewer for site-generated waste confinement and management (Chapter 14) to determine whether proposed operations of waste management systems have adequately addressed facilitation of decommissioning. The reviewer should consult with the radiation protection reviewer (Chapter 11) to determine whether proposed health physics surveys and recordkeeping will facilitate decommissioning.

13.5.3 Decommissioning Plan

The review has three major elements: (a) a determination of overall plan adequacy and completeness, including proposed decontamination and decommissioning activities, (b) the decommissioning cost estimate, and (c) the financial assurance mechanism.

13.5.3.1 General Provisions

In preparing for the review of the proposed decommissioning plan, the reviewer should consult the general review procedures contained in Policy and Guidance Directive FC-91-2, "Standard Review Plan: Evaluating Decommissioning Plans for Licensees Under 10 CFR Parts 30, 40, and 70." However, those review procedures apply to final plans submitted in support of license termination. The reviewer should also consult Regulatory Guide 3.65, but it also applies to plans prepared before license termination.

In determining the acceptability of the level of detail, the reviewer should consider the fact that plans submitted with license applications are prospective in nature and do not have the benefit of knowledge gained over the course of facility operation (e.g., detailed knowledge of the types, extent, and precise locations of contamination). Thus, it is not reasonable to expect plans submitted with applications to have the same level of detail as final plans, especially with respect to elements such as planned decontamination activities and the final radiation survey. As described later in this section, the consideration regarding level of detail does not apply to the decommissioning funding plan.

The reviewer should first determine that the decommissioning plan includes each of the elements required by 10 CFR 72.30. In addition to the identification and discussion of design features that facilitate decontamination and decommissioning (described in Section 13.5.1), the reviewer should ensure the plan includes a decommissioning funding plan, a cost estimate for decommissioning, and a financial assurance mechanism. The reviewer should verify that the plan is consistent with the objective of "timely removal of the facility from service and reducing residual radioactivity to a level that permits release of the property for unrestricted use and license termination."

Although the decommissioning plan specifically applies to activities licensed under Part 72, there may be interrelationships with other licensed activities, including co-located Part 50 facilities. The reviewer should evaluate any proposed provisions intended to accommodate conditions associated with the co-location of facilities. For example, the reviewer should consider a case in which a spill from reactor operations occurred underground in an area beneath a proposed ISFSI pad location and the licensee proposes to delay decommissioning this contaminated soil because of concerns with compromising ISFSI pad integrity. In this example, the reviewer should determine whether (a) such a condition was adequately addressed as part of designing the ISFSI for decommissioning, and (b) it is acceptable to include such interrelated activities as part of ISFSI decommissioning.

13.5.3.2 Cost Estimate

The cost estimate for decommissioning is expected to be a major review area that requires independent staff calculations in most cases. The reviewer should ensure that the cost estimate is based on "total project costs," including all applicable direct and indirect costs. The reviewer should ensure that the cost estimate covers the complete scope of the decommissioning plan, including:

- Planning and preparation of the facility for decommissioning
- Decontamination and dismantling of structures, systems, and components
- Packaging, transportation, and disposal of radioactive wastes
- Final radiation survey.

The reviewer should evaluate the applicant's methods and assumptions for estimating costs. The reviewer should verify that estimates are based on available technologies, practices, and disposal capacity. The reviewer should ensure that conservative adjustments have been applied to account for uncertainties in the cost estimate and that a contingency amount has also been applied.

The reviewer should consider the following items that could result in underestimating the decommissioning cost:

- Low estimates of volume of low-level waste that will probably require stabilization, containerization, transportation, and disposal

- Uncertainty in per-cubic-foot costs of disposal of low-level waste, and processing and disposal costs of pool coolant

- Need to transfer stored materials to other casks for transportation

- Low estimates of time for design and planning, obtaining regulatory approvals, and procurement of services

- Low estimate of staff and physical infrastructure costs during planning for engineering, procurement, and performance of decontamination and decommissioning operations.

The reviewer should ensure that the basis year for the dollar estimate (e.g., 1997 dollars) is identified. This year should not be for a year earlier than that in which the cost estimate is prepared. The reviewer should ensure that the plan includes provisions for updating the cost estimate, including periodic updates, as well as making changes necessitated by contamination events (e.g., spills and other accidents), new regulations, etc. The reviewer should ensure that the cost estimate does not include costs for activities not necessary to terminate the NRC license (e.g., dismantling of non-radioactive and non-contaminated structures, systems and components).

The reviewer should validate the cost estimate by performing independent calculations. The reviewer should clearly identify any methods or assumptions that differ from those in the applicant's cost estimate and discuss differences in results. The reviewer should evaluate nuclear facility cost experience available to the NRC and discuss trends that indicate how cost estimates have compared with actual costs.

13.5.3.3 Financial Assurance Mechanism

The review of the applicant's proposed financial assurance mechanism should use the specific guidance provided in NUREG/CR-1337, Rev. 1, "Standard Review Plan (SRP) for the Review of Financial Assurance Mechanisms for Decommissioning Under 10 CFR Parts 30, 40, 70, and 72." The reviewer should verify that the proposed mechanism conforms to the prescribed format and content. In reviewing the contents, the reviewer should consult with the Division of Waste Management and possibly the Office of the General Counsel for technical and legal assistance in this area.

13.6 Evaluation Findings

The reviewer should prepare evaluation findings on satisfaction of the regulatory requirements related to planning and providing for decommissioning, as identified at Section 13.3. If the documentation submitted with the application fully supports positive findings for each of the regulatory requirements, the statements of findings should be as follows (numbering is for convenience in referencing the FSRP section):

F13.1   The staff has reviewed the proposed decommissioning plan documentation submitted by the applicant for the [ISFSI/MRS] facility in accordance with the standard review plan for spent fuel dry storage facilities, and the description of the plan in the SAR. The staff has determined that the decommissioning plan submitted by the applicant sufficiently provides reasonable assurance that decommissioning issues for the [ISFSI/MRS] facility have been adequately characterized, so that the site will ultimately be available for unrestricted use for any private or public purpose. The staff, therefore, concludes that the proposed decommissioning plan complies with 10 CFR Part 72.

F13.2   The staff has reviewed the decommissioning funding plan documentation submitted by the applicant for the [ISFSI/MRS] facility in accordance with the standard review plan for spent fuel dry storage facilities. The staff has determined that the decommissioning funding plan submitted by the applicant is sufficient to provide reasonable assurance that costs related to decommissioning as characterized by the proposed decommissioning plan have been adequately estimated. The staff, therefore, concludes that the cost estimate in the decommissioning funding plan complies with 10 CFR Part 72.

F13.3   The staff has reviewed the financial assurance documentation submitted by the applicant, as part of the decommissioning funding plan for the [ISFSI/MRS] facility, in accordance with the standard review plan for spent fuel dry storage facilities. The staff has determined that the financial assurance mechanisms submitted by the applicant are sufficient to provide reasonable assurance that adequate funds will be available to decommission the facility so that the site will ultimately be available for unrestricted use for any private or public purpose. The staff, therefore, concludes that the financial assurance mechanisms in the decommissioning funding plan comply with 10 CFR Part 72.

13.7 References

NRC documents referenced are identified at Consolidated References, Section 17.

14 WASTE CONFINEMENT AND MANAGEMENT EVALUATION

14.1 Review Objective

The purpose of the review is to ensure that the design and proposed operation of the Independent Spent Fuel Storage Installations (ISFSI) or Monitored Retrievable Storage (MRS) provide for safe confinement and management of any radioactive waste generated as a result of facility operations. This review specifically concerns radioactive wastes generated by the handling and storage of spent fuel or high-level waste (HLW) at the site. These include (a) gaseous effluents from treatment and ventilation systems, (b) liquid wastes from laboratory, cask washdown, and decontamination activities, and (c) solid or solidified wastes. Neither the actual spent fuel or HLW being stored, nor the waste generated by eventual decommissioning of the facility fall within the scope of this review.

Radiation protection considerations for other waste management activities are addressed in Chapter 11, and monitoring of radioactivity in effluents is addressed under Chapter 9 of this Standard Review Plan (SRP).

Figure 14.1 presents an overview of the confinement evaluation review process. The figure shows that the confinement review draws information from both the application and the results of technical reviews of operations systems. The figure also shows that the results of the confinement review are used by other technical review areas (confinement evaluation and radiation protection evaluation).

14.2 Areas of Review

The following outline shows the areas of review addressed in Section 14.4, Acceptance Criteria, and Section 14.5, Review Procedures:

Waste Sources
Off-Gas Treatment and Ventilation
Liquid Waste Treatment and Retention
Solid Wastes
Radiological Impact of Normal Operations

14.3 Regulatory Requirements

This section identifies and presents a high-level summary of Title 10 of the Code of Federal Regulations (CFR) Part 72 relevant to the review areas addressed by this chapter. The NRC staff reviewer should read the exact regulatory language. The regulatory requirements affecting radiation protection at an ISFSI or MRS are extensive. Virtually the entire contents of 10 CFR 20, "Standards for Protection Against Radiation," are also applicable to this review. A matrix at the end of this section matches the regulatory requirements identified in this section to the areas of review identified in the previous section.

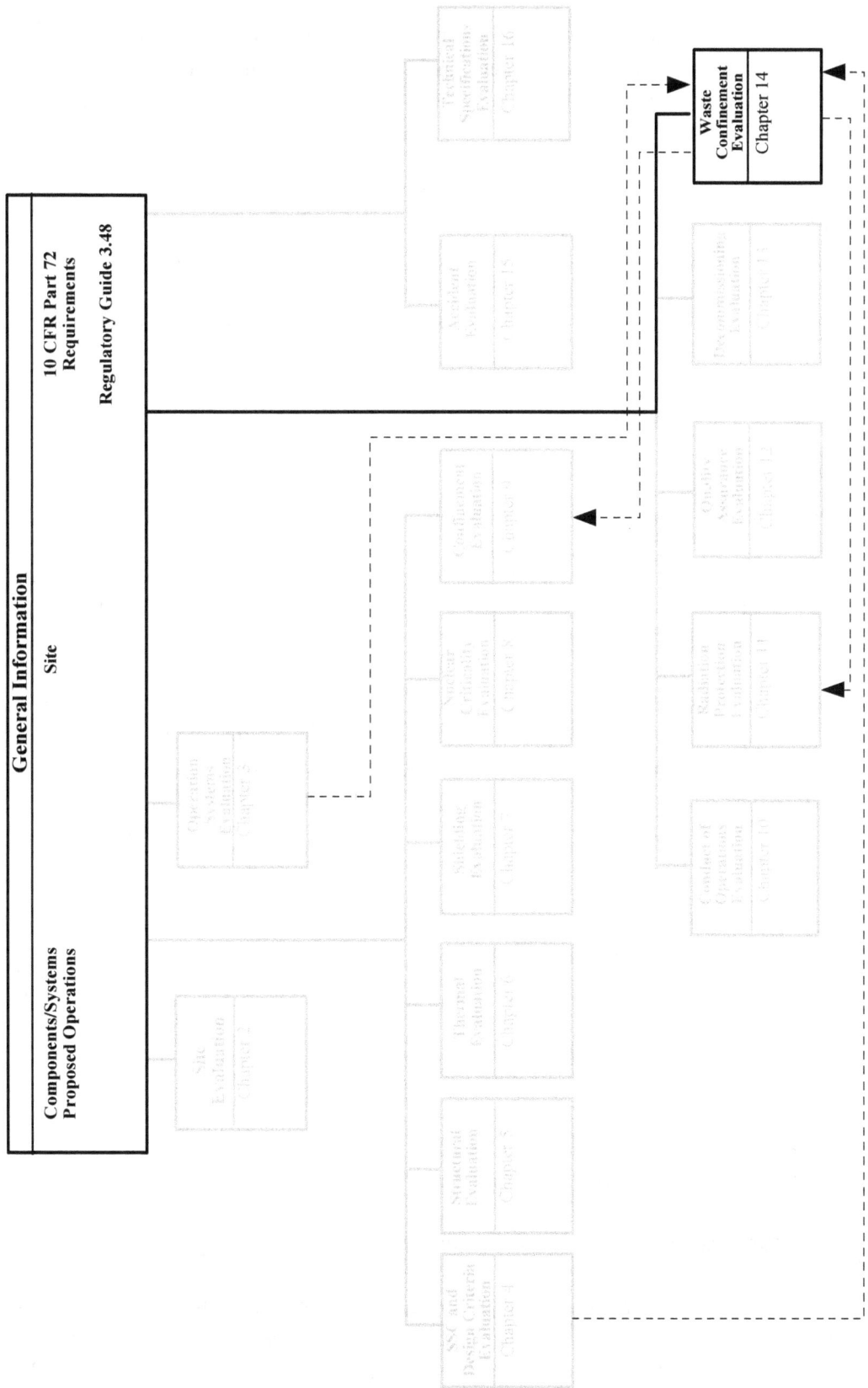

Figure 14.1 Overview of Waste Containment Operations Evaluation

20.1101 Radiation protection programs.

(d) "A constraint on air emissions of radioactive material to the environment . . . shall be established by licensees other than those subject to Section 50.34a, such that the individual member of the public likely to receive the highest dose will not be expected to receive a total effective dose equivalent in excess of 10 mrem (0.1 mSv) per year from these emissions."

20.1301 Dose limits for individual members of the public.

(a) "Each licensee shall conduct operations so that

    (1) The total effective dose equivalent to individual members of the public from the licensed operation does not exceed 0.1 rem (1 milliSievert) in a year

    (2) The dose in any unrestricted area from external sources does not exceed 0.002 rem (0.02 mSv) in any one hour."

(d) "In addition to the requirements of this part, a licensee subject to the provisions of EPA's generally applicable environmental radiation standards in 40 CFR part 190 shall comply with those standards."

20.1302 Compliance with dose limits for individual members of the public.

(b) "A licensee shall show compliance with the annual dose limit in 20.1301 by-

    (1) Demonstrating by measurement or calculation that the total effective dose equivalent to the individual likely to receive the highest dose from the licensed operation does not exceed the annual dose limit; or

    (2) Demonstrating that

        (i) The annual average concentrations of radioactive material released in gaseous and liquid effluents at the boundary of the unrestricted area do not exceed the values specified in Table 2 of Appendix B to Part 20; and

        (ii) If an individual were continuously present in an unrestricted area, the dose from external sources would not exceed 0.002 rem (0.02 mSv) in an hour and 0.05 rem (0.5 mSv) in a year."

20.2001 General requirements.

(a) "A licensee shall dispose of licensed material only

    (1) By transfer to an authorized recipient . . .

    (2) By decay in storage; or

    (3) By release in effluents within the limits in 20.1301; or

    (4) As authorized under 20.2002, 20.2003, 20.2004, or 20.2005."

20.2003 Disposal by release into sanitary sewerage.

(a) " A licensee may discharge licensed material into sanitary sewerage if each of the following conditions is satisfied:

    (1) The material is readily soluble (or is readily dispersible biological material) in water; and

    (2) The quantity of licensed or other radioactive material that the licensee releases into the sewer in 1 month divided by the average monthly volume of water released into the sewer by the licensee does not exceed the concentration listed in Table 3 of Appendix B to Part 20; and

(3) If more than one radionuclide is released, the following conditions must also be satisfied:

> (i) The licensee shall determine the fraction of the limit in Table 3 of Appendix B to Part 20 represented by discharges into sanitary sewerage by dividing the actual monthly average concentration of each radionuclide released by the licensee into the sewer by the concentration of that radionuclide listed in Table 3 of appendix B to 20.1001-20.2401; and

> (ii) The sum of the fractions for each radionuclide required by paragraph (a)(3)(i) of this section does not exceed unity; and

(4) The total quantity of licensed and other radioactive material that the licensee releases into the sanitary sewerage system in a year does not exceed 5 curies (185 GBq) of hydrogen-3, 1 curie (37 GBq) of carbon-14, and 1 curie (37 GBq) of all other radioactive materials combined."

72.24 Contents of application: Technical information [Contents of SAR]

(f) "The features of ISFSI or MRS design and operating modes to reduce to the extent practicable radioactive waste volumes generated at the installation."

(l) "A description of the equipment to be installed to maintain control over radioactive materials in gaseous and liquid effluent produced during normal operations and expected operational occurrences...The description must include:

> (1) An estimate of the quantity of each of the principal radionuclides expected to be released annually,

> (2) A description of the equipment and processes used in radioactive waste systems; and

> (3) A general description of the provisions for packaging, storage, and disposal of solid wastes containing radioactive materials resulting from treatment of gaseous and liquid effluents and from other sources."

72.40 Issuance of license.

(a) "Except as provided in paragraph (c) of this section, the Commission will issue a license under this part upon a determination that the application for a license meets the standards and requirements of the Act and the regulations of the Commission, and upon finding that

> (13) There is reasonable assurance that:

> > (i) The activities authorized by the license can be conducted without endangering the health and safety of the public."

72.104 Criteria for radioactive materials in effluents and direct radiation from an ISFSI or MRS.

(a) "During normal operations and anticipated occurrences, the annual dose equivalent to any real individual who is located beyond the controlled area must not exceed 25 mrem to the whole body, 75 mrem to the thyroid and 25 mrem to any other organ as a result of exposure to:

> (1) Planned discharges of radioactive materials . . .to the general environment,

> (2) Direct radiation from ISFSI or MRS operations, and

> (3) Any other radiation from uranium fuel cycle operations within the region."

(b) "Operational restrictions must be established to meet as low as is reasonably achievable objectives for radioactive materials in effluents."

(c) "Operational limits must be established for radioactive materials in effluents and direct radiation levels associated with ISFSI or MRS operations to meet the limits given in paragraph (a) of this section."

72.122 Overall requirements

(b) "Protection against environmental conditions and natural phenomena."

> (4) "If the ISFSI or MRS is located over an aquifer which is a major water resource, measures must be taken to preclude the transport of radioactive materials to the environment through this potential pathway."

(h) "Confinement barriers and systems."

> (3) "Ventilation systems and off-gas systems must be provided where necessary to ensure the confinement of airborne radioactive particulate materials during normal or off-normal conditions."

72.126 Criteria for radiological protection.

(c) "Effluent and direct radiation monitoring."

> (1) "As appropriate for the handling and storage system, effluent systems must be provided. Means for measuring the amount of radionuclides in effluents...must also be provided."

(d) "Effluent control. The ISFSI or MRS must be designed to provide means to limit to levels ALARA."

72.128 Criteria for spent fuel, high-level radioactive waste, and other radioactive waste storage and handling.

(a) "Spent fuel and high-level radioactive waste storage and handling systems. Spent fuel storage, high-level radioactive waste storage, and other systems that might contain or handle radioactive materials associated with spent fuel or high-level radioactive waste, must be designed to ensure adequate safety under normal and accident conditions. These systems must be designed with

> (5) Means to minimize the quantity of radioactive wastes generated."

(b) "Waste treatment. Radioactive waste treatment facilities must be provided. Provisions must be made for the packing of site-generated low-level wastes in a form suitable for storage onsite awaiting transfer to disposal sites."

A matrix that shows the primary relationship of these regulations to the specific areas of review associated with this SRP chapter is given in Table 14.1. The NRC staff reviewer should verify the matching of regulatory requirements to the areas of review presented in the matrix to ensure that no requirements are overlooked as a result of unique applicant design features.

Table 14.1 Relationship of Regulations and Areas of Review

Areas of Review	10 CFR Part 20 Regulations				
	20.1101	20.1301	20.1302	20.2001	20.2003
Waste Sources					
Off-gas treatment and Ventilation				●	
Liquid Waste Treatment and Retention				●	●
Solid Wastes					
Radiological Impact of Normal Operations	●	●	●		

Table 14.1 Relationship of Regulations and Areas of Review (continued)

Areas of Review	10 CFR Part 72 Regulations					
	72.24	72.40	72.104	72.122	72.126	72.128
Waste Sources	●		●	●		●
Off-gas treatment and Ventilation	●		●	●	●	●
Liquid Waste Treatment and Retention	●		●	●	●	●
Solid Wastes	●		●	●		●
Radiological Impact of Normal Operations		●				

14.4 Acceptance Criteria

The principal acceptance criteria that apply to confinement and management of site-generated waste are based on meeting the following regulations:

- 10 CFR 72.104, as it relates to sufficient information being provided to demonstrate that the proposed ISFSI or MRS waste storage and management system has been designed and will be operated so that during normal operations and anticipated occurrences, the annual dose equivalent to any real individual who is located beyond the controlled area does not exceed 25 mrem to the whole body, 75 mrem to the thyroid and 25 mrem to any other organ.

- 10 CFR 20.1302, as it relates to maximum levels of radioactivity in effluents to unrestricted areas.

- 10 CFR 20.1101, as it relates to constraints on air emissions of radioactive material to the environment such that no individual member of the public is likely to receive a total effective dose equivalent in excess of 10 mrem (0.1 mSv) per year from these emissions. Additional acceptance criteria apply to the descriptions provided in the SAR of waste sources and management systems, waste characteristics, operations, and monitoring. The SAR must describe the design bases for systems and equipment that maintain control over radioactive material in gaseous and liquid effluents, and identify the equipment and facilities important to safety. The SAR must also include the design objectives and the means to be employed to keep the levels of radioactivity in effluents as low as is reasonably achievable (ALARA) and to minimize the generation of waste. Waste operations, from generation and collection to final disposal offsite, must be described in the narrative descriptions and flowsheets.

Specific requirements are addressed in the following sections as they relate separately to waste sources, off-gas treatment and ventilation, liquid waste treatment and retention, and solid wastes. NUREG-0800, "Standard Review Plan," can also be used to identify requirements that apply to acceptance criteria for these categories.

14.4.1 Waste Sources

Radioactive wastes that result from an ISFSI or MRS can be separated into two main categories:

- Effluents -- radioactivity discharged to the environment in gaseous or liquid form. The activity content of these effluents must comply with regulatory limits and ALARA criteria.

- Wastes -- radioactive materials that are of sufficient hazard or regulatory concern that they require special care before final disposal. The generation of such wastes must be ALARA.

All actual and potential sources of site-generated radioactive waste must be identified in the SAR. Waste sources described must include activities that give rise to potentially radioactive wastes that would require treatment or special handling. The identification of sources must be comprehensive.

Anticipated radioactive wastes must be described and classified with respect to source, chemical and radiological composition, method and design for treatment and handling, and storage mode before disposal. Sources of non-radioactive waste such as combustion products and chemical wastes must also be identified to the extent that the reviewer can ascertain whether site activities can result in radioactive materials being added to such sources.

The total volume of liquid waste discharged to the environment must be estimated to provide the bases for determining concentrations of radionuclides in liquid effluents. Total sanitary sewer flow may be needed to determine concentrations of radionuclides in waste disposed to the sanitary sewer.

14.4.2 Off-Gas Treatment and Ventilation

Off-gas treatment and ventilation systems are typically provided for removing radioactive and non-radioactive hazardous materials from the atmosphere within a confinement barrier before being released in the environment. The SAR must provide flowsheet and narrative descriptions of off-gas treatment and ventilation system operations. It must also identify design criteria and applicable regulatory limits. General design criteria must be based on site conditions and accident-level and off-normal analyses, design objectives, and projected volumes of gaseous (or airborne) waste.

The SAR must also indicate those radioactive wastes that will be produced as a result of off-gas treatment. The applicant must show that system capacity is consistent with the confinement system requirements during normal and off-normal conditions.

The descriptions must also address replacement and disposal of items such as filters and scrubber solutions, as well as any transfers of wastes to other waste treatment systems. The design must address the potential for personnel exposures and contamination that can result from handling operations.

Continuous monitoring systems must be provided to detect effluent radioactivity and to alarm on high effluent activity. Monitoring systems are addressed in SRP Sections 9.4.2 and 11.4.2.

14.4.3 Liquid Waste Treatment and Retention

The SAR must provide flowsheet and narrative descriptions of the liquid waste treatment system and associated design criteria and regulatory limits.

14.4.3.1 Design Objectives

Basic liquid waste treatment concepts include volume reduction, immobilization of radioactive elements, change of composition, and removal of radioactive elements from the waste stream. The description of the facility liquid waste treatment and retention systems must identify the design objectives and demonstrate that the system can handle the expected volume of potentially radioactive and non-radioactive hazardous wastes generated during normal and off-normal operations.

In general, engineered features should be emphasized over procedures to meet protective requirements.

14.4.3.2 Equipment and System Description

The SAR must describe the features, systems, and special handling techniques that are important to safety. Drawings must include location of equipment, flow paths, piping, valves, instrumentation, and other physical features. Seismic and quality group classification must conform to the guidelines of Regulatory Guide 1.1.43, "Design Guidance for Radioactive Waste Management Systems, Structures and Components in Light-Water-Cooled Nuclear Reactor Power Plants." Where feasible, gravity flow must be used to reduce pressure and to avoid or minimize contamination of pumping and pressure system equipment. Measurement capability must be provided to determine the volume, concentration, and radioactivity of wastes fed into collection tanks.

The SAR must identify the sources of all liquid wastes generated and their flow into and out of the liquid treatment systems. Measurement capability must be provided to determine the volume and radioactivity of wastes fed to the collection system. Individual lines must be used for each waste stream fed to the central collection system, where necessary, to prevent chemical reactions or introduction of contaminants such as complexing agents that can interfere with waste decontamination. Individual lines outside confinement (and liquid containment) barriers must be designed not to rupture in the event of frost heave, earth or structure settlement, or earth-structure motions during design basis earthquakes. A separate confinement barrier (e.g., drained outer pipe or drained tunnel) must be provided for these lines.

Spills, overflows, or leakage from storage vessels must be collected or retained within a suitable secondary confinement structure (e.g., secondary vessel, elevated threshold, or dike). A capability must exist to transfer liquid from the secondary confinement to a suitable storage location. All transfer lines must have individual identification.

The piping must be designed to minimize entrapment and buildup of solids in the system. Bypasses which route waste streams around collection tanks must be avoided. Provisions must be made for clean out or decontamination of liquid waste piping, as necessary, to clear potential blockages, perform maintenance or repair, or maintain occupational doses ALARA.

Volume reduction or solidification methods may be used to process liquid wastes. Redundancy and other special features may be incorporated to safely confine the wastes. Adequate shielding must be provided for radioactive liquid waste system components, as necessary.

The SAR must describe how influents to radioactive liquid waste systems are controlled (as necessary, depending on the sources) to prevent introduction of material that may adversely affect system performance. Such materials include, but are not limited to, oils, other organics, insoluble solids, solvents, and hazardous wastes.

14.4.3.3 Operating Procedures

The flowsheets and narrative descriptions of operations must describe the design features and procedures that minimize generation of liquid waste and the possibility of spills, and they must provide for control and containment of spills. The SAR must state whether the procedures include performance tests, action levels, actions to be taken under normal and off-normal conditions, and methods for testing to ensure functional operation.

14.4.3.4 Characteristics, Concentrations, and Volumes of Solidified Wastes

The physical, chemical, and thermal characteristics of solidified (extracted or residue of liquid) wastes must be described. These wastes, or "characteristics," must be compatible with estimates of concentrations and volumes generated. They must also be compatible with the design ratings of the selected liquid waste treatment and retention systems.

14.4.3.5 Packaging

The SAR must describe the packaging for solidified wastes. The package information must show the materials of construction, and include welding information. It must also show the maximum temperatures for waste and container at the highest design heat loads, the homogeneity of the waste contents, the corrosive interactions of the waste on materials of construction, the means for preventing over pressurization of the package, and the confinement provided by the package under off-normal conditions.

If standard low-level waste containers (e.g., DOT-approved drums) are to be used for packaging, they must be identified. Otherwise, packaging details including vendor, make, model, and full manufacturer's catalog information must be provided in the SAR or the supporting documentation. Suitability of packaging for holding and storage of wastes onsite at the designated location must be demonstrated.

Aspects of the operating quality assurance program that specifically apply to solidified waste packaging must also be described.

14.4.3.6 Storage Facilities

The SAR must describe the storage facilities for site-generated liquid or solidified waste. Movement of containers into and out of storage, and monitoring must be described. Equipment, waste routing, and spare storage volume must be installed and available to transfer the contents of one tank to another.

The minimum spare volume must exceed the maximum liquid content of any one tank. Provisions must be made so that liquids can be analyzed before transfer. Agitators must be included in storage vessels, when necessary, to promote mixing of the waste to ensure uniform decay heat distribution, minimize settling, or provide representative waste samples.

If liquid wastes are to be held until site decommissioning or for radioactive decay, the SAR must demonstrate (by analyses or relevant experiential data) that the storage capability is appropriate for the duration of the ISFSI or MRS life and the chemistry of the contents.

14.4.4 Solid Wastes

The SAR must describe the solid wastes produced during ISFSI or MRS operations. The wastes must be listed and characterized (see Section 14.4.4.4), and systems used to treat, package, and contain these wastes must be described in terms of radionuclide content, container size, and generation rate.

14.4.4.1 Design Objectives

The SAR must identify the design objectives and demonstrate that the system can handle the expected volume of potentially radioactive solid wastes generated during normal and off-normal operations. The design objectives must reflect waste minimization as well as safe management. If the design basis includes regulatory limits, these limits must be identified.

14.4.4.2 Equipment and System Description

The SAR must describe the features, systems, and special handling techniques that are important to safety. Drawings must identify locations of equipment and associated features that will be used for volume reduction, confinement, packaging, storage, and disposal. The SAR must identify the source of all solid wastes generated and their flows into and out of the solid waste treatment systems.

Fundamental solid waste treatment concepts include volume reduction, immobilization of radioactive material, change of composition, and removal of radioactive material from the waste stream. Solid waste management systems must include provisions for shielding, confinement, handling, and decontamination, as necessary, to ensure that occupational doses are maintained ALARA.

14.4.4.3 Operating Procedures

The SAR must describe the procedures associated with solid waste system or equipment operations. The procedures must identify performance or functional testing, process limits, action levels, and actions to be taken under normal and off-normal conditions. The means for monitoring and controlling limits must also be described.

14.4.4.4 Characteristics, Concentrations, and Volumes of Solid Wastes

The SAR must describe the physical, chemical, and thermal characteristics of the solid wastes, and provide estimates of the waste volume and radionuclide concentrations. Those estimates must be consistent with design ratings of the solid waste treatment and retention systems.

14.4.4.5 Packaging

The SAR must describe the packaging for solid wastes (as for solidified liquid waste, described in Section 14.4.3.5). Aspects of the operating quality assurance program that specifically apply to solid waste packaging must also be described.

If a laundry is to be used (e.g., to minimize solid waste generation), the containers for transferring the used items must be described. If the laundry is offsite, it must be identified and must be licensed to possess radioactive material of the type and quantity to be generated at the ISFSI or MRS. (Note: An offsite laundry is not licensed under 10 CFR 72, but an on-site laundry capability to support the ISFSI or MRS can be included in the installation license.)

14.4.4.6 Storage Facilities

The SAR must describe the solid waste storage facilities. Movement of packages into and out of storage and monitoring to be performed must be described. Corrosive aspects of the wastes and monitoring of the confinement barrier must be addressed.

Planned disposal of the wastes must be described. If solid wastes are to be held until site decommissioning or for radioactive decay, the SAR must demonstrate (by analyses or relevant operating experience) that the storage containers/confinement are appropriate for the duration of the ISFSI or MRS life or the projected decay holding time. The SAR must also show how the wastes will be handled at the time the installation is permanently decommissioned.

14.4.5 Radiological Impact of Normal Operations

Regulatory Guide 3.48 requires that the SAR provide a summary of the radiological impacts of wastes generated during normal site operations. Information must include:

- A summary identifying each effluent and waste type

- The amount of each waste type generated per metric ton of spent fuel or high-level waste handled and stored per unit of time (e.g., per year)

- The quantity and concentration of each principal radionuclide in each waste stream

- Identification of the locations beyond the restricted areas (as defined in 10 CFR 20.1003) and beyond the controlled area (as defined in 10 CFR 72.3) that are potentially affected by radioactive materials in effluents

- The estimated concentrations of principal radionuclides at the locations identified and the collective (person-rem) dose to human occupants at these locations, including the contribution of each principal radionuclide to the dose

- Sample calculations and a discussion of the reliability of the concentration and dose estimates

- For each effluent, a summary of the constraints imposed on process systems and equipment to ensure safe operation.

When combined with other site effluents (e.g., those emitted from the material being stored), the radionuclide concentrations in gaseous or liquid effluents must not exceed the concentration limits specified in Table 2 of Appendix B to Part 20, and the resultant doses must not exceed the applicable criteria of 10 CFR 72.104. Constraints on air emissions must ensure that no member of the public receives a total effective dose equivalent in excess of 10 mrem (0.1 mSv) per year from those emissions.

14.5 Review Procedures

The review procedures in this section are based on the information required by Section 6 of Regulatory Guide 3.48, supplemented by NUREG-0800 guidance that may also apply to site-generated waste.

14.5.1 Waste Sources

The reviewer must determine that the SAR demonstrates that all waste materials generated as a result of facility operations will be safely contained until disposal.

Review the general description and operating features of the facility as discussed in SAR Section 1. Verify that the features of the facility design and operations reduce, to the extent practicable, the quantity of radioactive waste generated at the installation. If applicable, compare flowsheets and facility diagrams to ensure that the waste confinement and management systems are designed to minimize the quantity of radioactive wastes generated. Verify that the types of waste generated, the method and design for treatment and handling, and the mode of storage before disposal are generally accurate and are acceptable.

Ensure that all sources of waste have been identified. Consider the following list of sources that can exist at an ISFSI or MRS that has an on-site transfer capability:

- Wastes associated with normal operations
  - filters and membranes (for liquids and from the Heating, Ventilation, and Air Conditioning [HVAC] systems of the pool and waste management facilities)
  - wipes
  - material and liquids separated from pool water by the cleanup system (radioactive materials, corrosion products, and impurities)
  - material and liquid collected by skimming the pool surface (to provide clear water)
  - pool coolant seeps, minor leaks, and piping system flushing fluid
  - pool facility condensate on interior surfaces
  - pool coolant
  - HVAC duct flushing fluid
  - laboratory samples
  - decontamination station effluent
  - disposable (one-time use) and reusable personnel protective clothing and equipment
  - laundry effluent (from washing personnel protective clothing, clothing bags, etc.)
  - contaminated equipment and tools
  - radioactive waste containers and bags

- Wastes associated with off-normal events and conditions (which may be radioactive or handled as possibly radioactive):
  - confinement area sprinkler runoff
  - earth contaminated by spills or from other causes

Review the waste analysis, and check for potential interactions between non-radioactive chemical wastes or combustion products and radioactive materials. If applicable, review the method and design for treatment, handling, and disposal of chemical wastes.

14.5.2 Off-Gas Treatment and Ventilation

Review the flowsheet and narrative descriptions of off-gas treatment and ventilation system operations for consistency with the design and selection of equipment and facilities, general design criteria, and regulatory limits. Ensure that the description of the facility off-gas, waste treatment and ventilation systems identifies the driving regulatory requirements, design objectives, and general design criteria. Compare the proposed performance with similar facilities reviewed and approved by the NRC.

Verify that the design adequately addresses site conditions and includes reasonable estimates of airborne waste generation rates for normal, off-normal, and accident-level conditions. Verify that system capacity is consistent with the confinement system requirements during normal and off-normal conditions.

Ensure that the design addresses replacement and disposal of items such as filters and scrubber solution, as well as any transfers of wastes to other waste treatment systems. Verify that the design addresses potential personnel exposure and contamination that could result from handling operations. Review with the radiation protection and confinement reviewers the off-gas monitoring systems and verify that these systems provide adequate detection and alarm capability.

14.5.3 Liquid Waste Treatment and Retention

Review the flowsheet and narrative descriptions of liquid waste system operations for consistency with the design and selection of equipment and facilities, general design criteria, and regulatory limits. Compare the proposed performance with similar facilities reviewed by the NRC.

14.5.3.1 Design Objectives

Review the design objectives and verify that the system can handle the expected volume of potentially radioactive liquid wastes generated during normal and off-normal operations. Ensure that the design objectives clearly identify which waste streams will be processed to achieve volume reduction or solidification. Verify that all sources of liquid waste have been identified. Assess the applicant's estimates of expected inventories for each stream, and determine whether they are reasonable for design purposes.

14.5.3.2 Equipment and System Description

Verify that the SAR describes the features, systems, and special handling techniques that are important to safety, and that pressure vessels tank and piping systems important to safety will be constructed in accordance with the appropriate quality standard(s).

Review the design to ensure that (a) adequate measurement is provided (to determine liquid waste volume and radioactivity concentration), (b) the system is not vulnerable to contamination build-up, (c) influents to the liquid waste systems do not include materials (e.g., oils, insoluble solids, solvents, hazardous wastes, etc.) that may adversely affect system performance, (d) secondary confinement is provided for waste lines outside of the confinement barriers, and (e) provisions are made, as necessary, for component shielding and clean-out or decontamination of piping.

14.5.3.3 Operating Procedures

Review the flowsheets and narrative descriptions of operations to verify that proposed design features and procedures will minimize liquid waste generation and the possibility of spills, and provide for control and containment of spills. Verify that provisions are made for ensuring functional operation, including testing procedures, action levels, and associated actions.

14.5.3.4 Characteristics, Concentrations, and Volumes of Solidified Wastes

Review the applicant's description of physical, chemical, and thermal characteristics of solidified wastes, and determine whether they are consistent with estimates of liquid waste concentrations and volumes generated. Also determine whether they are consistent with the design ratings of the selected liquid waste treatment and retention systems.

14.5.3.5 Packaging

Review the descriptions of solidified waste packaging and verify that the container specifications are compatible with the forms of waste to be packaged. This determination should consider materials of construction (including welding information, if appropriate), heat load, potential corrosive interactions of waste and packaging, prevention of over-pressurization, and confinement provided by the package under off-normal conditions.

14.5.3.6 Storage Facilities

Review the description of storage facilities, and determine whether the storage capacity is consistent with estimates of liquid and solidified waste volumes to be generated and stored over the life of the facility. Review proposed operations to ensure that movement of containers into and out of storage, monitoring, equipment, waste routing, and spare storage volume (for liquid transfers) have been taken into account, as necessary. Ensure that provisions exist for spills, overflows, or leakage.

14.5.4 Solid Wastes

Review the process flow diagram and narrative descriptions of solid waste operations for consistency with the design and selection of equipment and facilities, general design criteria, and regulatory limits. If applicable, compare the proposed performance with similar facilities reviewed by the NRC.

14.5.4.1 Design Objectives

Verify that the SAR identifies all sources of wastes generated and that the system is capable of handling the expected volume of potentially radioactive and non-radioactive hazardous wastes generated during normal and off-normal operations. Specifically review waste generated from use of personal protective clothing and equipment that is to be treated as potentially contaminated, because they typically constitute a major fraction of total volume.

14.5.4.2 Equipment and System Description

Review the equipment and systems descriptions for solid waste collection and treatment to ensure that (a) features, systems, and special handling techniques that are important to safety have been identified, (b) locations of equipment and associated features that are used for volume reduction, confinement, or packaging, storage, and disposal are identified, and (c) provisions exist for shielding, confinement, handling, and decontamination, as necessary, to ensure that occupational doses are maintained ALARA.

14.5.4.3 Operating Procedures

Review the procedures associated with solid waste system or equipment operations, and verify that performance testing, process limits, and means for monitoring and controlling limits are adequately addressed. Ensure that action levels and associated response actions for normal and off-normal conditions are provided.

14.5.4.4 Characteristics, Concentrations, and Volumes of Solid Wastes

Review the applicant's description of physical, chemical, and thermal characteristics of solid wastes, and determine whether the estimates of concentrations and volumes generated are consistent with the design ratings of the solid waste treatment and storage systems.

14.5.4.5 Packaging

Review the descriptions of solid waste packaging, and verify that the container specifications are appropriate for forms of waste to be packaged. As with solidified wastes (but to a lesser extent), this determination must consider materials of construction, heat load, potential corrosive interactions of waste and packaging, and confinement provided by the package under off-normal conditions.

If an onsite laundry is to be used, verify that the containers and procedures for transfer of potentially contaminated items are adequately described. If the laundry is offsite, ensure that applicable regulatory requirements for offsite transportation are addressed.

14.5.4.6 Storage Facilities

Review the description of storage facilities, and determine whether the storage capacity is consistent with estimates of solid and solidified waste volumes to be generated and stored over the life of the facility. Review proposed operations to ensure that monitoring, equipment, and the movement of containers into and out of storage, monitoring, and equipment have been adequately addressed, as necessary. Verify that long-term storage options are reasonable in light of ultimate disposal plans and availability.

14.5.5 Radiological Impact of Normal Operations

Verify that the SAR contains a listing identifying (a) each effluent and waste type (b) the amount of each waste type generated per metric ton of spent fuel or HLW handled and stored per unit of time (e.g., per year), and (c) the quantity and concentration of each principal radionuclide in each waste stream. Check that these releases apply to operational occurrences (off-normal events and conditions). Verify that a summary of the constraints is imposed for each effluent process system to ensure safe operation.

Coordinate the evaluation of offsite impacts of site-generated waste effluents with the radiation protection reviewer, who determines whether these impacts, when combined with other effluent or dose impacts as required, meet both the effluent concentration limits of Part 20, Appendix B, Table 2, and the dose criteria of 10 CFR 72 and 10 CFR 20 (described in SRP Chapter 11 of this SRP).

14.6 Evaluation Findings

The reviewer prepares evaluation findings upon satisfaction of the regulatory requirements for waste confinement and measurement, as identified in Section 14.3. If the documentation submitted with the application fully supports positive findings for each of the regulatory requirements, the statements of findings must be substantially as follows (finding numbering is for convenience in referencing within the SRP and Safety Evaluation Report):

F14.1   The SAR adequately describes acceptable features of the [ISFSI or MRS] design and operating modes that reduce to the extent practicable the radioactive waste volumes generated by the installation, in compliance with 10 CFR 72.24(f) and 10 CFR 72.128(a)(5).

F14.2   The SAR adequately describes acceptable equipment to be installed to maintain control over radioactive materials in gaseous and liquid effluent produced during normal operations and expected operational occurrences; estimated radionuclide releases; and provisions for packaging, storage, and disposal of solid wastes containing radioactive materials resulting from treatment of gaseous and liquid effluents and from other sources, in compliance with 10 CFR 72.24(l).

F14.3   The SAR provides reasonable assurance that the waste confinement and management activities to be authorized by the license will not endanger public health and safety, in compliance with 10 CFR 72.40(a)(13).

F14.4   [If the [ISFSI or MRS] is located over an aquifer that is a major water resource (which may be interpreted as over any ground water)] The [ISFSI or MRS] design and procedures provide acceptable measures to preclude the transport of radioactive materials to the environment through the aquifer, in compliance with 10 CFR 72.122(b)(4).

F14.5   [If appropriate] The effects of operation of the proposed [ISFSI or MRS] combined with those of other nuclear facilities at the site will not constitute an unreasonable risk to the health and safety of the public, in compliance with 10 CFR 72.122(e).

F14.6   [If appropriate] The design of the [ISFSI or MRS] provides acceptable ventilation and off-gas systems to ensure the confinement of airborne radioactive particulate materials during normal or off-normal conditions, in compliance with 10 CFR 72.122(h)(3).

F14.7   [If appropriate] The design of the [ISFSI or MRS] provides [an] acceptable effluent system[s], which includes means for measuring the amount of radionuclides in effluents during normal operations and under accident conditions, and for measuring the flow of the diluting medium, in compliance with 10 CFR 72.126(c).

F14.8   The design of the [ISFSI or MRS] acceptably provides means to limit the release of radioactive materials in effluents during normal operation and to control the release of radioactive materials under accident conditions to levels in compliance with 10 CFR 72.126(d).

F14.9   The design of the [ISFSI or MRS] includes radioactive waste treatment facilities that include capability for packing site-generated, low-level wastes in a form suitable for storage onsite while awaiting transfer to disposal sites, in compliance with 10 CFR 72.128(b).

15 ACCIDENT ANALYSIS

15.1 Review Objective

The objective of this chapter is to provide guidance to the staff for a systematic evaluation of the applicant's identification and analysis of hazards for both off-normal and accident or design basis events involving structures, systems, and components (SSCs) important to safety.

Off-normal events are those expected to occur with moderate frequency or once per calendar year. ANSI/ANS 57.9 refers to these events as Design Event II.

Accident events are considered to occur infrequently, if ever, during the lifetime of the facility. ANSI/ANS 57.9 subdivides this class of accidents into Design Event III, a set of infrequent events that could be expected to occur during the lifetime of the ISFSI, and Design Event IV, events that are postulated because they establish a conservative design basis for SSCs important to safety. For purposes of this chapter of the Facilities Standard Review Plan (FSRP), no distinction is made between these two classes of events. The effects of natural phenomena, such as earthquakes, tornadoes, hurricanes, floods, tsunami, and seiches, are considered to be accident events.

A minimum list of events to be considered is in Section 15.2.

The interrelationship between the accident analysis review and other areas of review is illustrated in Figure 15.1. The figure shows that the evaluation draws on information in the application as well as results of other technical reviews. The figure also shows that the results of this review are used in other technical reviews.

15.2 Areas of Review

The following outline provides a minimum list of accident scenarios to be analyzed. The acceptance criteria for those events are in Section 15.4, and the review procedure for those events are in Section 15.5.

15.2.1 Off-Normal Events
> 15.2.1.1 Cask Drop less than Design Allowable Height
> 15.2.1.2 Partial Vent Blockage (if applicable)
> 15.2.1.3 Operational Events
> 15.2.1.4 Off-normal Ambient Temperatures
> 15.2.1.5 Off-normal Events Associated with Pool Facilities

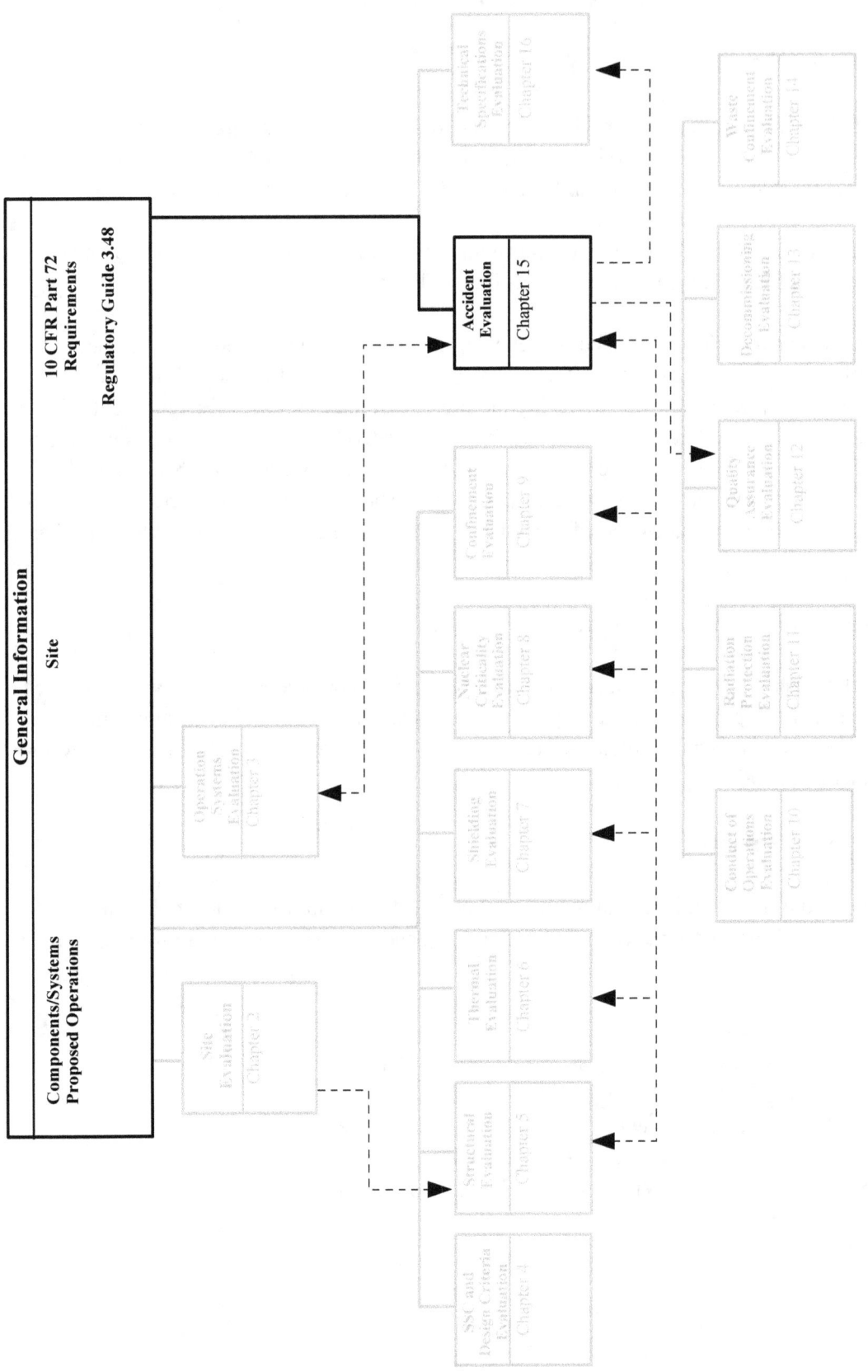

Figure 15.1 Overview of Accident Evaluation

ACCIDENT ANALYSIS

15.2.2 Accidents

    15.2.2.1 Cask Tipover
    15.2.2.2 Cask Drop
    15.2.2.3 Flood
    15.2.2.4 Fire and Explosion
    15.2.2.5 Lightning
    15.2.2.6 Earthquake
    15.2.2.7 Loss of Shielding
    15.2.2.8 Adiabatic Heatup
    15.2.2.9 Tornadoes and Missiles Generated by Natural Phenomena
    15.2.2.10 Accidents at Nearby Sites
    15.2.2.11 Accidents Associated with Pool Facilities
    15.2.2.12 Building Structural Failure onto SSCs

15.2.3 Other Non-Specified Accidents

This category of accidents covers all other accident scenarios that do not fall into the categories identified in Sections 15.2.1 and 15.2.2, as described above. These accidents may be postulated in the SAR, or they may arise as a result of the NRC evaluation of the accident analyses.

15.3 Regulatory Requirements

This section identifies and presents a high-level summary of the regulatory requirements that are applicable to the review areas addressed by this chapter. The NRC staff reviewer should read the exact regulatory language. At the end of this section is a matrix that maps the regulatory requirements identified in this section with the areas of review identified in the previous section.

72.24 Contents of application: Technical information [Contents of the SAR].
(a) "A description and safety assessment of the site with appropriate attention to design bases for external events"
(d)(2) [The SAR should include information on] the adequacy of structures, systems, and components provided for the prevention of accidents and the mitigation of the consequences of accidents including natural and manmade phenomena and events."
(m) "An analysis of the potential dose equivalent or committed dose equivalent to an individual outside the controlled area from accidents ..."

72.90 General considerations.
(b) "The ISFSI or MRS must be examined with respect to...external and man-induced events"

15-3

NUREG-1567

72.92   Design basis external natural event.
(a) "Natural phenomena that may exist or that can occur in the region of a proposed site must be identified and assessed."
(b) "The records of the occurrence and severity of those important natural phenomena [must be collected and recorded]."
(c) "Appropriate methods must be adopted for evaluating the design basis external natural events based on the characteristics of the region and the current state of knowledge about such events."

72.94   Design basis external man-induced events.
(a) "The important potential man-induced events that affect the ISFSI or MRS design must be identified."
(b) "Information concerning the potential occurrence and severity of such events must be...evaluated."
(c) "Appropriate methods must be adopted for evaluating the design basis external man-induced events."

72.104 Criteria for radioactive materials in effluents and direct radiation...
(a) "During normal and anticipated occurrences the annual dose equivalent ... must not exceed 25 mrem to the whole body ..."
(b) "Operational restrictions must be established to meet [ALARA] objectives..."
(c) "Operational restrictions must be established for radioactive materials in effluents and direct radiation levels..."

72.106 Controlled area of an ISFSI.
(b) "Any individual...shall not receive a dose... greater than 5 rem to the whole body...from any design basis accident."

72.122 Overall requirements.
(b) Protection against environmental conditions and natural phenomena.
    (1) "Structures, systems, and components important to safety must be designed to... withstand postulated accidents."
    (2) "Structures, systems, and components important to safety must be designed to withstand the effects of natural phenomena such as earthquakes, tornadoes, lightning, hurricanes, floods, tsunami, and seiches, without impairing their capability to perform safety functions. The design bases for the structures, subsystems, and components must reflect"
        (ii) "Appropriate combinations of the effects of normal and accident conditions and the effects of natural phenomena."
(c) Protection against fires and explosions. "Protection against fires and explosions is required. [The facility should be] provided with sufficient capacity and capability to minimize the adverse effects of fires and explosions on structures, systems and components."
(i) Instrumentation and control systems. "Instrumentation and control systems must be provided to monitor systems that are important to safety over anticipated ranges for normal...off-normal...and accident conditions."
(h) Confinement barriers and systems.

(2) "For underwater storage...the water level must be designed so that any abnormal operations or failure...will not cause the water level to fall below safe limits."

72.124 Criteria for nuclear criticality safety.
(a) "Spent fuel handling, packaging, transfer, and storage systems must be designed to be maintained subcritical..."

72.126 Criteria for radiological protection.
(c) "Means for measuring the amount of radionuclides in effluents during normal operations and under accident conditions must be provided for these systems."
(d) "The ISFSI must be designed to provide means to control effluent to ALARA level under normal and accident conditions"

72.128 Criteria for spent fuel, high-level radioactive waste, and other radioactive waste storage and handling
(a) "[Systems associated with spent fuel, high-level waste, and other radioactive waste storage] must be designed to ensure adequate safety under normal and accident conditions. These systems must be designed with-"
        (2) "Suitable shielding for radioactive protection under normal and accident conditions."

72.236 Specific requirements for spent fuel storage cask approval.
(c) "Spent fuel is maintained in a subcritical condition..."
(d) "Radiation shielding and confinement features must...meet...72.104 and 72.106."
(l) "[The casks] will reasonably maintain confinement...under normal, off-normal, and credible accident conditions."

A matrix that shows the primary relationship of these regulations to the specific areas of review associated with this FSRP chapter is given in Table 15.1. The NRC staff reviewer should verify the association of regulatory requirements with the areas of review presented in the matrix to ensure that no requirements are overlooked as a result of unique applicant design features.

15.4 Acceptance Criteria

This section identifies the acceptance criteria used for each accident in the accident analysis review. The SAR must include complete information for each event, analysis of the safety performance of the system, and demonstration of compliance with all applicable regulations. Each evaluation should include (1) the cause of the event, (2) means of detecting the event, (3) a summary of the analysis of the event, including estimated consequences and comparison to regulatory limits, and (4) a corrective course of action.

All events must meet the following acceptance criteria regarding criticality, confinement, retrievability, and instrumentation. Dose limit criteria are discussed in specific subsections.

Table 15.1 Relationship of Regulations and Areas of Review

Areas of Review	10 CFR Part 72 Regulations										
	72.24	72.90	72.92	72.94	72.104	72.106	72.122	72.124	72.126	72.128	72.236
Off-Normal Events	●	●	●	●	●	●	●	●	●	●	●
Accidents	●	●	●	●		●	●	●	●	●	●
Other Non-Specified Accidents	●					●	●	●	●	●	●

Criticality

10 CFR 72.124(a) and 72.236(c) require that the spent fuel must be maintained in a subcritical condition (i.e., k_{eff} equal to or less that 0.95) under credible conditions. At least two unlikely, independent and concurrent or sequential changes must be postulated to occur in the conditions essential to nuclear criticality safety before a nuclear criticality accident is possible (double contingency).

Confinement

10 CFR 72.128(a)(3) and 72.236(d) and (l) require that the systems important to safety must be evaluated, using appropriate tests or by other means acceptable to the Commission, to demonstrate that they will reasonably keep radioactive material confined under credible accident conditions. A breach of a confinement barrier is not acceptable for any accident event. A confinement system is defined in 10 CFR 72.3 as a system, including ventilation, that acts as a barrier between areas containing radioactive substances and the environment.

Retrievability

10 CFR 72.122(l) requires that ISFSI storage systems allow ready retrieval of the stored spent fuel or high-level waste for normal and off-normal design conditions. Retrievability is the capability of returning the stored radioactive material to a safe condition without endangering public health and safety or causing additional exposure to workers. Any potential release of radioactive materials during retrieval operations must not exceed the radioactive exposure limits in 10 CFR Part 20 or 10 CFR 72.122(h).

Instrumentation

10 CFR 72.122(h) through (j) and 72.128(a)(1) require that the SAR identify all instruments and control systems that must remain operational under accident conditions.

15.4.1 Off-Normal Events

In addition to the acceptance criteria stated in Section 15.4, 10 CFR 72.104 requires that the following criteria be met regarding dose limits for off-normal events. During an off-normal event, the annual dose equivalent to any individual located beyond the controlled area must not exceed 25 mrem to the whole body, 75 mrem to the thyroid, and 25 mrem to any other organ as a result of exposure to the following sources:

- planned discharges to the general environment of radioactive materials (with the exception of radon and its decay products)

- direct radiation from operations of the ISFSI

- any other cumulative radiation from uranium fuel cycle operations (i.e., nuclear power plant) in the affected area

All off-normal events listed in Section 15.2.1 must be considered.

15.4.2 Accidents

In addition to the criteria stated in Section 15.4, 10 CFR 72.106(b) requires that any individual located at or beyond the nearest controlled area boundary must not receive a dose greater than 5 rem to the whole body or any organ from any design basis accident.

All accidents listed in Section 15.2.2 must be considered.

15.4.3 Other Non-Specified Accidents

In addition to all accidents given above, the applicant must list and evaluate other accident events that are specific to his design. If these other non-specified accidents have results that are enveloped by the accidents previously considered, the applicant must provide the basis for this evaluation, and no further consideration is required. It is expected that events such as human errors, operational errors, material aging, etc., may be enveloped by the required accidents listed in Sections 15.2.1 and 15.2.2.

15.5 Review Procedures

This section provides review guidance for each accident event evaluation. The review guidance varies in complexity within each evaluation. In general, the evaluations review the operating environment, the physical parameters, the methodology used, and the actual analysis performed by the licensee.

Items of unique or special safety significance should receive special emphasis. Refer to Section 4 of this FSRP for a discussion of the SSCs important to safety.

The effects of various accidents may be interrelated, and some degree of overlap is expected to occur during the accident analysis review process. An example of such overlap would be a tornado missile accident, reviewed according to Section 15.5.2.9, lead ing to a loss of shielding, an accident reviewed according to Section 15.5.2.7. If two or more accidents are interrelated, the reviewer should assess the probability of the event and the consequences in determining the bounding event.

The reviewer should evaluate all credible accidents and focus the review on those accidents with potential consequences resulting in the failure of the confinement boundary. Such events should be evaluated against the requirements of 10 CFR 72.106 and 72.122(b). Recovery methods or the need for over-packs or dry transfer systems to maintain safe storage conditions would then not be considered and evaluated as part of the licensing process.

15.5.1 Off-Normal Events

This section discusses the review of off-normal conditions that may include malfunctions of systems, minor leakage, limited loss of external power, and operator error. The consequences of these events should not have a significant effect beyond the cask storage area or fuel pool storage building. Each event should include (a) a discussion of the cause of the event, (b) the means of detection of the event, (c) an analysis of the effects and consequences, and (d) actions required to return the system to a normal situation. Radiological impact from off-normal operations should be assessed.

15.5.1.1 Cask Drop Less than Design Allowable Height

The drop of the confinement cask at less than design allowable height is one of the hypothetical off-normal scenarios that the applicant must evaluate. The evaluation must show that the cask integrity and fuel spacing geometry are not compromised if the cask is dropped from a relatively low height. It must also show that the cask will continue to store fuel safely after such a drop. The following steps are intended to provide the reviewer with an outline of the methodology for evaluating the off-normal event. The steps are representative of a typical cask but are not intended to cover every aspect of every possible design.

Define the Operating Environment

The reviewer should verify that the SAR identifies the operating environment of the off-normal event, including:

- the operational configuration of the confinement cask (i.e., the stand-alone confinement cask with no other SSC; the confinement cask inside the transfer or transport cask, with or without impact limiters; etc.)

- the confinement cask orientation at the moment of impact (i.e., end drop on top or bottom, side drop at various azimuths, and corner drop at various azimuths and inclinations)

- the physical location of the drop event (i.e., outside the spent fuel pool building or inside the spent fuel pool building).

Define the Physical Parameters

The reviewer should verify that the SAR defines the physical parameters associated with the off-normal event including:

- the receiving surface upon which the confinement cask impacts (i.e., the storage pad materials, dimensions and properties and the foundation properties; or the spent fuel pool or building floor materials, dimensions, and properties) (The surface should be sufficiently characterized to quantify the maximum deceleration levels.)

- the design of the confinement cask and associated SSCs (i.e., material properties, dimensions, and weights)

- the drop height of the confinement cask onto the receiving surface for each orientation.

Review the Analysis Methodology

The reviewer should verify that the SAR defines the analysis methodology used in the evaluation, such as:

- reference to the design basis accident drop analysis as given in the SAR

- a basis for the validity of the results for the off-normal event, if based on an accident event analysis

- static equivalent deceleration with appropriate dynamic load factors

- dynamic modeling with appropriate test data to benchmark deceleration versus time of impact

- specific analysis modeling tools such as manual techniques or computer codes (ANSYS has been accepted by the NRC for previous applications.)

Off-Normal Event Analysis

The reviewer should verify that the SAR presents the off-normal event analysis and design criteria and design codes and standards, such as:

- deceleration level for each case considered

- design code for evaluation (the ASME B&PV Code Section III, Service Level C is acceptable to the NRC)

- calculated stress intensity should be in elastic range for Service Level C

- evaluation of calculated stress intensity level against the allowable stress intensity level at the design temperature for each component member of the confinement cask

- evaluation of buckling stability for each component member of the confinement cask subjected to compressive loading

- evaluation of deformation of cask internal members that contribute to spacing geometry of the spent fuel assemblies or high-level waste materials that are subject to criticality safety as given in Section 8 of this FSRP (no plastic deformation is permitted for an off-normal event)

- calculation of accidental dose

15.5.1.2 Partial Vent Blockage (if applicable)

For confinement systems, such as natural convection cooling systems that are subject to a temperature rise from a partial vent blockage an evaluation of the event should be made. The purpose of the evaluation is not to establish a surveillance frequency, as in the case of the adiabatic heatup accident, but rather to establish that no critical temperature limits will be reached for an extended time period.

Define the Operating Environment and Physical Parameters

The reviewer should verify that the SAR identifies the operating environment of the off-normal event, including:

- the operational configuration of the confinement system
- the fraction of vent blockage
- the ambient temperature
- the design basis decay heat load

Review the Analysis Methodology

The reviewer should verify that the SAR defines the analysis methodology used in the evaluation, including:

- assumptions and calculational models or experimental testing

Off-Normal Event Analysis

The reviewer should verify the vent flow area and revised vent flow loss coefficients associated with a blockage of one-half of the normal air inlet vent flow area. If an even number of air inlet vents are part of the ISFSI design, it is permissible to assume that one-half of these vents are completely blocked.

The reviewer should verify the air outlet temperature and all key ISFSI unit internal material maximum temperatures. The reviewer should use the flow areas and flow loss coefficients calculated in the above paragraph, assuming a normal ambient air temperature (usually assumed to be 21 °C [70 °F]). The reviewer should also use maximum design basis decay heat, and the identical thermal models and computer codes that were used in the normal conditions thermal analysis of the ISFSI.

The reviewer should compare the calculated maximum material temperatures with their respective short-term temperature limits. The reviewer should coordinate with the reviewer responsible for the structural integrity to ensure that these temperatures are used to determine the appropriate allowable stress intensity levels.

The reviewer should evaluate the worker dose required to clear debris that is blocking air inlet(s) using the design basis calculated dose rate at the air inlets and the estimated time necessary to clear the vents. This worker dose should be compared to the dose associated with loading one ISFSI unit and the worker dose limits in 10 CFR 20.101.

15.5.1.3 Operational Events

The SAR should define any off-normal events based on an estimate of the frequency of occurrence. An off-normal event would be expected to occur approximately once per year or at least several times during the initial period of the 20-year license.

A typical off-normal condition resulting in a radiological release is evaluated in the remainder of this subsection.

Define the Operating Environment and Physical Parameters

The reviewer should verify that the SAR describes the maximum allowable container surface contamination, based on applicable technical specifications and/or health physics procedures. This contamination is usually expressed in terms of either counts per minute, per unit surface area or μCi/square centimeter, and different values are provided for alpha contamination and beta/gamma contamination.

The reviewer should verify the calculation of the total surface area of the spent nuclear fuel or high-level waste container.

The reviewer should verify the calculation of the total container surface contamination by multiplying the values of surface contamination (in terms of curies of activity) and surface area.

Review the Analysis Methodology

The reviewer should verify that the SAR contains the 95% probability value for the atmospheric dispersion factor from the ISFSI cask to the nearest public boundary from the ISFSI site. The technical basis and applicability of the atmospheric dispersion value should be included. Regulatory Guide 1.145, "Atmospheric Dispersion Models for Potential Accident Consequence

Assessments at Nuclear Power Plants," provides detailed directions on acceptable methods for calculations of site-specific values of dispersion parameters. The NRC has previously accepted Regulatory Guide 1.4, "Assumptions Used For Evaluating the Potential Radiological Consequences of a Loss of Coolant Accidents for Pressurized Water Reactors," or 1.25, "Assumptions Used For Evaluating the Potential Radiological Consequences of a Fuel Handling Accident in the Fuel Handling and Storage Facility for Boiling and Pressurized Water Reactors," for conservative generic values of atmospheric dispersion factors in the absence of site-specific meteorological data.

The reviewer should verify that the SAR contains appropriate external, inhalation, and ingestion dose conversion factors for surface contamination, principally Cobalt-60, that is postulated to be released from the ISFSI cask. Acceptable values of these factors are available in Regulatory Guide 1.109, "Calculation of Annual Doses to Man from Routine Releases of Reactor Effluents For the Purpose of Evaluating Compliance with 10 CFR Part 50, Appendix I," or Federal Guidance Report No. 11.

Accident Analysis

Using the acceptable adult breathing rate presented in Section 7 of NUREG-1536, "Standard Review Plan for Dry Cask Storage Systems," and a Cobalt-60 surface contamination radionuclide release fraction of 1.0 along with the data compiled in the previous steps, the reviewer should verify the calculation of the whole body and individual organ dose at the ISFSI site public boundary.

15.5.1.4 Off-Normal Ambient Temperatures

Off-normal ambient temperatures are expected to occur during the operational life of the facility. The numerical values of off-normal ambient temperatures are expected to be greater than the normal ambient temperature but less than the accident ambient temperature. The higher probability of occurrence of off-normal ambient temperatures as compared to the accident temperatures requires that calculated material temperatures due to off-normal ambient temperatures meet the normal operational material temperature limits.

Define the Operating Environment

The reviewer should verify that appropriate maximum and minimum off-normal ambient temperatures have been specified in the SAR. An example of previously accepted conditions have been maximum and minimum ambient temperature values of 52°C (125°F) and -40°C (-40°F). For previously licensed facilities, a typical annual average ambient temperature has been 24°C (75°F). The maximum and minimum ambient temperature values should equal the 99% values in Table 1, Climatic Conditions for the United States, in the ASHRAE Fundamentals Handbook. Alternatively, the data from NUREG/CR-1390, "Probability Estimates of Temperature Extremes for the Contiguous United States," for the 0.99 probability level may be used. If the site does not correspond with the location cited in the reference, technical justification for using the same climatic data should be supplied.

Similarly, the site-specific or generic value of solar insolation or heat flux should be verified. This value should be used in conjunction with the normal and off-normal maximum ambient temperature, but a value of zero solar heat flux should be used with the minimum ambient air temperature scenario.

Review the Analysis

The reviewer should calculate the steady state temperature distribution within the ISFSI using the same methodology and computer codes that were used for the normal ambient air temperature scenario.

The reviewer should evaluate the calculated temperature distribution in terms of material temperature limits (e.g., fuel cladding, concrete, and proprietary neutron shielding materials) and thermal stresses. The material temperature limits should be consistent with the acceptable limits identified in the thermal analysis evaluation.

15.5.1.5 Off-Normal Events Associated with Pool Facilities

Pool facilities must be able to withstand the effects of a set of off-normal events that can be expected to occur with a frequency approximately once per storage year, or at least several times during the initial period of the twenty-year license. Section 5 of this FSRP identifies two NRC-approved ANSI/ANS standards dealing with wet spent fuel storage and handling facilities. The SAR should specify if ANSI/ANS 57.2 for light-water reactor spent fuel storage facilities or ANSI/ANS 57.7 for a water pool IFSFI is to be used for the pool facility. Depending on the standard specified in the SAR, the pool facilities should be designed to withstand the effects of off-normal events (e.g., the water level will not fall below safe limits established for worker exposure or fuel cladding temperature limits).

Define the Off-Normal Events

The reviewer should verify that the SAR defines the off-normal events associated with pool facilities. The following off-normal events are estimated to occur with a frequency of approximately once per year of storage operation and must be evaluated regardless of which ANSI standard is cited in the SAR; however, the list is intended to be representative and not all-inclusive:

•   failure of any single active component to perform its intended function on demand

•   spurious operation of certain active components such as a relief valve or a control valve

•   loss of external power supply for a limited duration (e.g., less than 8 hours) which could cause loss of cooling

•   single operator error followed by proper corrective action

•   minor leakage from component or pool liner

If ANSI/ANS 57.2 is cited in the SAR, then an off-normal event that must be considered in addition to the above events is a single failure in the electrical or control system.

Define the Operating Environment

The SAR should identify the operating environment and conditions of the off-normal events. Because each pool facility will be different, the reviewer should verify that the applicant has described, in detail, how each of the above off-normal events is or is not possible due to the specific design or operating environment.

Define the Physical Parameters

The reviewer should verify that the SAR defines the physical parameters associated with the off-normal events, including:

- level or temperature of water at time of failure or spurious operation of active components
- any protective devices designed to mitigate the consequences of the off-normal events
- alarms and response times for corrective action
- capacities of the storage or transfer pool, flow rates of the pumps and valves, time constants for the control systems, etc.

Review the Analysis Methodology

The reviewer should verify that the SAR defines the analysis methodology for evaluating the consequences of the off-normal events, including:

- modeling techniques for calculating the temperature rise of the storage pool or transfer pool water
- assumptions used as a part of the off-normal event

Off-Normal Event Analysis

The reviewer should verify that the SAR presents the analysis and design criteria and design codes and standards for each of the off-normal events that are defined in the SAR. The codes and standards given below are the primary design and construction codes acceptable to the NRC. For a more detailed listing of design codes and standards, the reviewer may consult ANSI/ANS 57.2 or 57.7.

Spent fuel storage and cask unloading/handling pools
    (a) ASME B&PVC, Section III
    (b) ACI 318 for reinforced concrete for ANSI/ANS 57.7 designs and ACI 349 for ANSI/ANS 57.2 designs
    (c) ANSI/AWS D1.1 for structural welding
    (d) ANSI/AWS D1.4 for welding reinforcing steel
    (e) AISI *Steel Products Manual*

Spent fuel storage racks
   (a) ASME B&PVC, Section III

Spent fuel pool water makeup, cooling and cleanup system
   (a) ANSI/IEEE C2 for electrical code and ANSI/NFPA 70
   (b) ANSI/ASME B31.1 for power piping
   (c) ASME B&PVC Section VIII Division I, and Section II for pressure vessels
   (d) ANSI/API 620, rules for design and construction for large, welded low-pressure
   storage tanks and ANSI/API 650 for welded steel tanks for oil storage

Spent fuel cask handling systems
   (a) 49 CFR 173.392 and 173.393
   (b) CMAA 70 for specifications for overhead traveling cranes
   (c) ANSI N14.6 for special lifting devices for shipping containers weighing more than
   10,000 pounds
   (d) ANSI/ANS B30.1 for overhead and gantry cranes for ANSI/ANS 57.2 designs

Spent fuel or waste form handling systems
   (a) ANSI/IEEE C2 for electrical code
   (b) IEEE-IPCEA, IEEE S-135 for power cable code
   (c) ANSI/NFPA 70 for electrical code
   (d) ANSI/ASME B30.16 for overhead hoists

Heating, ventilating and air-conditioning systems
   (a) ASHRAE Handbook
   (b) Air Movement and Control Association, Standards and Application Guides
   (c) ANSI/ASME N509, Nuclear Power Plant Air Cleaning Units and Components
   (d) Uniform Building Code Mechanical Code

Buildings
   (a) 10 CFR Part 73 for physical protection of plants and materials
   (b) ANSI/ACI 349 for reinforced concrete for ANSI/ANS 57.2 designs and ANSI/ACI
   318 for ANSI/ANS 57.7 designs
   (c) NFPA 78 for protection against lightning
   (d) AISI *Steel Products Manual*

Rad waste treatment
   (a) 10 CFR Part 71 for packaging and transportation
   (b) 49 CFR 173.391 for transportation
   (c) 10 CFR Part 20 for radiation protection

15.5.2 Accidents

The SAR should include a rigorous discussion of accident potential, both external natural events
and man-induced events, at the proposed ISFSI or MRS. Natural phenomena events to be

considered are presented in Chapter 2 of the SAR, and their review is discussed in the site characteristics Section of this FSRP. The accident analysis review focuses on the effects of the natural phenomena and man-induced events on SSCs important to safety. Analytical techniques, uncertainties, and assumptions should be presented.

The reviewer should verify that each event includes (a) a discussion of the cause of the event, (b) the means of detection of the event, (c) an analysis of the consequences (particularly any radiological consequences) and the protection provided by devices or systems designed to limit the extent of the consequences, and (d) any actions required of the operator. For each accident the applicant should provide and discuss the results of dose calculations.

15.5.2.1 Cask Tip-Over

The cask tip-over accident should be evaluated in the SAR. For this analysis, the NRC will accept cask tip-over about a lower corner onto a receiving surface from a position of balance with no initial velocity. The NRC has also accepted analysis of cask drops with the longitudinal axis horizontal that, together with the longitudinal axis vertical, could bound a non-mechanistic tip-over analysis.

Define the Operating Environment

The reviewer should verify that the SAR identifies the operating environment of the accident, including:

- the operational configuration of the confinement cask (i.e., a stand-alone cask in the pad; a cask inside a transfer cask suspended from a cable on a crane or hoist, with or without impact limiters; etc.)

- the physical location of the tip-over accident (i.e., outside the spent fuel pool building or inside the spent fuel pool building)

Define the Physical Parameters

The reviewer should verify that the SAR defines the physical parameters necessary to evaluate the accident, including:

- the receiving surface upon which the confinement cask slaps down (i.e., the storage pad materials, dimensions, and properties and the foundation properties; or the spent fuel pool or building floor materials) (The surface must be defined to quantify the maximum deceleration levels.)

- the design of the confinement cask and associated SSCs (i.e., material properties, dimensions, and weights)

Review the Analysis Methodology

The reviewer should verify that the SAR defines the analysis methodology used in the evaluation, such as:

- reference to horizontal and vertical analyses if the tip-over can be shown to be bounded by these two accidents

- specific analysis modeling tools such as closed-form manual techniques or computer codes (ANSYS has been accepted by the NRC for previous applications.)

Accident Analysis

The reviewer should verify that the SAR presents the accident analysis and design criteria and design codes and standards, such as:

- deceleration level

- design code for evaluation (The ASME B&PV Code Section III, Service Level D is acceptable to the NRC.)

- specification if elastic or elastic-plastic analysis is used and appropriate citation of design code

- evaluation of calculated stress intensity level against the allowable stress intensity level at the design temperature and pressure for each component in the confinement cask

- evaluation of buckling stability for each component member of the confinement cask subject to compressive loading

- evaluation of deformation of cask internal members that contribute to the spacing geometry for criticality safety

- calculation of the accidental dose

15.5.2.2 Cask Drop

The drop of the confinement cask is one of the hypothetical accident scenarios that must be evaluated by the applicant. The following steps are intended to provide the reviewer with an outline of the methodology that the applicant should provide in the SAR. The steps are representative of a typical cask but are not intended to cover every aspect of every possible design.

Define the Operating Environment

The reviewer should verify that the SAR identifies the operating environment of the accident, including:

- the operational configuration of the confinement cask (i.e., the stand-alone confinement cask with no other SSCs; the confinement cask inside the transfer or transport cask, with or without impact limiters; etc.)

- the confinement cask orientation at the moment of impact (i.e., end drop on top or bottom, side drop at various azimuths, and corner drop at various azimuths and inclinations)

- the physical location of the drop accident (i.e., outside the spent fuel pool building or inside the spent fuel pool building)

Define the Physical Parameters

The reviewer should verify that the SAR defines the physical parameters associated with the accident, including:

- the receiving surface upon which the confinement cask impacts (i.e., the storage pad materials, dimensions, and properties and the foundation properties; or the spent fuel pool or building floor materials, dimensions, and properties)(The surface should be sufficiently characterized to quantify the maximum deceleration levels.)

- the design of the confinement cask and associated SSCs (i.e., material properties, dimensions, and weights)

- the drop height of the confinement cask onto the receiving surface for each orientation (The maximum height above the impact surface to which the cask could be lifted should be used in the analysis.)

Review the Analysis Methodology

The reviewer should verify that the SAR defines the analysis methodology used in the evaluation, such as:

- static equivalent deceleration with appropriate dynamic load factors

- dynamic modeling with appropriate test data to benchmark deceleration

- specific analysis modeling tools such as manual techniques or computer codes (ANSYS and has been accepted by the NRC for previous applications.)

Accident Analysis

The reviewer should verify that the SAR presents the accident analysis and design criteria and design codes and standards, such as:

- deceleration level for each case considered

- design code for evaluation (The ASME B&PV Code Section III, Service Level D is acceptable to the NRC.)

- specification if elastic or elastic-plastic analysis is used and appropriate citation of design code

- evaluation of calculated stress intensity level against the allowable stress intensity level at the design temperature and pressure for each component member of the confinement cask

- evaluation of the buckling stability for each component member of the confinement cask subjected to compressive loading

- evaluation of the deformation of cask internal members that contribute to spacing geometry of the spent fuel assemblies or high-level waste materials that are subject to criticality safety as given in Section 8 of this FSRP.

- calculation of accidental dose

15.5.2.3 Flood

The flood accident is one of the accidents that must be evaluated as required by 10 CFR 72.122(b)(2). The flood evaluation in the SAR should be coordinated with the site characteristics. The following steps are suggested to provide the reviewer with an outline of the methodology that the applicant should provide in the SAR.

Define the Operating Environment

The reviewer should verify that the SAR identifies the operating environment of the accident, including:

- the operational configuration of the confinement cask or other SSCs important to safety (i.e., a confinement cask on a storage pad, a spent fuel pool building, a confinement cask in a shielding structure, etc.)

- the physical location of the SSCs important to safety at the time of the hypothetical flood

- the source of the flood water based on historical data for the site

Define the Physical Parameters

The reviewer should verify that the SAR defines the physical parameters associated with the flood condition, including:

- the quantity of flood water (i.e., the static head of water and the maximum flow velocity)

- any protection devices placed at the site to prevent ingress of water into the spent fuel pool building or to prevent casks from tip-over or sliding

Define the Analysis Methodology

The reviewer should verify that the SAR defines the analysis methodology used in the evaluation, such as:

- sliding and overturning
- evaluation of external pressure stress intensity

Accident Analysis

The reviewer should verify that the SAR presents the accident analysis and design criteria and design codes and standards, such as:

- the design basis flood conditions

- the determination of the maximum drag force acting on the confinement cask or other SSCs important to safety

- the determination of the pressure loading acting on the SSCs

- the determination of the external pressure stress intensity and comparison with the allowable stress as found in the ASME B&PV Code Section III, Service Level C

- determination that there is no sliding and overturning of the SSCs

- compliance with Regulatory Guides 1.59, "Design Basis Floods for Nuclear Power Plants," and 1.102, "Flood Protection for Nuclear Power Plants," where applicable

- calculation of accidental dose

15.5.2.4 Fire and Explosions

The fire and explosion accidents must be evaluated as required by 10 CFR 72.122(c). The evaluation of these accidents should be coordinated with the site characteristics, as defined in the SAR. The following steps are suggested to provide the reviewer with an outline of the methodology for evaluating the fire and explosion accidents.

Define the Operating Environment

The reviewer should verify that Chapters 2, 4, and 8 of the SAR identify the operating environment for a fire or explosion accident, including:

- the presence of materials that could accidentally burn or explode in the vicinity of the ISFSI or along the route of transfer at the site

- operational conditions that could accidentally initiate combustion or explosion

Define the Physical Parameters

The reviewer should verify that the SAR defines the physical parameters associated with the accidents, including:

- the quantity of combustible fuel present at the site
- the barriers in place to protect the SSCs from damage by heat or explosive overpressure
- the presence of a fire protection program (FPP)

Review the Evaluation Methodology

The reviewer should verify that the SAR defines the methodology by which the fire or explosion hazards are to be evaluated, including:

- modeling techniques for calculating temperature rise of SSCs

- assumptions and modeling techniques for predicting structural response of SSCs to external or internal pressure

- assumptions used to evaluate habitable areas such as the control room of a spent fuel pool (Regulatory Guide 1.78, "Assumptions for Evaluating the Habitability of a Nuclear Power Plant Control Room During a Postulated Hazardous Chemical Release," may be used by the reviewer.)

Accident Analysis

The reviewer should verify that the SAR presents the accident analysis and design criteria, and standards to:

- establish design criteria for temperature limits for temperature sensitive materials and SSCs such as concrete, fuel cladding, shielding materials, and confinement vessels subject to internal pressure rise or external pressure rise

- show that the maximum temperature due to the accidental fire does not reach the design limit and that the effect on the SSCs has been evaluated in the structural evaluation chapter

- show that the maximum internal pressure for a confinement cask is properly evaluated or bounded by another accidental event (e.g., the pressure should be based on the assumption that 100% of the fuel rods have failed)

- show that the maximum external pressure does not cause a breach of the confinement vessel and that the stress intensity level is below the stress limit (i.e., Service Level D of the ASME B&PVC)

- verify that an FPP provides assurance that a fire will not significantly increase the risk of radioactive releases to the environment (i.e., the FPP should consist of fire detection and extinguishing systems and equipment, administrative controls and procedures, and trained personnel)

- confirm that control room or control area ventilation system piping and instrumentation drawings show monitors located in the system intakes that can detect radiation, smoke, and toxic chemicals, if applicable

- confirm that monitors actuate alarms in the control room or other suitable locations, if applicable (The reviewer may consult NUREG-0800 BTP 9.5-1 for detailed guidance.), if applicable

- verify that areas storing flammable, combustible, and hazardous materials are located and protected so that a fire, explosion, or release of hazardous materials will not adversely affect any SSCs important to safety

- verify that materials that collect and contain radioactive materials such as spent ion exchange resins, charcoal filters, and HEPA filters are stored in closed metal tanks located away from ignition sources and combustible material

- confirm that any accidental release results in a dose less than 5 rem, in compliance with 10 CFR 72.106(b)

15.5.2.5 Lightning

Lightning is an event that must be evaluated in compliance with 10 CFR 72.122(b)(2). The following steps provide the reviewer with an outline of the methodology for evaluating the lightning accident.

Define the Operating Environment

The reviewer should verify that the SAR identifies the operating environment condition for a lightning strike, including:

- SSCs that are exposed to possible lightning strikes
- IFSFI facilities that are exposed to lightning strikes
- lightning protective devices included as a part of the design

Analysis of Effects and Consequences

The reviewer should verify that the SAR presents an analysis or discussion of the effects of lightning strikes on all SSCs important to safety and facility buildings, such as spent fuel pool storage facilities, including:

- a discussion of structural materials or components that might be damaged by heat or mechanical forces generated by passing current to ground

- a discussion of any equipment that is necessary for continued operation in a pool facility and the protective devices for such equipment (The "Lightning Protection Code" of the National Fire Protection Association, NFPA 78, may be used to determine if the protective devices are satisfactory.)

- any radiological consequences associated with the lightning strike

15.5.2.6 Earthquake

The earthquake accident is one of the accidents that must be evaluated as required by 10 CFR 72.122(b)(2). The evaluation in the SER should be coordinated with the site characteristics evaluation. The following steps are suggested to provide the reviewer with an outline of the methodology that the applicant should provide in the SAR.

Define the Operating Environment

The reviewer should determine the design ground motion according to the SAR. This parameter is discussed in Section 2.5.6 of this FSRP, where the rationale for its selection is evaluated.

The reviewer should verify the evaluation, performed in accordance with Section 2.5.6 of this FSRP, of the characteristics of the supporting media. Also, review the discussion regarding soil-structure interaction.

The reviewer should verify that the SAR has defined the configuration of the SSCs at the time of the seismic event (i.e., the cask on the storage pad, the cask in a transfer cask during transport, the cask suspended from a crane, the cask during operations in the pool, etc.)

Define the Physical Parameters

The reviewer should determine which components of the ISFSI or MRS must be designed to withstand the effects of the design-basis earthquake (DBE). General Design Criteria (GDC) 2, "Design Bases for Protection Against Natural Phenomena," of Appendix A, "General Design Criteria for Nuclear Power Plants," to 10 CFR 50, "Domestic Licensing of Production and Utilization Facilities," requires that nuclear power plant SSCs be designed to withstand the effects of earthquakes without loss of capability to perform their safety functions. Regulatory Guide 1.29, "Seismic Design Classification," describes a method for identifying those features of a light-water reactor (LWR) that should be designed to withstand the effects of the safe shutdown earthquake (SSE). The staff has interpreted this regulatory guide to mean that those SSCs identified as important to safety should be designed for the DBE. Refer to Section 4 of this FSRP for an evaluation of the identification of these components. Protection devices to mitigate effects of the event, such as a seismic sensor to trip power to overhead cranes or extra seismic supports to be installed during transfer operations, should be defined.

Review the Analysis Methodology

If an equivalent static load method is used, the reviewer should verify that the method produces conservative results and that the SSCs can be realistically represented by a simple model.

If a response spectrum analysis technique was used, the reviewer should verify that the response spectra meet the requirements of Regulatory Guide 1.60, "Design Response Spectra for Seismic Design of Nuclear Power Plants," and that damping ratios are in accordance with Regulatory Guide 1.61, "Damping Values for Seismic Design of Nuclear Power Plants."

If a time-history analysis has been performed, the reviewer should verify that the time-history acceleration is in compliance with ASCE 4-86, "Seismic Analysis of Safety-Related Nuclear Structures."

Accident Analysis

The reviewer should verify that the analysis has used the three components of earthquake motion and has combined them in accordance with NUREG-0800, Section 3.7.2, "Seismic System Analysis," subsection 6 and Regulatory Guide 1.92, "Combining Modal Responses and Spatial Components in Seismic Response Analysis."

In accordance with NUREG-0800, Section 3.7.2, "Seismic System Analysis," subsection 14, the reviewer should verify that a determination of Category I structure overturning moments has been considered. To be acceptable, the determination of the design overturning moment should incorporate three components of input motion and conservative consideration of vertical and lateral seismic forces. The reviewer should verify that the structure neither overturns nor slides due to the DBE.

The reviewer should verify that a summary of natural frequencies of the SSCs important to safety has been provided. If the direct integration method of analysis was used, the reviewer should verify that total responses of the SSCs have been calculated.

The reviewer should verify that any radiological consequences associated with the seismic event have been identified and that accidental dose rates have been calculated. SSCs are not required to survive accident-level earthquakes without permanent deformation; however, the reviewer should verify that the stress intensities are less than the stress allowables (i.e., Service Level D of the ASME B&PVC).

15.5.2.7 Loss of Shielding

The loss of shielding of any SSCs identified as important to safety must be evaluated to determine the dose to workers and the public. Loss of shielding can occur due to a tornado missile impact-induced penetration of concrete shielding, the reduction in hydrogen content of neutron shielding by high temperature exposure, loss of water or lowering of the water level by leakage from shields that are composed of water, structural failure or melting of shielding by fire or explosion, etc. The following steps are suggested to provide the reviewer with an outline of the methodology that the applicant should provide in the SAR.

Define the Operating Environment and Physical Parameters

The reviewer should verify that the SAR identifies the operating environment and the physical parameters of the accident, including:

- the operational configuration of the SCCs such as a cask design that uses a liquid shielding material

- the design threshold for safety pressure-relief valves or rupture discs for liquid shield tanks

Review the Analysis Methodology and Accident Event Analysis

The reviewer should determine the maximum reduction in ISFSI radiation shielding thickness, material shielding effectiveness, or loss of temporary shielding due to postulated accidents such as tornado missiles, explosions, fires, liquid shield tank leaks, and cask drop. All possible shielding areas should be evaluated.

The reviewer should perform a revised neutron and gamma dose rate shielding analysis with the accident-induced reduced shielding thickness or loss of shielding. The analysis should use computer codes and/or methodology identical to those of the design shielding calculations for the ISFSI.

The reviewer should evaluate the magnitude of worker dose rate due to a loss of shielding in terms of shielding repair efforts and ALARA.

15.5.2.8 Adiabatic Heatup

Adiabatic heatup is a key assumption for an evaluated accident because it ensures that the most conservative thermal transient response of the cask components has been evaluated. The transient temperature response of internal cask components, including spent nuclear fuel, is due solely to the spent fuel decay heat and the individual cask material heat capacity (i.e., mass and specific heat).

Define the Operating Environment

The reviewer should verify that the SAR defines the ambient temperature, including insolation, that is used in the analysis. The reviewer should verify that the configuration of the SSCs has been defined, (i.e., all inlets and outlets blocked (for casks) and cooling systems or pumps inoperable (for pools)). The reviewer should evaluate the highest design basis decay heat load of the ISFSI design that should be stated in the design criteria chapter of the ISFSI SAR.

Define the Physical Parameters

The reviewer should verify the minimum mass of each material that constitutes a component of the ISFSI. Such materials are typically uranium dioxide, zircalloy, stainless steel, anconal, carbon steel, boral, resin, polyethylene, and concrete. In general, the mass can be calculated by determining the volume of the material and using a value for density of the material that is obtained from an established reference of material properties. The density should be appropriate for the anticipated temperature range for this calculation.

The reviewer should determine the specific heat of each material from established references for the expected range of temperatures.

The reviewer should determine the maximum short-term accident temperature limit of each material within the ISFSI from established references.

Review the Methodology and Analysis

For casks having multiple air inlets and outlets, some cask designers have argued that, if all air inlets were completely blocked, one air outlet would become an air inlet while the other air outlet would continue to function as an outlet. This assumption has never been verified by experimental test data and has, therefore, been rejected by the staff. All casks that rely on natural air convection through internal labyrinthine passages are required to assume that all air inlet passages are completely blocked, and the thermal response must be calculated by assuming that no heat loss to the environment occurs.

The reviewer should calculate the sum of the product of mass and specific heat for each ISFSI material. This is denoted as the heat capacity of the ISFSI unit.

The reviewer should calculate the adiabatic heatup rate of the ISFSI unit by dividing the total ISFSI unit maximum decay heat load by the total ISFSI unit heat capacity.

Using the highest calculated temperature for each ISFSI material at normal operating ambient temperatures, the maximum short-term accident temperature limit for each material, and the ISFSI adiabatic heatup rate that was calculated in the above paragraph, the reviewer should determine the earliest time that a specific material temperature limit will be exceeded after the onset of an adiabatic heatup scenario. The reviewer should refer to Section 6.5.2.2 for temperature limits on fuel cladding and Section 6.5.2.3 for temperature limits on concrete.

The reviewer should report, as the key result, the minimum time to reach the first ISFSI material temperature limit during an adiabatic heatup event. A surveillance frequency must be incorporated into the technical specifications. The basis for this technical specification would be the material temperature limit during an adiabatic heatup.

The reviewer should verify that any radiological consequences associated with the adiabatic heatup have been identified and that accidental dose rates have been calculated.

15.5.2.9 Tornadoes and Missiles Generated by Natural Phenomena

The tornado and tornado-generated missile accidents must be evaluated as required by 10 CFR 72.122(b)(2). The evaluation in the SAR should be coordinated with the site characteristics. The following steps are suggested to provide the reviewer with an outline of the methodology that the applicant should provide in the SAR.

Define the Operating Environment and Physical Parameters

The reviewer should determine the design wind and tornado wind velocities according to the SAR. These parameters are discussed in Section 2.5.3 of this FSRP. The reviewer should verify that the design basis tornado characteristics given in Table 15.2, taken from Regulatory Guide 1.76, "Design Basis Tornado for Nuclear Power Plants," and NUREG-1503, "Final Safety Evaluation Report Related to the Certification of Advanced Boiling Water Reactor Design," have been used as design values; if less conservative values have been used, the reviewer should verify that the applicant has justified his selection.

Table 15.2 Design Basis Tornado Characteristics

| Region | Maximum Wind Speed[1] (mph) | Rotational Speed (mph) | Translational Speed (mph) | | Radius of Maximum Rotational Speed (feet) | Pressure Drop (psi) | Rate of Pressure Drop (psi/sec) |
			Maximum	Minimum			
I	360	290	70	5	150	3.0	2.0
II	300	240	60	5	150	2.25	1.2
III	240	190	50	5	150	1.5	0.6

[1]The maximum wind speed is the sum of the rotational speed and the maximum translational speed components.

The reviewer should verify that the applicant has used Spectrum I or Spectrum II in the analysis of missile impact. Spectrum I includes an 1800 Kg automobile, a 125 Kg 8-inch armor piercing artillery shell, and a 1-inch solid steel sphere, all impacting at 35% of the maximum horizontal windspeed of the design basis tornado. The first two missiles are assumed to impact at normal incidence, and the last is assumed to impinge upon barrier openings in the most damaging directions. Spectrum II missiles are given in Table 15.3. Refer to WASH-1300 for a definition of tornado regions.

Table 15.3 Spectrum II Tornado Missiles

Missile	Mass (Kg)	Dimensions (m)	Velocity (m/sec)		
			Region I[1]	Region II	Region III
A Wood Plank	52	.092 x .289 x 3.66	83	70	58
B 6-inch Sch 40 pipe	130	.168D x 4.58	52	42	10
C 1-inch Steel rod	4	.0254D x .915	51	40	8
D Utility pole	510	.343D x 10.68	55	48	26
E 12-inch Sch 40 pipe	340	.32D x 4.58	47	28	7
F Automobile	1810	5 x 2 x 1.3	59	52	41

[1]Tornado regions are defined in WASH-1300.

Vertical velocities of 70% of the postulated horizontal velocities are acceptable in both spectra except for the small missile in Spectrum I and Missile C above that should be assumed to have the same velocity in all directions. The artillery shell in Spectrum I and Missiles A, B, C, and E in Spectrum II are to be considered at all elevations. The automobile in Spectrum I and Missiles D and F in Spectrum II are to be considered at elevations up to 30 ft above all grade levels within one-half mile of the facility structure.

Review the Methodology and Analysis

The reviewer should verify the transformation of wind velocity into pressure. The procedures used to transform the wind velocity into an effective pressure to be applied to structures and parts and portions of structures, and found in ASCE 7, are acceptable. These procedures specify that the maximum velocity pressure, p (in psf), should be obtained from the formula, $p = 0.00256 V^2$, where V is in mph; the velocity pressure should be assumed constant with height; and the maximum pressure applies at the radius of the tornado funnel at which the maximum velocity occurs. ASCE Paper No. 3269, "Wind Forces on Structures," may be used to obtain the effective wind pressures for cases that ASCE 7 does not cover.

The reviewer should verify that all SSCs important to safety have been analyzed for damage from missiles that might be generated by the design basis tornado for that ISFSI or MRS. Also review the applicant's analysis of missile-impact on SSCs important to safety. In previous submittals, the NRC has accepted use of "A Review of Procedures for the Analysis and Design of Concrete Structures to Resist Missile Impact Effects," by R. P. Kennedy, "Topical Report, Design of Structures for Missile Impact," by R.B. Linderman et al., and "U.S. Reactor Containment Technology," by W.B. Cottrell and A.W. Savolainen.

The reviewer should verify that the most adverse combination of tornado wind, differential pressure, and missile load is calculated and used in combination with other loads. To obtain the most adverse combination, the combinations to be considered should include wind alone, differential pressure alone, missile alone, wind plus half of the differential pressure, wind plus missile, and wind plus missile plus half of the differential pressure.

ANSI/ANS 57.9 provides acceptable criteria for resistance to overturning and sliding.

The reviewer should verify that any radiological consequences associated with the tornado and tornado-generated missiles have been identified and that accidental dose rates have been calculated.

15.5.2.10 Accidents at Nearby Sites

The reviewer should verify that potential accidents at nearby sites and transportation routes have been considered. The procedures for reviewing these accidents are expected to have been covered in other sections of this accident chapter (i.e., a natural gas explosion at a nearby site may result in an explosive overpressure that is evaluated in Section 15.5.2.4, and the effects of a fire at a nearby site is also evaluated in Section 15.5.2.4). The reviewer should verify that the effects of nearby site accidents have been encompassed by other reviews.

15.5.2.11 Accidents Associated with Pool Facilities

Pool facilities must be capable of withstanding the effects of natural phenomena such as earthquakes, tornadoes, hurricanes, tornado driven missiles, lightning, and floods. These natural phenomena are discussed in other sections of the accident chapter and are not repeated here. In addition, pool facilities should be designed to withstand the effects of operational accidents such that:

- the loss of water from the pool does not uncover radioactive material or fall below a safe limit
- radioactive material is protected from mechanical damage
- offsite exposure due to release of radioactive materials is below 10 CFR Part 20 limits (See Regulatory Guide 1.13)

Section 5 of this FSRP identifies two NRC-approved ANSI/ANS standards dealing with wet spent fuel storage and handling facilities. The SAR should specify if ANSI/ANS 57.2 for light-water reactor spent fuel storage facilities or ANSI/ANS 57.7 for a water pool IFSFI is to be used for the pool facility. The accidents associated with pool facilities are dependent on the particular standard cited in the application.

Define the Accident

The following accidents are estimated to occur with a frequency of 10^{-1} to 10^{-2} per storage year, and the reviewer should verify that they have been evaluated regardless of which ANSI standard is cited in the SAR:

- a passive failure of a radioactive liquid retaining boundary

- a loss of external power for an extended interval (e.g., up to eight hours)

- overfilling the pool

- loss of non-seismic Category I portion of the spent fuel pool cooling system (if applicable to design)

- loss of air supply to seals on gates resulting in leakage from the pool (if applicable to design)

- drop of spent fuel cask from controlled normal height (not from maximum achievable height)

The following accidents are estimated to occur with a frequency of 10^{-2} to 10^{-6} per storage year for facilities designed to ANSI/ANS 57.2 standards, and the reviewer should verify that they have been addressed:

- rupture of all fuel rods in a spent fuel assembly

- inadvertent opening of a gate (e.g., cask loading isolation gate) when the adjoining area is empty, resulting in reduced shielding

- drop of a spent fuel cask from maximum achievable height

- loss of offsite power for up to seven days

The above accidents are suggested to the reviewer as representative of possible accidents. The list is not intended to cover all pool facilities, since each facility will be unique. Some pool facilities may be the existing pools at 10 CFR Part 50 licensed reactors, in which case the accident scenarios would correspond to guidelines in NUREG-0800 and ANSI/ANS 57.2.

Define the Operating Environment

The reviewer should verify that the SAR identifies the operating environment and conditions of the accidents. Because each pool facility will be different, the applicant must describe, in detail, how each of the above accidents is possible or not possible due to the specific design or operating environment.

Define the Physical Parameters

The SAR should define the physical parameters associated with the accidents, including:

- level or temperature of water at time of failure of external power

- any protective devices designed to mitigate the consequences of the accidents

- alarms and response times for corrective action

- capacities of the storage or transfer pool, flow rates of the pumps and valves, time constants for the control systems, etc.

•   height of cask drop and characterization of surface(s) impacted by the cask

Review the Analysis Methodology

The reviewer should verify that the SAR defines the analysis methodology for evaluating the consequences of the accidents, including:

•   modeling techniques for calculating the temperature rise of the storage pool or transfer pool water

•   assumptions used as a part of the accidents

Accident Analysis

The reviewer should verify that the SAR presents the accident analysis and design criteria and design codes and standards for each of the accidents that are defined in the SAR. The codes and standards given below are the primary design and construction codes that are acceptable to the NRC, and a complete reference for the publications can be found in Section 17. For a more detailed listing of design codes and standards, the reviewer may consult ANSI/ANS 57.2 or 57.7.

Spent fuel storage and cask unloading/handling pools
> (a) ASME B&PVC, Section III
> (b) ACI 318 for reinforced concrete for ANSI/ANS 57.7 designs and ACI 349 for ANSI/ANS 57.2 designs
> (c) ANSI/AWS D1.1 for structural welding
> (d) ANSI/AWS D1.4 for welding reinforcing steel
> (e) AISI *Steel Products Manual*

Spent fuel storage racks
> (a) ASME B&PVC, Section III

Spent fuel pool water makeup, cooling and cleanup system
> (a) ANSI/IEEE C2 for electrical code, and ANSI/NFPA 70
> (b) ANSI/ASME B31.1 for power piping
> (c) ASME B&PVC Section VIII Division I, and Section II for pressure vessels
> (d) ANSI/API 620, Rules for design and construction for large, welded low-pressure storage tanks and ANSI/API 650 for welded steel tanks for oil storage

Spent fuel cask handling systems
> (a) 49 CFR 173.392 and 173.393
> (b) CMAA 70 for specifications for overhead traveling cranes
> (c) ANSI N14.6 for special lifting devices for shipping containers weighing more than 10,000 pounds
> (d) ANSI/ANS B30.1 for overhead and gantry cranes for ANSI/ANS 57.2 designs

Spent fuel or waste form handling systems

(a) ANSI/IEEE C2 for electrical code

(b) IEEE-IPCEA, IEEE S-135 for power cable code

(c) ANSI/NFPA 70 for electrical code

(d) ANSI/ASME B30.16 for overhead hoists

Heating, ventilating and air-conditioning systems

(a) ASHRAE Handbook

(b) Air Movement and Control Association, Standards and Application Guides

(c) ANSI/ASME N509, Nuclear Power Plant Air Cleaning Units and Components

(d) Uniform Building Code Mechanical Code

Buildings

(a) 10 CFR Part 73 for physical protection of plants and materials

(b) ANSI/ACI 349 for reinforced concrete for ANSI/ANS 57.2 designs and ANSI/ACI 318 for ANSI/ANS 57.7 designs

(c) NFPA 78 for protection against lightning

(d) AISI *Steel Products Manual*

Rad waste treatment

(a) 10 CFR Part 71 for packaging and transportation

(b) 49 CFR 173.391 for transportation

(c) 10 CFR Part 20 for radiation protection

The reviewer should verify that any radiological consequences associated with pool facility accidents have been identified and that accidental dose rates have been calculated.

15.5.2.12 Building Structural Failure onto SSCs

A building that houses the SSCs (contains spent fuel casks, is used for transfer operations, or is used for temporary storage) must be designed to prevent massive collapse due to accident conditions. Although not considered an SSC important to safety, the building must be designed to withstand collapse due to the effects of flood, fire and explosion, lightning, earthquake, tornado and tornado-generated missiles, and accidents at nearby sites. Review procedures for these events have been presented in previous subsections of Section 15.5.2 of this FSRP for SSCs important to safety. The reviewer should verify that the applicant has analyzed the building structure to meet the applicable portions of these procedures. The applicant's analysis should provide evidence that, although equipment or structures may be damaged, the surviving equipment and structures will continue to protect the fuel and high-level waste and that the radiological consequences are within the acceptable levels.

15.5.3 Other Non-Specified Accidents

The reviewer should evaluate other accident scenarios included in the SAR but not identified in previous subsections. The reviewer should coordinate the accident analysis review with the evaluations of all technical chapters to verify that design characteristics of the ISFSI or MRS do not pose potential accidents that have not been identified by the applicant.

15.6 Evaluation Findings

NRC staff reviewers prepare evaluation findings regarding satisfaction of the regulatory requirements related to accident analysis. If the documentation submitted with the application fully supports positive findings for each of the regulatory requirements, then the findings should substantially be stated as follows (finding numbering is for convenience in referencing within this FSRP and would be of convenience in the SER):

F15.1   The SAR includes acceptable analyses of the design and performance of SSCs important to safety under off-normal and accident scenarios. Applicable off-normal accidents, analyzed in the SAR, included [reviewer to select from the following:] partial vent blockage, operational events [reviewer to list], off-normal ambient temperature scenarios, and off-normal events associated with the pool facilities. Applicable accident events, analyzed in the SAR, included [reviewer to select from the following:] cask tipover, cask drop, flood, fire and explosion, lightning, earthquake, loss of shielding [if applicable], adiabatic heatup of the cask, tornadoes and missiles generated by natural phenomena, accidents at nearby sites, accidents associated with pool facilities, building structural failure onto SSCs, and [other scenarios supplied by the applicant].

F15.2   The analyses of off-normal and accident events and conditions and reasonable combinations of these and normal conditions show that the design of the [ISFSI/MRS] will acceptably meet the requirements without endangering the public health and safety, in compliance with the overall requirements of 10 CFR 72.122.

F15.3   The analyses of off-normal and accident events and conditions and reasonable combinations of these and normal conditions show that the design of the [ISFSI/MRS] will acceptably meet the requirements of 10 CFR 72.124 regarding the maintenance of the spent fuel in a subcritical condition.

F15.4   The analyses of off-normal and accident events and conditions and reasonable combinations of these and normal conditions show that the design of the [ISFSI/MRS] will acceptably meet the requirements of 10 CFR 72.126 regarding criteria for radiological protection.

F15.5   The analyses of off-normal and accident events and conditions and reasonable combinations of these and normal conditions show that the design of the [ISFSI/MRS] will acceptably meet the requirements of 10 CFR 72.128 regarding handling, storage, and retrievability of the spent fuel and other radioactive material.

15.7 References

NRC documents referenced are identified at Consolidated References, Section 17.

ACI 318, "Building Code Requirements for Reinforced Concrete."

ACI 349, "Code Requirements for Nuclear Safety Related Concrete Structures," and ACI 349R, "Commentary."

AISI, *Steel Products Manual.*

ANSI/ANS 57.2-1983, "Design Requirements for Light-Water Reactor Spent Fuel Storage Facilities at Nuclear Power Plants."

ANSI/ANS 57.7-1988, "Design Criteria for an Independent Spent Fuel Storage Installation (Water Pool Type)."

ANSI/ANS 57.9-1984, "Design Criteria for an Independent Spent Fuel Storage Installation (Dry Storage Type)."

ANSI/ASME B30.2.0-1978, "Overhead and Gantry Cranes (Top Running Bridge, Multiple Girder)."

ANSI/ASME B30.16, "Overhead Hoists."

ANSI/ASME B31.1, "Power Piping."

ANSI/ASME N509, "Nuclear Power Plant Air Cleaning Units and Components."

ANSI/IEEE C2, "Electrical Code."

ANSI N14.6-1993, "American National Standard for Radioactive Materials - Special Lifting Devices for Shipping Containers Weighing 10,000 Pounds (4500 kg) or More," Institute for Nuclear Materials Management, 1993.

"ANSYS Basic Analysis Procedures Guide," Fourth Edition, ANSYS Release 5.6, ANSYS, Inc., Canonsburg, Pennsylvania, November 1999.

API 620, "Rules for Design and Construction for Large, Welded Low-Pressure Storage Tanks."

API 650, "Welded Steel Tanks for Oil Storage."

ASCE 4-86, "Seismic Analysis of Safety-Related Nuclear Structures," American Society of Civil Engineers, New York, NY, 1986.

ASCE 7 (formerly ANSI A58.1), "Minimum Design Loads for Buildings and Other Structures."

ASHRAE Handbook, 1989, Fundamentals, American Society of Heating Refrigeration and Air Conditioning Engineers, Inc.

ASME Boiler and Pressure Vessel Code, Section II, Division 1, "Material Specifications."

ASME Boiler and Pressure Vessel Code, Section III, Division 1, "Rules for Construction of Nuclear Power Plant Components."

ASME Boiler and Pressure Vessel Code, Section VIII, "Pressure Vessels."

AWS D1.1, "Structural Welding Code - Steel."

AWS D1.4, "Structural Welding Code - Reinforcing Steel."

CMAA 70, "Specifications for Electric Overhead Traveling Cranes," Crane Manufacturers Association of America.

DOE/RW-0184-R1, "Characteristics of Potential Repository Wastes," Office of Civilian Radioactive Waste Management, US Department of Energy, July 1992.

IEEE S-135, IEEE-IPCEA, "Power Cable Ampacities."

NFPA 70, "National Electric Code."

NFPA 78, "Lightning Protection Code."

UBC, "Uniform Building Code," International Conference of Building Officials.

Air Movement and Control Association, Standards and Application Guides.

Eckerman, K. F., Wolbarst, A. B., and Richardson, A. C. B., "Limiting Values of Radionuclide Intake and Air Concentration and Dose Conversion Factors for Inhalation, Submersion, and Ingestion," Federal Guidance Report No. 11, DE89-011065, ORNL, 1988.

Kennedy, R. P., "A Review of Procedures for the Analysis and Design of Concrete Structures to Resist Missile Impact Effects," ORNL-NSIC-5, Volume 1, Chapter 6.

Linderman, R. B., Rotz, J. V., Yeh, G. C. K., "Design of Structures for Missile Impact," Topical Report BC-TOP-9-A, Revision 2, Bechtel Power Corporation, September 1974.

Oak Ridge National Laboratory, "ORIGEN2.1: Isotope Generation and Depletion Code - Matrix Exponential Method," 1991.

WASH-1300, "Technical Basis for Interim Regional Tornado Criteria," U.S. Atomic Energy Commission, May 1974.

16 TECHNICAL SPECIFICATIONS

16.1 Review Objective

The objective of this review is to evaluate the applicant's proposed technical specifications, including their justification, to ensure that they are completely and appropriately defined and justified. The proposed technical specifications must be supported by the technical disciplines under review. The NRC also determines whether the Safety Evaluation Report (SER) should incorporate any additional requirements.

The technical specifications define the conditions that are deemed necessary and sufficient for safe ISFSI or MRS use. The technical specifications include functional and operating limits, monitoring instruments and limiting control settings, limiting conditions, surveillance requirements, design features, and administrative controls that ensure safe operation of the facility. Each specification must be clearly documented and justified in the technical review sections of the Safety Analysis Report (SAR) and associated SER section as necessary for safe facility operation. These technical specifications are included in the site-specific license.

An overview of the technical specifications review process is shown in Figure 16.1. The figure shows that the technical specification review draws information from all technical sections as well as the accident and operation systems chapters.

16.2 Areas of Review

This chapter provides guidance for use in evaluating the technical specifications that the applicant deems necessary for safe use of the ISFSI. A comprehensive review of the proposed specifications assesses the applicant's compliance with the regulatory requirements represented by the following five areas:

Functional/operating limits, monitoring instruments, and limiting control settings
Limiting conditions
Surveillance requirements
Design features
Administrative controls

16.3 Regulatory Requirements

This section identifies and presents a high-level summary of Title 10 of the Code of Federal Regulations (CFR) Part 72 relevant to the review areas addressed by this chapter. The NRC staff reviewer should read the exact regulatory language. A matrix at the end of this section matches the regulatory requirements identified in this section to the areas of review identified in the previous section.

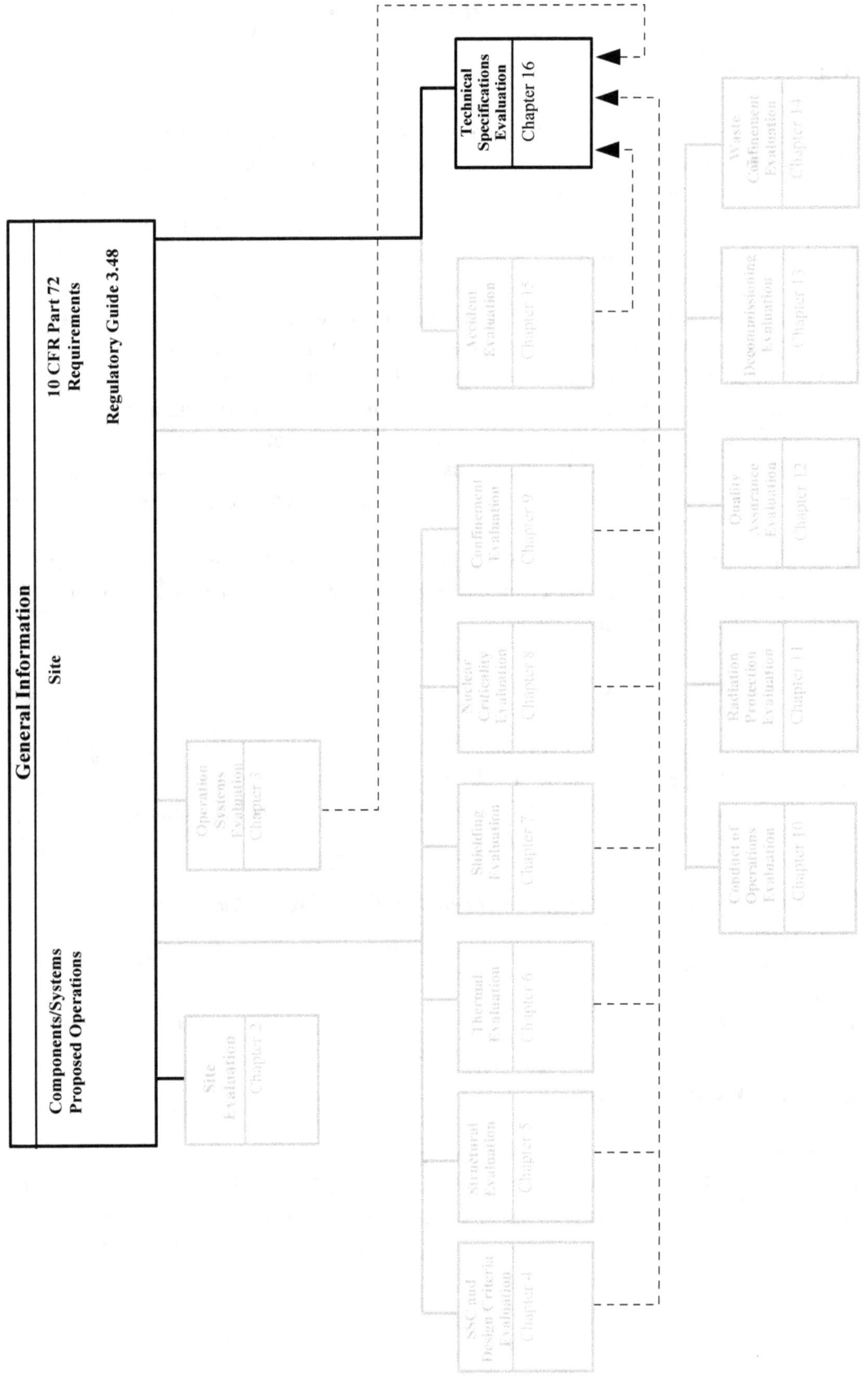

Figure 16.1 Overview of Technical Specifications Evaluation

72.24 Contents of application: Technical information [Contents of SAR].
(g) "An identification and justification for the selection of those subjects that will be probable license conditions and technical specifications"

72.26 Contents of application: Technical specifications.
"Each application under this part shall include proposed technical specifications"

72.44 License conditions.
(a) "Each license issued under this part shall include license conditions."
(c) "Each license issued under this part must include technical specifications. Technical specifications must include requirements in the following categories:"

    (1) Functional and operating limits and monitoring instruments and limiting control settings.
    (2) Limiting conditions.
    (3) Surveillance requirements.
    (4) Design features.
    (5) Administrative Controls.
(d) "Each license ... must include"
    (3) "An annual report...specifying the quantity of each of the principal radionuclides released to the environment."

A matrix showing the primary relationship of these regulations to the specific areas of review in this chapter is given in Table 16.1. The reviewer should independently verify the relationships in this matrix to ensure that no requirements are overlooked because of unique applicant design features.

Table 16.1 Relationship of Regulations and Areas of Review

Areas of Review	10 CFR Part 72 Regulations		
	72.24	72.26	72.44
Functional/operating limits, monitoring instruments, and limiting control settings	●	●	●
Limiting conditions	●	●	●
Surveillance requirements	●	●	●
Design features	●	●	●
Administrative control	●	●	●

16.4 Acceptance Criteria

The technical specifications are required by 10 CFR 72.44(c) to include functional/operating limits, monitoring instruments, limiting control settings, limiting conditions, surveillance requirements, design features, and administrative controls. The applicant must identify proposed technical specifications necessary to maintain subcriticality, confinement, shielding, heat removal, and structural integrity under normal, off-normal, and accident conditions. The applicant must identify the basis for each of the proposed technical specifications by reference to the analysis in the SAR.

Acceptance criteria for functional and operating limits, monitoring instruments, and limiting control settings include limits placed on fuel, waste handling, and storage conditions to protect the integrity of the fuel and container, to protect the employees against occupational exposures, and to guard against the uncontrolled release of radioactive materials. Acceptance criteria for limiting conditions are the lowest levels required for safe operation.

Acceptance criteria for establishing surveillance requirements include the frequency and scope of surveillance requirements to verify performance and availability of SSCs important to safety, and the verification of the bases for the proposed limiting conditions.

Acceptance criteria for administrative controls include organizational and management procedures, recordkeeping, review and audit systems, and reporting necessary to ensure that the ISFSI or MRS is managed in a safe and reliable manner. Administrative action that must be taken in the event of non-compliance with a limit or condition should be specified.

16.4.1 Code Exceptions

There is no existing American Society of Mechanical Engineers (ASME) Code for the design and fabrication of spent fuel dry storage casks. Therefore, the industry adopted, and NRC accepted, the use of ASME Code Section III as an acceptable standard for the design and fabrication of dry storage casks. However, since dry storage casks are not pressure vessels, ASME Code Section III cannot be implemented without allowing some exceptions to its requirements. Therefore, in the past, NRC has allowed specific exceptions to the code for those requirements that were not applicable or practical to implement for spent fuel dry cask storage systems.

Early spent fuel dry cask storage licenses and certificates of compliance were issued without documenting which specific exceptions to ASME Code Section III were approved. Poor quality assurance practices during design and fabrication sometimes led to significant deviations from the Code without appropriate certificate holder design review or NRC review and approval.

Therefore, the applicant should document commitments to ASME Code Section III, with proposed exceptions, in the application. Likewise the NRC should document these commitments in the 10 CFR Part 72 licenses, certificates of compliance, or technical specifications and its approval of the proposed exceptions in the SER. Also, the NRC should include a statement (in license, certificate of compliance, or technical specifications) which refers the reader to the SAR and applicable SERs for any exceptions to the codes. In addition, to ensure that similar problems do not exist in other areas, all other codes and standards applied to components important to safety should be identified in the SAR and should be included in the license, certificate of compliance, or technical specification.

Applicants should propose a condition to a license, certificate of compliance, or technical specification (Section 4, "Design Features) that describes commitments to specified codes. The condition or technical specification should also describe a process to address deviations from the applicable codes that may be necessary. In such cases, the licensee should request an exception to the requirements of the applicable code from the NRC. If the staff finds that the deviation does not adversely impact safety, it may authorize the requested exception in writing. The following is an example of a provision for allowing exceptions to applicable codes:

4.3   Codes and Standards

The American Society of Mechanical Engineers (ASME) Boiler and Pressure Vessel Code, Section III, 1992 Edition with Addenda through 1994, is the governing Code for the storage system.

4.3.1   Design Exceptions to Codes, Standards, and Criteria

SAR Table XXX lists all approved exceptions for the design of the ISFSI.

4.3.2   Construction/Fabrication Exceptions to Codes, Standards, and Criteria

Proposed alternatives to ASME Code Section III, 1992 Edition with Addenda through 1994, including exceptions referenced in Section 4.3.1, may be used when authorized by the Director of the Office of Nuclear Material Safety and Safeguards or designee.

The proposal to the NRC must demonstrate that the alternatives would provide an acceptable level of quality and safety, or that compliance with the specified requirements of ASME Code, Section III, 1992 Edition with Addenda through 1994, would result in hardship or unusual difficulty without a compensating increase in the level of quality and safety.

Request for exceptions should be submitted in accordance with 10 CFR 72.4.

(TO BE INCLUDED IN THE SAR)
LIST OF ASME CODE EXCEPTIONS FOR PLANT ISFSI
SAR Table XXX

Component	Reference ASME Code Section/Article	Code Requirement	Exception, Justification & Compensatory Measures
Cask-specific data to be added as applicable.			

16.5 Review Procedures

The review of the technical specifications is based on information presented in the various technical design chapters and the technical specifications chapter of the applicant's SAR. The variability of designs and operations, make it impossible to define each instance for which a technical specification is necessary. For this reason, it is important that the staff conduct a detailed, thorough, and independent evaluation of each technical section of the SAR. Since the basis for a specification may have been extensively discussed in the technical sections, the applicant may use an abbreviated format in the technical specifications section of the SAR. All technical disciplines conduct the technical specification review under the coordination of the project manager. The various technical disciplines should review the results of their specific evaluations that identify technical specifications necessary for safe ISFSI or MRS operation. These technical specifications identified by the NRC staff should be compared to those identified by the applicant.

Review procedures are given in NUREG-1536, Chapter 12, Section V. In addition to providing guidance to the reviewers, Section V provides a listing of areas for which the staff have previously required technical specifications. In addition to these example areas, others topics may include:

*   frequency and scope proposed for the surveillance requirements

*   administrative controls which include organization and administrative systems and procedures, record keeping, review, and audit systems that are required to ensure that the ISFSI or MRS is managed in a safe and reliable manner

*   administrative action that must be taken in the event of non-compliance with a limit or condition.

The reviewer should identify any additional technical specifications deemed necessary, using the recommended format from Regulatory Guide 3.48, "Standard Format and Content for the Safety Analysis Report for an Independent Spent Fuel Storage Installation, (Dry Storage)," for presenting technical specifications.

Upon completion of the review, a separate table, chapter, or appendix to the SER should designate explicitly those technical specifications determined to be necessary.

16.5.1 Code Exceptions

The reviewer should verify that the applicant included a written description in a condition to the license, certificate of compliance, or technical specification (Section 4, "Design Features) which documents the codes committed to. The condition or technical specification should also describe a process to address any deviations from the ASME Code or other codes that may be needed. Likewise, the reviewer should verify that these commitments are documented in the 10 CFR Part 72 license, certificate of compliance, or technical specifications. A list of proposed exceptions to code requirements should also be provided in the SAR; this list should be revised, as necessary, to reflect all NRC-authorized exceptions.

16.6 Evaluation Findings

NRC staff reviewers prepare evaluation findings regarding satisfaction of the regulatory requirements related to technical specifications. Evaluation findings developed or included in all SER sections relating to technical specifications are also listed in this section. These findings are presented as technical specifications that state staff recommendations relating to satisfaction of license conditions as required by 10 CFR 72.44. These statements should be similar to the following models (finding numbering is for convenience in referencing within the SRP and SER):

F16.1   The staff concludes that the conditions for use at the [ISFSI name] identify necessary technical specifications to satisfy 10 CFR Part 72 and that the applicable acceptance criteria have been satisfied. The proposed technical specifications provide reasonable assurance that the ISFSI will allow safe storage of spent fuel. This finding is based on the regulation itself, appropriate regulatory guides, applicable codes and standards, and accepted practices. The technical specifications identified by the applicant include the following: [Reviewer to specify]

F16.2   In addition to the applicant's proposed technical specifications, the staff finds that the following additional technical specifications are required: [Reviewer to specify]

17 CONSOLIDATED REFERENCES

17.1 NRC Documents Cited

17.1.1 Title 10 Code of Federal Regulations, Chapter I - Nuclear Regulatory Commission

Part 20, "Standards for Protection Against Radiation."

Part 30, "Rules of General Applicability to Domestic Licensing of Byproduct Material."

Part 40, "Domestic Licensing of Source Material."

Part 50, "Domestic Licensing of Production and Utilization Facilities."

Part 70, "Domestic Licensing of Special Nuclear Material."

Part 71, "Packaging and Transport of Radioactive Material."

Part 72, "Licensing Requirements for the Independent Storage of Spent Nuclear Fuel and High-Level Radioactive Waste."

Part 73, "Physical Protection of Plants and Materials."

Part 100, "Reactor Site Criteria."

17.1.2 Regulatory Guides (RG)

RG 1.4, "Assumptions Used for Evaluating the Potential Radiological Consequences of a Fuel Handling Accident in the Fuel Handling And Storage Facility for Boiling and Pressurized Water Reactors," Revision 2, 1972.

RG 1.8, "Qualification and Training of Personnel for Nuclear Power Plants." Revision 2, April 1987.

RG 1.13, "Spent Fuel Storage Facility Design Basis." Revision 1, December 1975.

RG 1.21, "Measuring, Evaluating, and Reporting Radioactivity in Solid Wastes and Releases of Radioactive Materials in Liquid and Gaseous Effluents from Light-Water-Cooled Nuclear Power Plants." Revision 1, June 1974.

RG 1.23, "Onsite Meteorological Programs." February 1972.

RG 1.25, "Assumptions Used for Evaluating the Potential Radiological Consequences of a Fuel Handling Accident in the Fuel Handling and Storage Facility for Boiling and Pressurized Water Reactor (Safety Guide 25)." March 1972.

RG 1.26, "Quality Group Classifications and Standards for Water-, Steam-, and Radioactive-Waste-Containing Components of Nuclear Power Plants." Revision 3, February 1976.

RG 1.28, "Quality Assurance Program Requirements (Design and Construction)." Revision 3, August 1985.

RG 1.29, "Seismic Design Classification." Revision 3, September 1978.

RG 1.33, "Quality Assurance Program Requirements." Revision 2, February 1978.

RG 1.59, "Design Basis Floods for Nuclear Power Plants." Revision 2, August 1977.

RG 1.60, "Design Response Spectra for Seismic Design of Nuclear Power Plants." Revision 1, December 1973.

RG 1.61, "Damping Values for Seismic Design of Nuclear Power Plants." October 1973.

RG 1.76, "Design Basis Tornado for Nuclear Power Plants." April 1978.

RG 1.78, "Assumptions for Evaluating the Habitability of a Nuclear Power Plant Control Room During a Postulated Hazardous Chemical Release." June 1974.

RG 1.86, "Termination of Operating Licenses for Nuclear Reactors." June 1974.

RG 1.92, "Combining Modal Responses and Spatial Components in Seismic Response Analysis." Revision 1, February 1976.

RG 1.97, "Instrumentation for Light-Water-Cooled Nuclear Power Plants to Assess Plant and Environs Conditions During and Following an Accident." Revision 3, May 1983.

RG 1.102, "Flood Protection for Nuclear Power Plants." Revision 1, September 1976.

RG 1.109, "Calculation of Annual Dose to Man from Routine Releases of Reactor Effluents for the Purpose of Evaluating Compliance with 10 CFR 50, Appendix I." Revision 1, October 1977.

RG 1.115, "Protection Against Low-Trajectory Turbine Missiles." Revision 1, July 1977.

RG 1.117, "Tornado Design Classification." Revision 1, April 1978.

RG 1.120, "Fire Protection Guidelines for Nuclear Power Plants." Revision 1, November 1977.

RG 1.122, "Development of Floor Design Response Spectra for Seismic Design of Floor-Supported Equipment or Components." Revision 1, February 1978.

RG 1.140, "Design, Testing, and Maintenance Criteria for Normal Ventilation Exhaust System Air Filtration and Adsorption Units of Light-Water-Cooled Containment Isolation Provisions for Fluid Systems." Revision 1, October 1979.

RG 1.143, "Design Guidance for Radioactive Waste Management Systems, Structures, and Components Installed in Light-Water-Cooled Nuclear Power Plants." Revision 1, October 1979.

RG 1.145, "Atmospheric Dispersion Models for Potential Accident Consequence Assessments at Nuclear Power Plants." Revision 1, November 1982.

RG 3.48, "Standard Format and Content for the Safety Analysis Report for an Independent Spent Fuel Storage Installation, (Dry Storage)" Revision 1, August 1989.

RG 3.54, "Spent Fuel Heat Generation in an Independent Spent Fuel Storage Installation." Revision 1, September 1999.

RG 3.60, "Design of an Independent Spent Fuel Storage Installation (Dry Storage)." March 1987.

RG 3.65, "Standard Format and Content of Decommissioning Plans for Licensees Under 10 CFR Parts 30, 40, and 70." August 1989.

RG 3.66, "Standard Format and Content of Financial Assurance Mechanisms Required for Decommissioning Under 10 CFR 30, 40, 70, and 72." June 1990.

RG 3.67, "Standard Format and Content for Emergency Plans for Fuel Cycle and Materials Facilities." January 1992.

RG 4.1, "Programs for Monitoring Radioactivity in the Environs of Nuclear Power Plants." Revision 1, April 1975.

RG 4.16, "Measuring, Evaluating, and Reporting Radioactivity in Releases of Radioactive Materials in Liquid and Airborne Effluents from Nuclear Fuel Processing and Fabrication Plants." Revision 1, December 1985.

RG 5.20, "Training, Equipping, and Qualifying of Guards and Watchmen." January 1974.

RG 5.55, "Standard Format and Content of Safeguards Contingency Plans for Fuel Cycle Facilities." March 1978.

RG 7.6, "Design Criteria for the Structural Analysis of Shipping Cask Containment Vessels." Revision 1, March 1978.

RG 7.10, "Establishing Quality Assurance Programs for Packaging Used in the Transport of Radioactive Material, Revision 1." Revision 1, June 1986.

RG 7.11, "Fracture Toughness Criteria of Base Material for Ferritic Steel Shipping Cask Containment Vessels with a Maximum Wall Thickness of 4 Inches (0.1 m)." June 1991.

RG 7.12, "Fracture Toughness Criteria of Base Material for Ferritic Steel Shipping Cask Containment Vessels with a Wall Thickness Greater than 4 Inches (0.1m) But Not Exceeding 12 Inches (0.3m)." June 1991.

RG 8.4, "Direct-Reading and Indirect-Reading Pocket Dosimeters." February 1973.

RG 8.5, "Criticality and Other Interior Evacuation Signals." Revision 1, March 1981.

RG 8.6, "Standard Test Procedures for Geiger-Mueller Counters." May 1973.

RG 8.8, "Information Relevant to Ensuring that Occupational Radiation Exposures at Nuclear Power Stations Will Be As Low As Is Reasonably Achievable." Revision 3, June 1978.

RG 8.9, "Acceptable Concepts, Models, Equations, and Assumptions for a Bioassay Program." Revision 1, July 1993.

RG 8.10, "Operating Philosophy for Maintaining Occupational Radiation Exposures As Low As Is Reasonably Achievable." Revision 1-R, May 1997..

RG 8.13, "Instruction Concerning Prenatal Radiation Exposure." Revision 3, June 1999.

RG 8.14, "Personnel Neutron Dosimeters." Revision 1, August 1977.

RG 8.15, "Acceptable Programs for Respiratory Protections." Revision 1, October 1999.

RG 8.25, "Air Sampling in the Workplace." Revision 1, June 1992.

RG 8.26, "Applications of Bioassay for Fission and Activation Products." September 1980.

RG 8.27, "Radiation Protection Training for Personnel at Light-Water-Cooled Nuclear Power Plants." March 1981.

RG 8.28, "Audible Alarm Dosimeters." August 1981.

RG 8.29, "Instruction Concerning Risks from Occupational Radiation Exposure." Revision 1, February 1996.

RG 8.34, "Monitoring Criteria and Methods to Calculate Occupational Radiation Doses." July 1992.

RG 8.38, "Control of Access to High and Very High Radiation Areas of Nuclear Power Plants." June 1993.

RG 8.39, "Control of Access to High and Very High Radiation Areas of Nuclear Plants." April 1997.

17.1.3 NUREG

NUREG-0612, "Control of Heavy Loads at Nuclear Power Plants," July 1980.

NUREG-0800 "Standard Review Plan for the Review of Safety Analysis Reports for Nuclear Power Plants," draft report, June 1996.

NUREG-1337, Rev. 1, "Standard Review Plan (SRP) for the Review of Financial Assurance Mechanisms for Decommissioning under 10 CFR Parts 30, 40, 70, and 72." December 1988.

NUREG-1503, "Final Safety Evaluation Report Related to the Certification of Advanced Boiling Water Reactor Design." July 1994. Supplement 1, May 1997.

NUREG-1536, "Standard Review Plan for Dry Cask Storage Systems," Final Report, January 1997.

17.1.4 NUREG/CR

NUREG/CR-0200, "SCALE: A Modular Code System for Performing Standardized Computer Analyses for Licensing Evaluation," ORNL. March 1997.

NUREG/CR-0722, "Fission Product Release from Highly Irradiated LWR Fuel," by R.A.Lorenz, J.L. Collins, A.P. Malinauskas, O. Kirkland, and R.L. Towns, ORNL, February 1980.

NUREG/CR-0781, "SKYSHINE-II Procedure: Calculation of the Effects of Structure Design on Neutron, Primary Gamma-Ray, and Secondary Gamma-Ray Dose Rate in Air." January 1981.

NUREG/CR-1390, "Probability Estimates of Temperature Extremes for the Contiguous United States." May 1980.

NUREG/CR-1815, "Recommendations for Protecting Against Failure by Brittle Fracture in Ferritic Steel Shipping Containers Up to Four Inches Thick," LLNL. August 1981.

NUREG/CR-4554, "SCANS (Shipping Cask Analysis System): A Microcomputer Based Analysis System for Shipping Cask Design Review," LLNL. March 1998.

NUREG/CR-6007, "Stress Analysis of Closure Bolts for Shipping Casks," LLNL. January 1993.

NUREG/CR-6242, "CASKS (Computer Analysis of Storage Casks): A Microcomputer Based Analysis System of Storage Casks Design Review," LLNL. February 1995.

NUREG/CR-6322, "Buckling Analysis of Spent Fuel Baskets," LLNL. May 1995.

NUREG/CR-6407, "Classification of Transportation Packaging and Dry Spent Fuel Storage System Components According to Importance to Safety," INEL. February 1996.

NUREG/CR-6451, "A Safety and Regulatory Assessment of Generic BWR and PWR Permanently Shutdown Nuclear Power Plants," by R.J. Travis, R.E. Davis, E.J. Grove, and M.A. Azarm, Brookhaven National Laboratory. August 1997.

17.1.5 Information Notices, Bulletins, and other Publications

Information Notice No. 91-26, "Potential Nonconservative Errors in the Working Format Hansen-Roach Cross-Section Set Provided with the Keno and Scale Codes," April 15, 1991.

NRC Bulletin 96-04, "Chemical, Galvanic, or other Reactions in Spent Fuel Storage and Transportation Casks."

17.2 Codes, Standards, and Specifications

American Concrete Institute (ACI)
ACI 318, "Building Code Requirements for Reinforced Concrete."

ACI 349, "Code Requirements for Nuclear Safety Related Concrete Structures," and ACI 349R, "Commentary."

ACI 359, "Code for Concrete Reactor Vessels and Containments" (also designated as ASME Boiler and Pressure Vessel Code, Section III, "Rules for Construction of Nuclear Power Plant Components," Division 2), ACI and ASME (Joint Committee).

American Institute of Steel Construction (AISC)
"Code of Standard Practice for Steel Buildings and Bridges," published in the AISC "Manual of Steel Construction."

"Specification for Structural Steel Buildings, Allowable Stress Design and Plastic Design," published in the AISC "Manual of Steel Construction."

American Iron and Steel Institute (AISI)
AISI, *Steel Products Manual.*

American National Standards Institute/American Nuclear Society (ANSI/ANS)
ANSI/ANS 2.8-1981, "Determining Design Bases Flooding at Power Reactor Site."

ANSI/ANS 3.1-1993, "Selection, Qualification, and Training of Personnel for Nuclear Power Plants."

ANSI/ANS 6.1.1, "American National Standard for Neutron and Gamma-Ray Fluence to Dose Factors," Institute for Nuclear Materials Management/American Nuclear Society, 1991.

ANSI/ANS-HPSSC-6.8.1-1981, "Location and Design Criteria for Area Monitoring Systems for Light Water Reactors."

ANSI/ANS 8.1-1983, "Nuclear Criticality Safety In Operations with Fissionable Materials Outside Reactors," 1985.

ANSI/ANS 8.17-1984, "Criticality Safety Criteria for the Handling, Storage, and Transportation of LWR Fuel Outside Reactors."

ANS 8.20, "Nuclear Criticality Safety Training."

ANSI/ANS 8.21-1995, "Use of Fixed Neutron Absorbers in Nuclear Facilities Outside Reactors," 1995.

ANSI/ANS N13.1, "Guide to Sampling Airborne Radioactive Material."

ANSI/ANS N13.2, "Guide to Administrative Practices in Radiation Monitoring."

ANSI/ANS N13.6, "Practice for Occupational Radiation Exposure Record Systems."

ANSI/ANS-HPS N13.30-1996, "Performance Criteria for Radiobioassay."

ANSI/ANS-HPS-N13.32-1995, "Performance Testing of Extremity Dosimeters."

ANSI/ANS-HPS-N13.41-1997, "Criteria for Performing Multiple Dosimetry."

ANSI/ANS-HPS-N13.42-1997, "Internal Dosimetry Program for Mixed Fission and Activation Products."

ANSI N14.5-1987, "American National Standard for Radioactive Materials - Leakage Tests on Packages for Shipment," Institute for Nuclear Materials Management, January 1987.

ANSI N14.6-1993, "American National Standard for Radioactive Materials - Special Lifting Devices for Shipping Containers Weighing 10,000 Pounds (4500 kg) or More," Institute for Nuclear Materials Management, 1993.

ANSI/ANS N45.2.11-1974, "Quality Assurance Requirements for the Design of Nuclear Power Plants," 1974.

ANSI N210-1976/ANS-57.2, "Design Objectives for Light Water Reactor Spent Fuel Pool Storage Facilities at Nuclear Power Stations," [Referenced in NUREG-0800, Draft Revision 4, 1996].

ANSI/ANS 57.2-1983, "Design Requirements for Light Water Reactor Spent Fuel Storage Facilities at Nuclear Power Plants."

ANSI/ANS 57.7-1988, "Design Criteria for an Independent Spent Fuel Storage Installation (Water Pool Type)."

ANSI/ANS 57.9-1984, "Design Criteria for an Independent Spent Fuel Storage Installation (Dry Storage Type)" [Referenced to the extent that ANSI/ANS 57.9-1984 is stated as suitable in Regulatory Guide 3.60].

ANSI/ASME B30.2.0-1978, "Overhead and Gantry Cranes (Top Running Bridge, Multiple Girder)."

ANSI/ASME B30.16, "Overhead Hoists."

ASME/ANSI B31.1, "Power Piping."

ANSI/ASME N509, "Nuclear Power Plant Air Cleaning Units and Components."

ANSI/IEEE C2, "Electrical Code."

American Petroleum Institute (API)
API 620, "Rules for Design and Construction for Large, Welded Low-pressure Storage Tanks."

API 650, "Welded Steel Tanks for Oil Storage."

American Society for Nondestructive Testing (ASNT)
Recommended Practice No. SNT-TC-1A, December 1992.

American Society for Testing and Materials (ASTM)
A 36, "Standard Specification for Structural Steel."

A 53, "Standard Specification for Welded and Seamless Steel Pipe."

A 82, "Standard Specification for Cold-Drawn Steel Wire for Concrete Reinforcement."

A 184, "Standard Specification for Fabricated Deformed Steel Bar Mats for Concrete Reinforcement."

A 185, "Standard Specification for Welded Steel Wire Fabric for Concrete Reinforcement."

A 242, "Standard Specification for High-Strength Low-Alloy Structural Steel."

A 416, "Standard Specification for Uncoated Seven-Wire Stress-Relieved Steel Strand for Prestressed Concrete."

A 421, "Standard Specification for Uncoated Stress-Relieved Steel Wire for Prestressed Concrete."

A 441, "Standard Specification for High-Strength Low-Alloy Structural Manganese Vanadium Steel."

A 496, "Standard Specification for Deformed Steel Wire for Concrete Reinforcement."

A 497, "Standard Specification for Welded Deformed Steel Wire Fabric for Concrete Reinforcement."

A 500, "Standard Specification for Cold-Formed Welded and Seamless Carbon Steel Structural Tubing in Rounds and Shapes."

A 501, "Standard Specification for Hot-Formed Welded and Seamless Carbon Steel Structural Tubing."

A 572, "Standard Specification for High-Strength Low-Alloy Columbium-Vanadium Steels of Structural Quality."

A 588, "Standard Specification for High-Strength Low-Alloy Structural Steel with 50,000 psi Minimum Yield Point to 4 inches Thick."

A 615, "Standard Specification for Deformed and Plain Billet-Steel Bars for Concrete Reinforcement."

A 706, "Standard Specification for Low-Alloy Steel Deformed Bars for Concrete Reinforcement."

A 772, "Standard Specification for Uncoated High-Strength Steel Bar for Prestressing Concrete."

C 31, "Standard Method of Making and Curing Concrete Test Specimens in the Field."

C 33, "Standard Specification for Concrete Aggregates."

C 39, "Standard Method of Test for Compressive Strength of Cylindrical Concrete Specimens."

C 42, "Standard Method of Obtaining and Testing Drilled Cores and Sawed Beams of Concrete."

C 88, "Standard Method of Test for Soundness of Aggregates by Use of Sodium Sulfate or Magnesium Sulfate."

C 94, "Standard Specification for Ready-Mixed Concrete."

C 109, "Standard Method of Test for Compressive Strength of Hydraulic Cement Mortars (Using 2-inch or 50-mm Cube Specimens)."

C 131, "Standard Test Method for Resistance to Degradation of Small-Size Coarse Aggregate by Abrasion and Impact in the Los Angeles Machine."

C 144, "Standard Specification for Aggregate for Masonry Mortar."

C 150, "Standard Specification for Portland Cement."

C 172, "Standard Method of Sampling Fresh Concrete."

C 192, "Standard Method of Making and Curing Concrete Test Specimens in the Laboratory."

C 260, "Standard Specification for Air-Entraining Admixtures for Concrete."

C 289, "Standard Method of Test for Potential Reactivity of Aggregates (Chemical Method)."

C 441, "Standard Method of Test for Effectiveness of Mineral Admixtures in Preventing Excessive Expansion of Concrete, Due to the Alkali-Aggregate Reaction."

C 494, "Standard Specification for Chemical Admixtures for Concrete."

C 496, "Standard Method of Test for Splitting Tensile Strength of Cylindrical Concrete Specimens."

C 595, "Standard Specification for Blended Hydraulic Cements."

C 618, "Standard Specification for Fly Ash and Raw or Calcined Natural Pozzolan for Use as a Mineral Admixture in Portland Cement Concrete."

C 637, "Standard Specification for Aggregates for Radiation-Shielding Concrete."

C 685, "Standard Specification for Concrete Made by Volumetric Batching and Continuous Mixing."

C 1017, "Standard Specification for Chemical Admixtures for Use in Producing Flowing Concrete."

E 1167, "Guide for Radiation Protection Program for Decommissioning Operations."

E 1168, "Guide for Radiation Protection Training for Nuclear Facility Workers."

American Society of Civil Engineers (ASCE)
ASCE 4-86, "Seismic Analysis of Safety-Related Nuclear Structures," American Society of Civil Engineers, New York, NY, 1986.

ASCE 7 (formerly ANSI A58.1), "Minimum Design Loads for Buildings and Other Structures."

American Society of Mechanical Engineers (ASME)
ASME Boiler and Pressure Vessel Code, Section II, Division 1, "Material Specifications."

ASME Boiler and Pressure Vessel Code, Section III, Division 1, "Rules for Construction of Nuclear Power Plant Components."

ASME Boiler and Pressure Vessel Code, Section V, "NDE Specifications and Procedures."

ASME Boiler and Pressure Vessel Code, Section VIII, "Pressure Vessels."

ASME Boiler and Pressure Vessel Code, Section IX, "Welding and Brazing Qualifications."

ASME NQA-1, "Quality Assurance Program Requirements for Nuclear Facilities, 1983 Edition."

American Welding Society (AWS)
AWS A2.4, "Standard Symbols for Welding, Brazing and Nondestructive Examination."

AWS D1.1, "Structural Welding Code - Steel."

AWS D1.4, "Structural Welding Code - Reinforcing Steel."

American Society of Heating Refrigeration and Air Conditioning Engineers, Inc. (ASHRAE)
ASHRAE Handbook, 1989, Fundamentals.

Crane Manufacturing Association of America (CMAA)
CMAA 70, "Specifications for Electric Overhead Traveling Cranes."

Department of Energy (DOE)
DOE/RW-0184-R1, "Characteristics of Potential Repository Wastes," Office of Civilian Radioactive Waste Management, U. S. Department of Energy, July 1992.

International Conference of Building Officials (ICBO)
"Uniform Building Code" (UBC).

Institute of Electrical and Electronic Engineers, Inc. (IEEE)
IEEE S-135, IEEE-IPCEA, "Power Cable Ampacities."

National Council on Radiation Protection (NCRP)
NCRP Report No. 59-1978, "Operational Radiation Safety Program."

NCRP Report No. 71-1983, "Operational Radiation Safety Training."

National Fire Protection Association (NFPA)
NFPA, "Handbook of Fire Protection."

NFPA 1, "Fire Prevention Code" (including but not limited to)
>   NFPA 70, "National Electric Code"
>   NFPA 78, "Lightning Protection Code"
>   NFPA 101, "Code for Safety to Life from Fire in Buildings and Structures"
>   NFPA 512, "Standard for Truck Fire Protection"
>   NFPA 801, "Recommended Fire Protection Practice for Facilities Handling Radioactive Materials."

17.3 Manuals and Texts

Alcouffe, R.E., et al., "User's Guide for TWODANT: A Code Package for Two-Dimensional, Diffusion Accelerated, Neutral Particle Transport," LA-10049-M Rev., Los Alamos National Laboratory, April 1992.

"ANISN/PC-Multigroup One-Dimensional Discrete Ordinates Transport Code System with Anisotropic Scattering," CCC-J14 Micro, Oak Ridge, Tennessee, 1990.

"ANSYS Basic Analysis Procedures Guide," Fourth Edition, ANSYS Release 5.6, ANSYS, Inc., Canonsburg, Pennsylvania, November 1999.

Hoerner, S.F., "Fluid Dynamic Drag," 1965, Hoerner Fluid Dynamics, P.O. Box 342, Brick Town, New Jersey 08723.

Marker, B.R., et al., "NIKE3D-A Nonlinear, Implicit, Three-Dimensional Finite Element Code for Solid and Structural Mechanics-User's Manual," UCRL-MA-105268, LLNL, January 1991.

"MCBEND - A Monte Carlo Program for Shielding Calculations; User Guide for Version 6," issue 4, Answers (MCBEND) 2: UKAEA Winfrith, UK, March 1990.

National Safety Council, "Accident Prevention Manual for Industrial Operations."

O'Dell, R.D., et al., "Revised User's Manual for ONEDANT: A Code for One-Dimensional, Diffusion Accelerated, Neutral Particle Transport," LA-9184-M, Rev., LANL, December 1989.

"QAD-CGGP, A Combinatorial Geometry Version of QAD-P5A, A Point Kernel Code System for Neutron and Gamma-Ray Shielding Calculations Using the GP Buildup Factor," CCC-493, Oak Ridge, Tennessee, 1994.

"RANKERN, A Point Kernel Integration Code for Complicated Geometry Probelems; User Guide to Version 12," Issue 2, ANSWERS (RANK)2, UKAEE, UK, September 1987.

"SCALE-PC Version 4.1," CCC-619, ORNL, Radiation Shielding Information Center, December 1993.

TRW Environmental Safety Systems, Inc., "DOE Characteristics Data Base, User Manual for the CDB-R," November 16, 1992.

U.S. Environmental Protection Agency, Federal Guidance Report No. 11, "Limiting Values of Radionuclide Intake and Air Concentration and Dose Conversion Factors for Inhalation, Submersion, and Ingestion," DE89-011065, 1988.

U.S. Environmental Protection Agency, "Manual of Protective Action Guides and Protective Action for Nuclear Incidents," EPA 520/1-75-001-A, January 1990.

Worku, G., et al., "MicroShield Version 4.2 User's Manual," Grove Engineering, Inc., Rockville, Maryland, 1995.

Young, W.C., "Roark's Formulas for Stress and Strain," McGraw-Hill. 1989.

17.4 Technical Reports

Gulf General Atomic
Archibald, R., Lathrop, K.D., and Mathews, D., "1DFX-A Revised Version of the IDF (DTF-IV) Sn Transport Theory Code," Gulf-GA-10820, September 1971.

Walti, P., and Kock, P., "MICROX- A Two Region Flux Spectrum Code for the Efficient Calculation of Group Cross Section," Gulf-GA-A10827, April 1972.

Lawrence Livermore National Laboratory (LLNL)
Lloyd, W.R., "Determination and Application of Bias Values in the Criticality Evaluation of Storage Cask Designs," UCID-21830, Lawrence Livermore National Laboratory, January 1990.

Schwartz, M.W. and Witte, M.C., "Spent Fuel Cladding Integrity During Dry Storage," UCID-21181, LLNL, September 1987.

Slemmons, D.B., et al. "Assessment of Active Faults for Maximum Credible Earthquakes of the Southern California-Northern Baja Region," University of California, LLNL publication No. UCID 19125, 48 p., 1982.

Thomas, G.R., and Carlson, R.W., "Evaluation of Use of Homogenized Fuel Assemblies in the Thermal Analysis of Spent Fuel Storage Casks," LLNL, Publication No. UCRL-ID-134567, July 1999.

Los Alamos National Laboratory (LANL)
"MCNP - A General Monte Carlo Code for Neutron and Photon Transport," LA-7396-M, LANL, July 1978.

"MCNP 4A, Monte Carlo N-Particle Transport Code System," LANL, December 1993.

Manteufel, R.D., and Todreas, N.E., "Effective Thermal Conductivity and Edge Configuration Model for Spent Fuel Assembly," Nuclear Technology, Vol. 105, pp. 421–440, March 1994.

Oak Ridge National Laboratory (ORNL)

"Assessment of Shielding Analysis Methods, Codes, and Data for Spent Fuel Transport/Storage Applications," ORNL/CSD/TM-246, July 1988.

Cottrell, W.B., and A.W. Savolainen, "U.S. Reactor Containment Technology," ORNL-NSIC-5, Vol. 1, Chapter 6, ORNL. August 1965.

Eckerman, K.F., Wolbarst, A.B., and Richardson, A.C.B., "Limiting Values of Radionuclide Intake and Air Concentration and Dose Conversion Factors for Inhalation, Submersion, and Ingestion," Federal Guidance Report No. 11, DE89-011065, ORNL, 1988.

Kennedy, R.P., "A Review of Procedures for the Analysis and Design of Concrete Structures to Resist Missile Impact Effects," ORNL-NSIC-5, Volume 1, Chapter 6. September 1975.

"ORIGEN2: Isotope Generation and Depletion Code-Matrix Exponential Method," 1991.

Petrie, L.M., Cross, N.F., "KENO-IV, An Improved Monte Carlo Criticality Program," ORNL-4938, ORNL, November 1975.

Rhoads, W.A., "The TORT Three-Dimensional Discrete Ordinates Neutron/Photon Transport Code," ORNL-6268, ORNL, November 1987.

Pacific Northwest Laboratory (PNL)

Cunningham, M.E., et al., "Control of Degradation of Spent LWR Fuel During Dry Storage in Inert Atmosphere," PNL-6364, PNL, October 1987.

Johnson, A.B., and Gilbert, E.R., "Technical Basis for Storage of Zircaloy-Clad Spent Fuel in Inert Gases," PNL-4835, PNL, September 1983.

Levy, I.S., et al., "Recommended Temperature Limits for Dry Storage of Spent Light Water Zircalloy Clad Fuel Rods in Inert Gas," PNL-6189, PNL, May 1987.

PNL-4835, "Technical Basis for Storage of Zircalloy-Clad Spent Fuel in Inert Gases," PNL, September 1983.

Sandia National Laboratories (SNL)

Sandoval, R.P., et al., "Estimate of CRUD Contribution to Shipping Cask Containment Requirements," Sandia Report, SAND88-1358, TTC-0811, UC-71, SNL, January 1991.

Shaffer, Clinton J., "Radiological Source Term Analysis for PWR Spent Fuel Transportation Accidents," Sandia Report SEA NO. 97-4121-A:1, SNL, June 24, 1997.

Wilmot, Edwin L., "Transportation Accident Scenarios for Commercial Spent Fuel," Sandia Report SAND80-2124, SNL, February 1981.

Science Applications International Corporation (SAIC)
Stokley, J.R. and Williamson, D.H., "Structural Integrity of Spent Nuclear Fuel Storage Casks Subjected to Drop," Nuclear Technology, American Nuclear Society, April 1996.

Other Sources
Bonilla, M.G., Mark, R.K., and Lienkaemper, J.J., "Statistical Relations among Earthquake Magnitude, Surface Rupture Length, and Surface Fault Displacement," Bulletin of the Seismological Society of America, vol. 74, pp. 2379-2411, 1984.

Campbell, K.W., and Bozorgnia, Y., "Near Source Attenuation Peak Horizontal Acceleration From Worldwide Accelerograms Recorded from 1975 to 1993," Fifth U.S. National Conference on Earthquake Engineering, Chicago, IL, July 10-14, 1994.

Hill, M.D., Simmonds, J.R., and Jones, J.A., "NRPB Methodology for Assessing the Radiological Consequences of Accidental Releases of Radionuclide to Atmosphere - MARC-1," NRPB-R224, September 1989.

Kennedy, R.P., "A Review of Procedures for the Analysis and Design of Concrete Structures to Resist Missile Impact Effects," Holmes and Narver, Inc., September 1975.

Linderman, R.B., Rotz, J.V., Yeh, G.C.K., "Design of Structures for Missile Impact," Topical Report BC-TOP-9-A, Revision 2, Bechtel Power Corporation, September 1974.

National Oceanic and Atmospheric Administration, Technical Report NWS-23, "Meteorological Criteria for the Standard Project Hurricane and Probable Maximum Hurricane Windfields, Gulf and East Coasts of the United States." 1979.

Slemmons, D.B., "State-of-the-Art for Assessing Earthquake hazards in the United States: Report 6, Faults and Earthquake Magnitude," Miscellaneous Paper S-73-1, U.S. Army Engineer Waterways Experiment Station, Corps of Engineers, Vicksburg, Mississippi, 1977.

WASH-1300, "Technical Basis for Interim Regional Tornado Criteria," U.S. Atomic Energy Commission, May 1974.

18 BIBLIOGRAPHY

18.1 U.S. Code of Federal Regulations and NRC Documents

10 CFR Part 19, "Notices, Instructions and Reports to Workers: Inspection and Investigations."

10 CFR Part 21, "Reporting of Defects and Noncompliance."

10 CFR Part 51, "Environmental Protection Regulations for Domestic Licensing and Related Regulatory Functions."

10 CFR Part 61, "Licensing Requirements for Land Disposal of Radioactive Wastes."

29 CFR 1910.119, Appendix A, "List of Highly Hazardous Chemicals, Toxics and Reactives (Mandatory)."

40 CFR 69, "Chemical Accident Prevention Provisions."

40 CFR 190, "Environmental Radiation Protection Standards for Nuclear Power Operations."

40 CFR 191, "Environmental Radiation Protection Standards for the Management and Disposal of Spent Nuclear Fuel, High-Level and Transuranic Radioactive Wastes."

RG 1.22, "Periodic Testing of Protection System Actuation Functions (Safety Guide 22)." Revision 0, February 1972.

RG 1.52, "Design, Testing, and Maintenance Criteria for Postaccident Engineered-Safety-Feature Atmosphere Cleanup System Air Filtration and Adsorption Units of Light-Water-Cooled Nuclear Power Plants." Revision 2, March 1978.

RG 1.91, " Evaluation of Explosions Postulated to Occur on Transportation Routes Near Nuclear Power Plants." Revision 1, February 1976.

RG 1.111, "Methods for Estimating Atmospheric Transport and Dispersion of Gaseous Effluents in Routine Releases from Light-Water-Cooled Reactors." Revision 1, July 1977.

RG 1.112, "Calculation of Releases of Radioactive Materials in Gaseous and Liquid Effluent from Light-Water- Cooled Power Reactors."   Revision 0-R, May 1977.

RG 1.113 "Estimating Aquatic Dispersion of Effluents from Accidental and Routine Reactor Releases for the Purpose of Implementing Appendix I." Revision 1, April 1977.

RG 1.125, "Physical Models for Design and Operation of Hydraulic Structures and Systems for Nuclear Power Plants." Revision 1, October 1978.

RG 1.132, "Site Investigations for Foundation of Nuclear Power Plants." Revision 1, March 1979.

RG 1.138, "Laboratory Investigations of Soils for Engineering Analysis and Design of Nuclear Power Plants." Revision 0, April 1978.

RG 3.44, "Standard Format and Content for the Safety Analysis Report for an Independent Spent Fuel Storage Installation (Water-Basis Type)." Revision 2, January 1989.

RG 3.49, "Design of an Independent Spent Fuel Storage Installation (Water-Basin Type.)" Revision 0, December 1981.

RG 3.50, "Guidance on Preparing a License Application to Store Spent Fuel in an Independent Spent Fuel Storage Installation." Revision 1, September 1989.

RG 3.61, "Standard Format and Content for a Topical Safety Analysis Report for Spent Fuel Storage Casks." Revision 0, February 1989.

RG 3.62, "Standard Format and Content for the Safety Analysis Report for Onsite Storage of Spent Fuel Storage Casks." Revision 0, February 1989.

RG 4.13, "Performance, Testing, and Procedural Specifications for Thermoluminescence Dosimetry: Environmental Applications." Revision 1, July 1977.

RG 4.15, "Quality Assurance for Radiological Monitoring Programs   (Normal Operations) - Effluent Streams and the Environment." Revision 1, February 1979.

RG 5.7, "Entry/Exit Control for Protected Areas, Vital Areas, and Material Access Areas." Revision 1, May 1980.

RG 5.12, "General Use of Locks in the Protection and Control of Facilities and Special Nuclear Materials." Revision 0, November 1973.

RG 5.14, "Use of Observation (Visual Surveillance) Techniques in Material Access Areas." Revision 1, May 5, 1980.

RG 5.25, "Design Considerations for Minimizing Residual Holdup of Special Nuclear Material in Equipment for Wet Process Operations." Revision 0, June 1974.

RG 5.43, "Plant Security Force Duties." Revision 0, January 1975.

RG 5.44, "Perimeter Intrusion Alarm System." Revision 3, October 1997.

RG 8.1, "Radiation Symbol." February 1973.

NUREG-0654/FEMA-REP-1, "Criteria for Preparation and Evaluation of Radiological Emergency Response Plans and Preparedness in Support of Nuclear Power Plants," November 1980.

NUREG-0761, "Radiation Protection Plans for Nuclear Power Reactor Licensees," March 1981.

NUREG-1101, "Onsite Disposal of Radioactive Waste." Volume 1, March 1986. Volume 2, February 1987. Volume 3, November 1986.

NUREG-1140, "A Regulatory Analysis on Emergency Preparedness for Fuel Cycle and Other Radioactive Material Licensees." January 1988.

NUREG/CR-4775, "Guide for Preparing Operating Procedures for Shipping Packages," LLNL. December 1988.

18.2 Codes, Standards, and Specifications

American Institute of Steel Construction (AISC)
ANSI/AISC N690, "Nuclear Facilities - Steel Safety-Related Structures for Design Fabrication and Erection."

AISC, "Manual of Steel Construction - Allowable Stress Design," Ninth Edition, 1989.

American National Standards Institute/American Nuclear Society
ANSI/ANS 6.4, "Guidelines on the Nuclear Analysis and Design of Concrete Radiation Shielding for Nuclear Power Plant," 1985.

ANSI/ANS 6.4.2, "Specification for Radiation Shielding Materials," 1985.

ANS 8.19, "Administrative Practices for Nuclear Criticality Safety."

ANSI/ANS N13.10, "Specification and Performance of On-Site Instrumentation for Continuously Monitoring Radioactivity in Effluents."

ANSI/ANS 40.35, "Volume Reduction of Low-Level Waste," American Nuclear Society, 1984.

ANSI N 323, "Radiation Protection Instrumentation Test and Calibration."

American Society for Testing and Materials (ASTM)
A 312, "Standard Specification for Seamless and Austenitic Stainless Steel Pipe."

C 992, "Specification for Boron-Based Neutron Absorbing Material Systems for Use in Nuclear Spent Fuel Storage Racks."

E 1281, "Guide for Nuclear Facility Decommissioning Plans."

American Society of Mechanical Engineers (ASME)
ASME/ANSI NOG-1, "Rules for Construction of Overhead and Gantry Cranes (Top Running Bridge, Multiple Girder)."

ASME/ANSI B96.1, "Specification for Welded Aluminum-Alloy Field Erected Storage Tanks."

ASME N510, "Testing of Nuclear Air Cleaning ."

American Water Works Association (AWWA)
AWWA, "Manual of Water Quality and Treatment."

American Welding Society (AWS)
ANSI/AWS D1.2, "Structural Welding Code - Aluminum."

ANSI/AWS D1.3, "Structural Welding Code - Sheet Steel."

Institute of Electrical and Electronic Engineers, Inc. (IEEE)
IEEE 141, "Recommended Practice for Electric Power Distribution for Industrial Plants."

IEEE 142, "Recommended Practice for Grounding of Industrial and Commercial Power Systems."

IEEE 336, "Standard Installation, Inspection, and Testing Requirements for Power, Instrumentation and Control Equipment at Nuclear Facilities."

IEEE 446, "Recommended Practice for Emergency and Standby Power for Industrial and Commercial Applications."

IEEE 498, "Standard Requirements for the Calibration and Control of Measuring and Test Equipment Used in Nuclear Facilities."

IEEE 603, "Standard Criteria for Safety Systems for Nuclear Power Generating Stations."

IEEE C 37, "Circuit Breakers, Switchgear, Relays, Substations and Fuses - Series."

18.3 Technical Literature

"Assessment of Shielding Methods, Codes, and Data for Spent Fuel Transport/Storage Applications," ORNL/CSD/TM-246, July 1988.

Knoll, R.W., et al., "Evaluation of Cover Gas Impurities and Their Effects on the Dry Storage of LWR Spent Fuel," PNL-6365, DE88003983, PNL, November 1987.

U.S. Environmental Protection Agency, Federal Guidance Report No. 11, "Limiting Values of Radionuclide Intake and Air Concentration and Dose Conversion Factors for Inhalation, Submersion, and Ingenstion," DE89-011065, 1988.